Hazmatology
The Science of Hazardous Materials

Hazmatology: The Science of Hazardous Materials,
Five-Volume Set
9781138316072

Volume One - Chronicles of Incidents and Response
9781138316096

Volume Two - Standard of Care and Hazmat Planning
9781138316768

Volume Three - Applied Chemistry and Physics
9781138316522

Volume Four - Common Sense Emergency Response
9781138316782

Volume Five - Hazmat Team Spotlight
9781138316812

Chronicles of Incidents and Response

Robert A. Burke

CRC Press
Taylor & Francis Group
Boca Raton London New York

CRC Press is an imprint of the
Taylor & Francis Group, an **Informa** business

CRC Press
Taylor & Francis Group
6000 Broken Sound Parkway NW, Suite 300
Boca Raton, FL 33487-2742

© 2021 by Taylor & Francis Group, LLC
CRC Press is an imprint of Taylor & Francis Group, an Informa business

No claim to original U.S. Government works

Printed on acid-free paper

International Standard Book Number-13: 978-1-138-31609-6 (Hardback)

This book contains information obtained from authentic and highly regarded sources. Reasonable efforts have been made to publish reliable data and information, but the author and publisher cannot assume responsibility for the validity of all materials or the consequences of their use. The authors and publishers have attempted to trace the copyright holders of all material reproduced in this publication and apologize to copyright holders if permission to publish in this form has not been obtained. If any copyright material has not been acknowledged, please write and let us know so we may rectify in any future reprint.

Except as permitted under U.S. Copyright Law, no part of this book may be reprinted, reproduced, transmitted, or utilized in any form by any electronic, mechanical, or other means, now known or hereafter invented, including photocopying, microfilming, and recording, or in any information storage or retrieval system, without written permission from the publishers.

For permission to photocopy or use material electronically from this work, please access www.copyright.com (http://www.copyright.com/) or contact the Copyright Clearance Center, Inc. (CCC), 222 Rosewood Drive, Danvers, MA 01923, 978-750-8400. CCC is a not-for-profit organization that provides licenses and registration for a variety of users. For organizations that have been granted a photocopy license by the CCC, a separate system of payment has been arranged.

Trademark Notice: Product or corporate names may be trademarks or registered trademarks, and are used only for identification and explanation without intent to infringe.

Visit the Taylor & Francis Web site at
http://www.taylorandfrancis.com

and the CRC Press Web site at
http://www.crcpress.com

Dedication

Kevin Saunders

Kevin Saunders would be considered a miracle by some, to others a survivor, to many a hero, but to me he is an inspiration. After meeting Kevin and getting to know him I would be hard pressed to ever complain about anything again. You can read his story under the Corpus Christi, TX Grain Elevator Explosion. Kevin is a miracle. He is a survivor, he is a hero, but most of all he is an inspiration to everyone he talks to, whether personally or during his motivational speaking engagements.

Contents

Preface..xv
Acknowledgements xvii
Author...xxi

Chronicles of Incidents and Response
Origins of Hazardous Materials Response 1
Beginnings of Hazardous Materials Transportation Regulations............4
Dust and Boiler Explosions 5
Petroleum Industry is Born 5
Nebraska Firefighters Museum Hazardous Materials Display 7
Hastings, Nebraska "Gasoline & Oil Clean-Up Squad"............................8
Milwaukee, Wisconsin's Petroleum Dispersal Unit 1...............................9
Evolution of Present Day Hazmat Teams...10
Jacksonville, FL's First Emergency Services Hazmat Team in the
United States 11
   Chief Yarbrough's Vision..11
   The "Godfather of Hazmat" 11
   America's First Emergency Services Hazmat Team.............................12
   My Visits to Jacksonville 16
   Captain Ron Gore Retires 17
Martin County Florida: First Volunteer Hazmat Team in
United States 18
   Fire Department History..18
   Hazmat Team History 19
Houston, Texas Pioneer Hazmat Team 20
   Beginnings of Hazmat Response in Houston.......................................20
   Houston Distribution Warehouse Complex...23
Memphis, Tennessee Pioneer Hazmat Team 24
   Hazmat Team History 24
   Hazardous Materials Exposure 24
   Incidents 25
   Drexel Chemical Company Fire & Explosion 25
   Pro-Serve Fire (Brooks Road) 28

vii

viii   *Contents*

Selected Historical Hazmat Incidents

The 1800s 30
First Incident Found During Research 31
   Syracuse, NY August 30, 1841 The Great Gunpowder Explosion 31
Minneapolis, MN May 2, 1878 Washburn "A" Flour Mill Explosion........ 32
Chester, PA February 16, 1882 Jackson Fireworks Plant Explosion........... 35
Cavan Point, NJ May 11, 1883 Standard Oil Works Fire Started by
Lightning.. 38
Oil City, Titusville, and Meadville, PA June 1892 Flood and Explosion .. 41
Bradford, PA, May 14, 1894 Refinery Fire... 43
Butte, MT January 15, 1895 Warehouse District Dynamite Explosion
and Fire... 45
   "The Great Disaster of 1895".. 45
Chicago, IL August 5, 1897 Grain Elevator Fire 51
Firemen That Made The Supreme Sacrifice 54
Philadelphia, PA August 5, 1897 Naphtha Explosion 54
The 1900s 56
New York, NY November 2, 1900 Tarrant & Co Drug House Explosion ... 56
   (Greenwich Street Volcano FDNY) 56
Sheridan, PA Naphtha Car Explosions May 12, 1902................................. 59
Milwaukee, WI February 4, 1903 Schwab Stamp & Seal Acid Spill 62
   Firemen That Made The Supreme Sacrifice.. 62
      Injured.. 63
Avon, CT September 15, 1905 Fuse Co. Explosion.. 63
Portland, OR June 27, 1911 Oil Company Fire.. 64
   Fireman That Made The Supreme Sacrifice 65
Eddystone, PA April 10, 1917 Ammunition Works Disaster 65
Morgan, NJ October 4, 1918 T.A. Gillespie & Company Explosions......... 67
Boston, MA January 15, 1919 'Molasses Flood' Tank Explosion 71
   Fireman That Made The Supreme Sacrifice 73
Long Island City, NY September 14-15, 1919 Oil Tank Blaze 73
   Firemen That Made The Supreme Sacrifice.. 75
      Unknown.. 75
New York, NY July 18, 1922 Chemical Explosion 75
   Fireman Made the Ultimate Sacrifice... 76
Pekin, IL January 3, 1924 Corn Products Refining Corn Starch
Plant Explosion... 77
Milwaukee, WI April 22, 1926 Sawdust Explosion...................................... 78
   Firemen That Made The Supreme Sacrifice.. 79
Dover, NJ July 10, 1926 Picatinny Arsenal Munitions Explosion 79
Chicago, IL March 11, 1927 Chemical Plant Explosion 80
   Firemen Who Made the Supreme Sacrifice 81
   Seriously Injured Firefighter.. 81
San Diego, CA August 8, 1927 Acetylene Factory Explosion..................... 81

Contents

Heidelberg, PA December 25, 1928 Refinery Fire and Explosions 82
    Firemen Who Made The Supreme Sacrifice 83
Lakehurst, NJ May 3, 1937 Hindenburg Disaster Hydrogen
Explosion & Fire 83
Atlantic City, NJ July 16, 1937 Gasoline Explosion 84
Hastings, NE September 15, 1944 Explosion Naval Ammunition Depot... 85
    There Were Occasional Explosions.. 86
    Air Force Presence.. 88
Cleveland, OH October 20, 1944 LNG Leak, Explosion, Fire..................... 88
Westport, CT May 3, 1946 Truck Explosion.. 89
    Firemen Who Made The Supreme Sacrifice 90
Atlanta, GA December 30, 1946 Drug Fire 90
Texas City, TX April 16, 1947 Ammonium Nitrate Ship Explosions......... 91
    The Day Texas City Lost Their Fire Department.................................... 91
    Texas City Firefighters Who Made The Supreme Sacrifice on
    April 16, 1947 97
        Privates... 97
Minot, ND July 22, 1947 Oil Storage Tank Explosion................................. 98
Reno, NV August 16, 1948 Massive Dynamite Explosion 99
    Firemen Who Made The Supreme Sacrifice 100
Perth Amboy, NJ June 24, 1949 Asphalt Plant Explosion 100
    Firemen Who Made The Supreme Sacrifice. 101
Philadelphia, PA October 10, 1954 Chemical Plant Explosion................. 101
    Firefighters That Made The Supreme Sacrifice.................................... 101
    Injured.. 102
Dumas, TX Oil July 29, 1956 Shamrock Oil & Gas Corporation
Tank Farm Explosion.. 102
    Firemen Who Made The Supreme Sacrifice 103
Brownfield, TX December 23, 1958 Butane Truck Explosion................... 104
    Firemen Who Made The Supreme Sacrifice 105
Schuylkill Haven, PA June 2, 1959 Propane Truck Explosion................. 105
    Sgt. Clouser Gave This Official Version of the Tragedy 106
    Fireman That Made The Supreme Sacrifice 107
Meldrim, GA June 28, 1959 Railroad Trestle Disaster............................... 107
    First Train Derailment Releasing Hazardous Materials Kills 23 107
    First Victims to Be Killed by a Hazardous Materials Release
    from a Train Derailment (ICC) 109
Roseburg, OR August 8, 1959 Explosives Truck Explosion......................110
    Firemen and Police Officer Who Made The Supreme Sacrifice.......... 112
Kansas City, KS August 18, 1959 Southwest Boulevard Fire 113
    Dedication of Memorial...117
    Firemen Who Made The Supreme Sacrifice 118
        Pumper 19...118
        Pumper 25...118

x Contents

Auburn, NY March 31, 1960 Service Station Explosion............................118
    Firemen Who Made the Supreme Sacrifice 119
Kingsport, TN October 5, 1960 Eastman Chemical Plant Explosion.......119
Philadelphia, PA April 17, 1961 Gasoline Station Explosion 120
    Firemen Who Made the Supreme Sacrifice 120
Houston, TX February 3, 1962 Gasoline Trucks Collide.........................121
Brandtsville, PA April 29, 1963 LPG Cars Explode At Wreck Site...........121
Cleveland, OH August 13, 1963 Avis Rent A Truck Propane Truck
Explosion 123
    Firemen That Made the Supreme Sacrifice...124
Hammond, IN December 23, 1963 Indiana Storage Company
Warehouse Explosion 124
    Fireman Who Made the Supreme Sacrifice...125
Anchorage, AK March 27, 1964 Great Alaskan Good Friday
Earthquake...125
Marshalls Creek, PA June 26, 1964 Explosives Fire and Explosion.........128
    50th Anniversary Memorial..130
    Firefighters That Made The Supreme Sacrifice..131
Houston, TX August 10, 1965 Chemical Plant Explosion 131
Chicago, IL Feb 7, 1968 Gasoline Tanker Explosion & Fire 132
    Firemen Who Made the Supreme Sacrifice 134
Richmond, IN April 6, 1968 Natural Gas and Explosives Fire and
Explosion 134
    Richmond Indiana's "Most Tragic Day and Finest Hour" 134
    The Explosion..134
    Finding Jack Bales 140
Davenport, NE June 10, 1968 Propane Tank Fire.......................................143
Crete, NE February 19, 1969 Derailment and Anhydrous
Ammonia Release 145
    Shelter in Place Effectiveness Substantiated 145
    Casualties 147
    Hatchetts 725 W. 13th Street 147
    Erdmans Between 1005 and 1045 W. 13th Street 148
    Hoesche 1005 W. 13th Street...148
    Safranek 905 W. 13th Street...148
    Svarc 813 W. 13th Street..148
    Kovar 907 W. 13th Street..149
    Svarc 915 Redwood Street 149
    Crete Fire Department Responds 149
Carpentersville, IL December 10, 1969 Chemical Plant
Explosion & Fire...152
Crescent City, IL June 21, 1970 Derailment and LPG Rail Car BLEVE... 154
Woodbine, GA February 3, 1971 Thiokol Factory Solid Rocket Fuel
Explosion 157

Contents

Waco, GA June 1971 Car-Truck Collision Sparks Dynamite Explosion.....158
    Firefighter Who Made The Supreme Sacrifice 159
Houston, TX October 19, 1971 Mykawa Road Rail Car BLEVE & Fire ... 159
    Firefighter Who Made The Supreme Sacrifice 161
Kingman, AZ July 5, 1973 Propane Rail Car BLEVE 161
    Doxol Gas Distribution Plant..162
    Kingman, Arizona Remembers 35 Years Later...................................... 166
    Remembering Those Who Lost Their Lives...167
    Eleven Kingman Firefighters Who Made the Supreme Sacrifice 168
        Donald G. Webb... 168
        Arthur A. Stringer 169
        Frank (Butch) Henry 169
        Christopher G. Sanders 170
        Alan H. Hansen.. 170
        John O. Campbell.. 170
        Joseph M. Chambers 170
        M.B. (Jimmy) Cox.. 171
        William L. Casson 171
        Roger A. Hubka 171
        Richard Lee Williams 171
Houston, TX September 21, 1974 Englewood Rail Yard Collision
Fire and Explosion 172
Philadelphia, PA August 17, 1975 Gulf Oil Refinery Fire 173
    Firefighters Who Gave the Supreme Sacrifice...................................... 177
    Philadelphia Pays Tribute to Gulf Oil Refinery Fire Fallen 177
Niagara Falls, NY December 15, 1975 Hooker Chemical Chlorine
Tank Explosion 179
Houston, TX May 11, 1976 Anhydrous Ammonia Tanker Accident....... 180
    Worst Accident in Houston History 180
Westwego, LA December 22, 1977 Grain Elevator Explosion.................. 181
Galveston, TX December 29, 1977 Grain Elevator Explosion.................... 183
Waverly, TN February 24, 1978 Derailment and LPG Explosion............. 185
    Those Emergency Responders Who Made the Supreme Sacrifice..... 190
Youngstown, FL February 27, 1978 Train Derailment with Chlorine
Gas Cloud... 190
Brownson, NE April 2, 1978 Train Derailment Phosphorus
Fire & Explosion 191
Gettysburg, PA March 22, 1979 Phosphorous Truck Fire & Explosion.... 193
Collinsville, IL August 7, 1978 Propane Tank Car Explosion 195
Mississauga, Ontario, Canada November 10, 1979 Derailment
Fires Explosions... 195
    The "Mother of All Hazmat Evacuation" 195
    Saturday November 10, 1979.. 197
    Sunday, November 11, 1979.. 198

Monday, November 12, 1979 201
Tuesday, November 13, 1979 202
Wednesday, November 14, 1979 202
Thursday, November 15, 1979 203
Friday, November 16, 1979 203
Somerville, MA April 3, 1980 Phosphorus Trichloride Release 204
Corpus Christi, TX April 7, 1981 Grain Elevator Blast 207
    Meet Kevin Saunders 210
Bellwood, NE April 7, 1981 Grain Elevator Explosion 213
Thermal, CA January 7, 1982 Derailment Radiation Container 214
Livingston, LA September 28, 1982 Train Derailment, Fire and Vinyl
Chloride Release 215
Denver, CO April 4, 1983 Railcar Leak Nitric Acid 221
    Executive Summary 221
    Probable Cause 222
Houston, TX December 11, 1983 Borden's Anhydrous Ammonia
Explosion 223
    By Bob Parry 223
Buffalo, NY December 27, 1983 Propane Gas Explosion 226
    Firefighters That Made the Supreme Sacrifice 227
        Firefighters Truck 5 227
Shreveport, LA September 17, 1984 Anhydrous Ammonia Fire
Incident 228
    Firefighter That Made The Supreme Sacrifice 229
Norfolk, VA September 4, 1984 Gas Tank Truck Fire 230
Bhopal, India December 2-3, 1984 Release of Methyl Isocyanate 231
    Background 231
    The Incident 232
    Lessons Learned 233
    Since 1984 234
Miamisburg, OH July 8, 1986 Derailment Phosphorus Fire 236
Ord, NE July 15, 1987 Anhydrous Ammonia Leak 239
Henderson, NV May 4, 1988 PEPCON Explosions 240
Kansas City, MO November 29, 1988 Ammonium
Nitrate Explosion 242
    Chronology of the Explosion 243
    Firefighters That Made The Supreme Sacrifice 244
    Lessons Learned 244
Houston, TX October 23, 1989 Phillips 66 Houston Chemical
Complex Explosion & Fire 245
    Initial Response 246
    Firefighting 247
    Search and Rescue 247
Philadelphia, PA February 23, 1991 One Meridian Plaza
High-Rise Fire 248

Contents

xiii

Firefighters Who Made The Ultimate Supreme Sacrifice 249
New York City, NY February 26, 1993 World Trade Center Bombing 249
Ste. Elisabeth de Warwick, Quebec, Canada June 27, 1993
Propane Explosion 251
Firefighters That Made The Supreme Sacrifice 251
Orrtanna, PA December 6, 1993 Ammonia Leak Knouse Foods 251
Plant Workers/Firefighters Who Lost Their Lives in the Incident 253
Port Neal, IA December 13, 1994 Ammonium Nitrate Explosion 253
Oklahoma City, OK April 19, 1995 Terrorist Bombing 255
Response to the Bombing 256
Oklahoma City Bombing National Memorial 257
Investigation 258
Oklahoma City Bombing 259
Delaware County, PA August 6, 1995 Ammonia Leak 259
Weyauwega, Wisconsin March 4, 1996 Train Derailment and Fire 261
Burnside, Illinois October 2, 1997 LPG Tank Explosion 268
Firefighters That Made the Supreme Sacrifice 270
Albert City, Iowa April 9, 1998 Propane Explosion 270
Key Findings 271
Firefighters That Made The Supreme Sacrifice 273
Springer, OK September 1998 Liquid Nitrogen Asphyxiation 273
2000s 274
Baltimore, MD July 18, 2001 Howard Street Tunnel Fire 274
Minot, ND January 18, 2002 Train Derailment & Anhydrous
Ammonia Release 277
Macdona, TX June 28, 2004 Train Derailment Chlorine Leak 283
Graniteville, SC March 24, 2005 Chlorine Disaster 286
A Tale of Two Hurricanes 291
New Orleans, LA August 29, 2005 Hurricane Katrina 291
Houston, TX August 24, 2017 Hurricane Harvey 297
Crosby, TX August 29, 2017 Chemical Plant Explosion &
Fire During Hurricane Harvey 300
St. Louis, MO June 24, 2005 Flammable Gas Cylinder
Fire & Explosion 303
Shepherdsville, KY January 16, 2007 Crude Oil Train Derailment &
Fire 303
Executive Summary 303
Probable Cause 304
Ghent, WV January 30, 2007 Propane Explosion and Fire 305
Emergency Responders Who Made the Supreme Sacrifice 307
Oneida, NY Monday, March 12, 2007 Tank Car Explodes 307
Dallas, TX July 25, 2007 Bottled Gas Plant Fire and Explosions 309
Jacksonville, FL December 19, 2007 Chemical Plant Explosion 310
Cherry Hill, IL June 19, 2009 Ethanol Train Derailment and Fire 311
Probable Cause 312

Norfolk, NE December 10, 2009 Propane Storage Tank Fire Protient
Propane Fire...316
Tiskilwa, IL October 7, 2011 Train Derailment Ethanol Fire 317
West, Texas April 17, 2013 Ammonium Nitrate Explosion.......................318
Journal of Hazardous Materials.. 320
   Explosions of Ammonium Nitrate Fertilizer in Storage or
   Transportation 320
  West Memorial... 320
  Emergency Responders that Made The Supreme Sacrifice................. 321
  Citizen Fatalities Not Firefighters.. 322
  Chemical Safety Board (CSB) Investigation 322
   West Volunteer Fire Department (WVFD)...................................... 322
  Abbott, Bruceville-Eddy, Mertens, and Navarro Mills Volunteer
  Fire Departments.. 323
  Ammonium Nitrate Hazards 324
  Key Contributing Factors to Emergency Responders' Fatality 324
  Lack of Incident Command System... 325
  Lack of Established Incident Management System 326
  Firefighter Training.. 326
  Lessons Learned.. 328
  Key Findings... 328
Lac-Megantic, Quebec, CA July 6, 2013 Train Derailment Crude
Oil Fire 329
Casselton, ND December 30, 2013 Crude Oil Train Derailment & Fire ...331
Lynchburg, VA April 30, 2014 Crude Oil Train Derailment & Fire 332
Heimdal, ND May 6, 2015 Crude Oil Train Derailment & Fire................ 333
Houston, TX July 16, 2017 Propane Tanker Crash 334
  RIMS Incident 334
Cambria, WI May 31, 2017 Didion Mill Explosion 335
Hyndman, PA August 2, 2017 Train Derailment & Fire........................... 337
Oklahoma City, OK August 16, 2018 Oil Well Blow Out......................... 339
Tilford, SD September 8, 2018 Propane Tank Explosion........................... 341
Deer Park, TX March 17, 2019 Petrochemical Storage Tank Fires 342
September 16, 2019 Farmington, ME Propane Explosion 342
  Firefighter that Made The Supreme Sacrifice.................................... 343
  Firefighters Injured 344
CSB Releases Call to Action on Combustible Dust Hazards................... 344
To Date, the CSB Has Issued Four Recommendations 346
Perceptions about Dust Vary... 347
Factors Influencing Dust Hazard Perceptions.. 348
Dust Incidents 2006–2017 349
  U.S. Chemical Safety Board (CSB).. 349

Bibliography.. 355
Index 365

Preface

When we are being called to the scene of a hazardous materials incident, it is because the situation is out of control. Our job is to stabilize the incident with a systematic approach based upon science and risk management. The first and most important task for emergency responders is to be able to determine if hazardous materials are present. If we cannot do that, nothing else matters. Once it is determined that a hazardous material is present, responders must consider all of the actors that are present. Actors include the hazardous material, container, physical state, nonhazardous materials, weather, and environment.

Following determination of the hazards, it is necessary to confirm who and what is at risk or in harm's way. What is at risk if we do nothing? Can we make the situation better than it would have been if we hadn't been there at all? What we decide to do should be based upon the risk to the public, responders, property, and environment, and not just because we can. Following the determination of hazard and risk and the decision to do something, we need to determine how to protect ourselves and the public.

Protection includes moving or sheltering the public, moving responders, utilizing personnel protective equipment, or letting the incident take its course. Protection requires proper numbers of personnel and the proper equipment to carry it out. Safety of the public and response personnel is the number one priority.

Once the safety and protection issues are addressed and the decision is made to address the hazards of the incident, tactical options need to be implemented. Tactical options are based upon the overall incident goals and stabilization of the incident scene. Clean-up is not a responsibility of emergency responders. There are clean up companies either procured by the spiller or on a list at the state Department of Environment to conduct clean up operations.

Options for stabilization can vary widely depending on the scope and size of the incident and hazards involved. Science and risk analysis play an important role in the successful outcome of the stabilization portion of the response. Following incident stabilization, the response portion of the

xv

xvi   *Preface*

incident operations can be scaled back. During this period, restocking and documentation occur. Critiques are generally conducted formally some time after the conclusion of the incident. Some documentation needs to be done while all is fresh in the minds of response personnel to be utilized at the formal critique. This portion of the incident, while not as exciting as the tactical phase, is very important to do thoroughly.

Acknowledgements

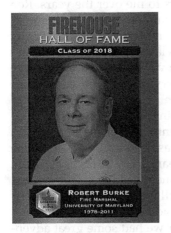

Thanks to the many fire departments and members across the United States and Canada that I have visited and become friends with during my visits to their departments over the years. Thanks to the firefighters from classes I have attended as a student, and then later taught, at the National Fire Academy, Maryland Fire and Rescue Institute and Community College of Baltimore County since 1988. Learning is a two-way street and I have learned much from the students as well. Thanks to the many friends I have met during the 40 plus years in the fire, EMS, hazardous materials, and emergency management fields. Many of them I have not seen for a while, some are no longer with us, but once a friend, always a friend.

Thanks to *Firehouse Magazine* for allowing me to write stories about hazmat for 32 years and counting. During those years, I have had the pleasure of writing under every editor of the magazine, including magazine founder Dennis Smith who gave me the chance to be published for the first time. I would also like to thank *Firehouse* editors, Janet Kimmerly, Barbara Dunleavy, Jeff Barrington, Harvey Eisner, Tim Sendelbach, and Peter Mathews for their support over the years. When I read my first copy of *Firehouse Magazine* in the late 1970s I was hooked. My dream was to someday go to Baltimore to attend a Firehouse Expo. Never did I dream I would not only attend an expo but teach at numerous expos, write for the magazine, and in 2018 be inducted into the Firehouse Hall of Fame Figure of Hall of Fame. To be placed in a fraternity with 16 of the people who had an enormous impact on the fire service in the United States and who I looked up to my entire career was very humbling.

Several people have been my mentors and have impacted my life and career. When I worked with the State Fire Marshall of Nebraska, Wally Barnett allowed me to accomplish things in the State Fire Marshal's Office

Brent Boydston, Chief Bentonville, AR Fire Department.

that I otherwise would not have. Because of his ability to let his employees reach their potential, I was able to write for *Firehouse Magazine* and become a contributing editor, teach for the National Fire Academy, and other things too numerous to mention. He was proud when I gave him a copy of my first book. I owe much of my success in the fire services to the opportunities Wally gave me. Jan Kuczma and Chris Waters at the National Fire Academy have been mentors to me over the years. Ron Gore, retired Captain from the Jacksonville, Florida Fire Department and Owner of Safety Systems, has had a large impact on my life and career. The Jacksonville Hazmat Team was the first emergency services hazmat team in the United States. Ron Gore is the Godfather of Hazmat response in the United States.

A former student of mine and the current Chief of the Bentonville, Arkansas Fire Department, Brent Boydston has been a great friend to me and has become a part of my family over the years. Rudy Rinas, Gene Ryan, and John Eversole of the Chicago Fire Department have been fellow classmates and students. Mike Roeshman and Bill Doty of the Philadelphia Fire Department are both former students and retired as hazmat Chief Officers. I used to ride with Bill and together we had some great adventures. Mike showed me Philadelphia's historical areas, like the spot where Ben Franklin flew his kite and his post office, which is so obscure today in downtown Philadelphia. I also stood on the spot where Rocky stood at the top of the steps in the movie. I had numerous adventures in Philadelphia which would not have happened without Bill and Mike.

Mike Roeshman Retired Hazmat Chief Philadelphia Fire Department.

Just outside of Philadelphia in Delaware County, Tom Micozzie, Hazmat Coordinator for Delaware County, was a former student and a great friend. We had many adventures together and I will never forget his introduction to me of the Galati at Rita's Italian Ice! Rita's was started by a retired Philadelphia firefighter and not long ago one opened up in Lincoln, Nebraska. Thanks to Richmond Indiana Chief Jerry Purcell, who I met during a visit to Richmond to do a Firehouse story on the 1968 explosion. As a result of the Richmond story being published, I was able

Acknowledgements

William, "Bill" Doty retired Hazmat Chief Philadelphia Fire Department.

to locate and become friends with survivor of the blast, Jack Bales. More recently visited to do another story on their hazmat team and propane training. Thanks to new friend Ron Huffman who traveled to Richmond to conduct the propane training utilizing water injection. This article appeared in the September 2019 issue of *Firehouse Magazine*.

Needless to say, I am grateful. Thanks to Tod Allen, Fire Chief in Crete Nebraska, who I met when I was researching for another friend Kent Anderson. Kent and I have worked on many anhydrous ammonia projects together and he has given me much information and introduced me to many experts in the field. Tod and I have become good friends. Tod is the apparatus operator on Truck One at Station One for the Lincoln Nebraska Fire Department. Tod invited me to come and ride with him, and many adventures later, I still go there a couple of times a month. Thanks also to all of my friends, both past and present, on "B" Shift at Station One for making me feel at home and showing me a good time whenever I am there. Thanks to my friend Captain Mark Majors for sharing his experiences with Nebraska Task Force 1 Urban Search and Rescue Team (USAR). I also thank Captain Francisco Martinez, Lincoln Hazmat. Thanks to Chief Michael Despain and Assistant Chief Patrick Borer for their friendship and hospitality while visiting the Lincoln Fire Department. This is only a short list, I would have to write a separate book to thank all of you I have met and for the impact you have had on my life over the past 40+ years. You know who you are, I appreciate your friendship and assistance, and consider yourselves thanked again.

Chief Jerry Purcell Richmond, IN Fire Department.

During my year long book writing adventure that led to *Hazmatology: The Science of Hazardous Materials*, I met and spoke to many people and made new friends. Thanks to my cousin Dustin Schroeder, Senior Captain at Station 68 and the FF, and others I met. Thanks to Kevin Okonski Hazmat at Houston Station 22. Thanks to Ludwig Benner, Bill Hand (Houston), Richard Arwood, Charles Smith (Memphis), Kevin Saunders (Motivator), Chief Jeff Miller (Butte, MT), and all of the Nebraska regional team leaders and members.

xx   *Acknowledgements*

I also want to thank my cousin Jeanene and her husband Randy for coming all the way from Montana to be with me at the Firehouse Hall of Fame induction. Thanks also to James Rey Milwaukee, Wilbur Hueser, Saskatoon in Canada for their hospitality and tour, and Captain Oscar Robles, Imperial, California. The list just goes on and on, there is not enough room here for everyone, but the rest of you know who you are and I want you to know how much I appreciate your assistance. You are all friends and I hope we will talk and meet again. Finally, thanks to librarians and historians across the country for all of their research assistance and willingness to help. To all of you, thanks for the memories!

Robert Burke

Author

Robert A. Burke, born in Beatrice and grew up in Lincoln, Nebraska, graduated high school in Dundee, Illinois, earned an A.A. in Fire Protection Technology from Catonsville Community College Baltimore County, Maryland (now Community College of Baltimore County), and a B.S. in Fire Administration from the University of Maryland. He has also completed graduate work at the University of Baltimore in Public Administration. Mr. Burke has attended numerous classes at the National Fire Academy in Emmitsburg, Maryland, and additional classes on firefighting, hazardous materials, and Weapons of mass destruction at Oklahoma State University; Maryland Fire and Rescue Institute; Texas A&M University, College Station, Texas; the Center for Domestic Preparedness in Anniston, Alabama; among others.

Mr. Burke has over 40 years of experience in emergency services as a career and volunteer firefighter, and he has served as a Lieutenant for the Anne Arundel County, Maryland Fire Department; an assistant fire chief for the Verdigris Fire Protection District in Claremore, Oklahoma; Deputy State Fire Marshal in the State of Nebraska; a private fire protection and hazardous materials consultant; an exercise and training officer for the Chemical Stockpile Emergency Preparedness Program (CSEPP) for the Maryland Emergency Management Agency; and retired as the Fire Marshal for the University of Maryland. He has served on several volunteer fire companies, including West Dundee, Illinois; Carpentersville, Illinois; Sierra Volunteer Fire Department, Chaves County, New Mexico; Ord, Nebraska; and Earleigh Heights Volunteer Fire Company in Severna Park, Maryland, which is a part of the Anne Arundel County, Maryland Fire Department.

Mr. Burke has been a Certified Hazardous Materials Specialist (CFPS) by the National Fire Protection Association (NFPA) and has been certified by the National Board on Fire Service Professional Qualifications as a Fire Instructor III; Fire Inspector; Hazardous Materials Incident.

Commander; Fire Inspector III; and Plans Examiner II. He served on the NFPA technical committee for NFPA 45 Laboratories Using Chemicals for 10 years. He has been qualified as an expert witness for arson trials as well.

Mr. Burke retired as an adjunct instructor at the National Fire Academy in Emmitsburg, Maryland in April 2018 after 30 years. He taught hazardous materials, weapons of mass destruction, and fire protection curriculums. He taught at his Alma Mater, Community College of Baltimore County, Catonsville Campus, and Howard County Community College in Maryland. He has published articles in various fire service trade magazines for the past 34 years. Mr. Burke is currently a contributing editor for *Firehouse Magazine*, with a bimonthly column titled "Hazmat Studies." soon to be remaned "Hazmatology:" He has also published numerous articles in *Firehouse, Fire Chief, Fire Engineering*, and *Nebraska Smoke Eater* magazines. He was inducted into the Firehouse Hall of Fame in October 2018 in Nashville, Tennessee. Mr. Burke has also been recognized as a subject matter specialist for hazardous materials and been interviewed by newspapers, radio, and television about incidents that have occurred in local communities, including Fox Television in New York City live during a tank farm fire on Staten Island.

Mr. Burke has been a presenter at Firehouse Expo in Baltimore, Maryland and Nashville, Tennessee numerous times, most recently in 2017. He presented at the EPA Region III SERC/LEPC Conference in Norfolk, Virginia in November 1994. He also presented at the 1996 Environmental and Industrial Fire Safety Seminar, Baltimore, Maryland on DOT ERG. He was a speaker at the 1996 International Hazardous Materials Spills Conference June 26, 1996, New Orleans, Louisiana, a speaker for the 5th Annual 1996 Environmental and Industrial Fire Safety Seminar, Baltimore, Maryland, sponsored by Baltimore City Fire Department and LEPC. He served as the instructor for Hazmat Chemistry August 1999 at Hazmat Expo 2000 in Las Vegas, Nevada. He also delivered a Keynote presentation at the Western Canadian Hazardous Materials Symposium Saskatoon, Saskatchewan, Canada in 2008.

Mr. Burke has developed several CD-ROM-based training programs, including the Emergency Response Guide Book; Hazardous Materials and Terrorism Awareness for Dispatchers and 911 Operators; Hazardous Materials and Terrorism Awareness for Law Enforcement; Chemistry of Hazardous Materials Course; Chemistry of Hazardous Materials Refresher; Understanding Ethanol; Understanding Liquefied Petroleum Gases; Understanding Cryogenic Liquids; Understanding Chlorine; and

Author xxiii

Understanding Anhydrous Ammonia. He has also developed the "Burke Placard Hazard Chart." He has published seven additional books titled *Hazardous Materials Chemistry for Emergency Responders*, 1st, 2nd, and 3rd Editions, *Counterterrorism for Emergency Responders*, 1st, 2nd, and 3rd editions, *Fire Protection: Systems and Response and Hazmat Teams Across America*.

Currently, Mr. Burke serves on the Homestead LEPC in Nebraska. He also manages a hazardous materials section at the Nebraska Firefighters Museum and periodically rides with friends on "B" shift Station 1, Lincoln Fire Department. He can be reached on the Internet at robert.burke@windstream.net. *Facebook* https://www.facebook.com/RobertAb8731 or via his website www.hazardousmaterialspage.com.

Volume One

Chronicles of Incidents and Response

> History is not just the chronicling of both major and minor events of lives or organizations. It's the basis and the lifeblood for tradition to be carried on by each succeeding generation.
>
> *From History of Houston Texas Fire Department*

Origins of Hazardous Materials Response

The wonders of this world we live in have been present since the beginning, that is, millions and millions of years ago, long before man arrived on the scene. Components of things that are a part of our everyday life today have been present from the beginning. However, man had to evolve to a point where technology enabled the lifestyle we enjoy today to become a reality. In the beginning, man focused on survival, that is, the basics, food, water, shelter, and clothing. All the things man needed for survival were present here on Earth. He just had to discover the means to procure them. It was just a matter of discovering them, many times through trial and error.

Maslow and his Hierarchy of Needs were thousands of years from entering the picture. Yet, in reality, the needs that Maslow identified were already in play. In addition to the aforementioned basics, man needed safety and security. Once all the basic needs were met, some men, often out of curiosity, started exploring the wonders of this planet, and slowly but surely discovered many of its wonders that ultimately made life on Earth more substantial. Motivation increases as needs are met. Since the discovery of fire, man has periodically unlocked some of the life-improving secrets and wonders hidden on our planet.

This volume is not a history of the world; many have already been there and done that. My motivation in bringing any of this up again is to illustrate how during man's journey through time he discovered hazardous materials that aided the progress but also caused some serious issues along the way. Unfortunately, the journey to modernization results in monetary gains, which clouds the view of the dangers posed by the hazardous materials needed to accomplish this modernization.

Scientists have given their lives during the discovery of hazardous materials on our planet, not out of thoughts of monetary gain but because of the ignorance of the dangers. Researchers lost their lives before they realized how to safely work with radiation by implementing protective measures. Today, scientists know how to conduct research safely with proper protective equipment and safety procedures when dealing with chemical and biological materials. Emergency responders also have that same opportunity to respond to hazardous materials incidents and wear proper protective equipment and follow safety procedures so that everyone goes home safely. Looking through all the incidents that have occurred since 1841, we have the opportunity to witness what our predecessors experienced to get us to the point we are at today.

Many times throughout history, we have tried to close the barn door on the dangers of hazardous materials after the horse had already been let loose. This happened in the 1960s and 1970s with the transportation of liquefied petroleum gases and again in the 2000s with ethanol and crude oil. It wasn't until civilians and emergency responders were killed that efforts were made to mitigate the issues with under-designed tank cars.

Knowledge of chemicals and their usefulness began with advanced civilizations such as the Romans and Egyptians. The discovery of the usefulness and economic value of chemicals is the only reason that hazardous materials we encounter today exist. Most of these materials do not exist naturally. They are made from known elements that exist on earth, and by combining them, useful substances are created. Of course, that brings up a whole other issue that is beyond the scope of the present discussion; hazardous waste. Because these compounds do not exist naturally, they do not always decompose and disappear. Again, the focus has often been on economics and not the environmental impacts of these waste materials. I am trying to make the point that through the study of hazardous materials and incidents that have occurred in the past, we can learn how to safely and effectively deal with hazardous materials in the present when responders are called upon to deal with them.

Responses to hazardous materials involve teams that are usually part of a fire department. My intent is not to portray the history of the fire department but to point out that there is a culture in the fire service that has been difficult, if not impossible, to change over the years in certain situations. Somewhere in my career, I heard someone say that "The fire service is 200 years of tradition totally unaffected by progress," and I have never forgotten this. We do have a rich legacy of traditions in the fire service and we should be proud of those.

However, these same traditions have led to unintended loss of life of firefighters at incident scenes or through illness that lingered for years

Chronicles of Incidents and Response

before death occurred. We are taught during training that dealing with fires tactics are pretty much the same for all fires. There are similarities in how fire behaves in a given circumstance and the actions needed to put it out. We are also taught, or the tradition has developed over years, that we need to act quickly on the scene of a fire to save public lives. Tradition also tells us that we often act quickly without regard for our own safety and are willing to give our lives to save others. Firefighters do not consider this as being heroic, but rather just a part of their job.

Organized hazardous materials response teams in the fire service started in the late 1970s with the Jacksonville, Florida, hazardous materials team being the first. Hazmat response in the fire service is still in its infancy compared to firefighting. There really hasn't been enough time or enough consistence in mission for traditions to form. Hazardous materials response is an ever-changing field compared to firefighting. Hazardous materials have the potential to kill and injure many more people, including responders. The fire service has evolved into many other fields beyond firefighting, including hazmat, Emergency Medical Service (EMS), and various types of rescue operations. Firefighting tactics have not undergone major changes over the years, and the incidences of structural fires are decreasing for many reasons. EMS and rescue have become a large part of the fire service today.

Compared to the other responsibilities of the fire service department, hazardous materials response is not a large part of responders' responsibilities. In terms of numbers, there are more rescue responses than fire responses, which has led to a trend in the fire service toward special operations teams that include hazmat and other technical rescue concentrations. It is too early to determine if this is a real trend or a fad. Only time will tell.

This volume deals with hazmat incident response and history. Looking at historical hazmat incidents provides us with lessons learned that can help us become safer now and in the future. Hazardous materials incidents were occurring for over 150 years before we even knew what hazardous materials were. They were encountered as fires and explosions and were addressed in much the same way. Fortunately, many of the explosions, which led to fires, were over before firefighters arrived at the scenes. What they faced was the aftermath, fire, massive deaths, and injuries among the general public.

The types of hazardous materials encountered have a direct connection to the technology of the times. Changes in types of hazardous materials have followed advancements in technology. The presence of hazardous materials and incidents associated with have also been driven by politics and wars. The safety of hazardous materials has also increased because of evolving safety concerns and regulations regarding their use, storage, and transportation. This increase in safety and regulation has reduced

4   *Hazmatology: The Science of Hazardous Materials*

the number of firefighters and hazardous materials personnel killed and injured during incidents. Now it is up to us to be less aggressive with hazardous materials and more scientific to determine when we should and when we should not intervene during an incident. If you are going to err, do so on the side of safety.

During my research on historical incidents involving a material that we would call hazardous today, the first incident I could find that happened and was recorded occurred in 1841. This incident involved gunpowder, an explosive material in common use at the time. During the 1800s, most of the hazardous materials incidents involved explosive materials or flammable liquids. Certain aspects of human survival, such as hunting for food, required guns, which, in turn, required gunpowder.

Then there was the Civil War, which required more gunpowder and other explosives for the war effort in addition to hunting for food. Gunpowder and munitions plants were located throughout the eastern part of the United States. Dynamite was also commonplace, and was sold to and used freely by anyone who wanted to buy it and for any use they wanted, such as stump blasting, rock blasting, and others. As the United States expanded west farmers creating new farm land required more explosives. Explosives were the technology of the time and were common place. Accidents happened as explosives were manufactured, transported and stored. They could be found almost anywhere. In the presence of zoning requirements that restricted where explosives could be manufactured, stored, and used, there were fewer accidents. Regulations and laws for safe handling and use resulted in less casualties for both the public and emergency responders.

Beginnings of Hazardous Materials Transportation Regulations

In 1866, the first law was passed regulating the transportation of what we know today as hazardous materials, specifically shipments of explosives and flammable materials such as nitroglycerin and glonoine oil. An 1871 statute established criminal sanctions against persons transporting specific hazardous commodities on passenger vessels in U.S. navigable waters in violation of Treasury Department Regulations. Rail shipments of explosives during and after the Civil War were addressed by unmodified statutes and contractual obligations between both shippers and carriers based on English common-law principles. Under the common law, common carriers were granted a public charter to operate and were obliged to provide service to anyone upon reasonable request, for reasonable cost, and without unjust discrimination. Carriers could, however, prescribe conditions under which certain freight would be accepted. A shipper was obligated to

Chronicles of Incidents and Response

identify the hazards of a dangerous commodity, use adequate packaging, and provide a clear warning to the carrier of the shipment's hazards.

The establishment of the Interstate Commerce Commission (ICC) in 1887 marked the beginning of a Federal effort to impose a degree of regulatory uniformity on all modes of transportation. ICC was the first independent regulatory agency of the U.S. government. Although ICC requirements were first developed for rail transportation, they were eventually extended to other modes of transportation. ICC was the primary regulatory agency with authority over transportation of hazardous materials until 1966. Hazardous Materials Transportation Regulations, Chapter 4, Princeton University.

Dust and Boiler Explosions

During the 1800s, dust and boiler explosions were quite common. Boiler explosions will not be covered in this volume because they do not involve a hazardous material. Combustible dusts are not a regulated hazardous material because they are ordinary materials, usually, but not always a by product of the use, storage or manufacturer of some other product. They are however a danger to employees and emergency responders and remain a problem to the present day. Dust explosions covered any commercial venture that involved the creation of dust or the use of particulate material such as starch, woodwork, grain storage, among others. One of the most tragic incidents occurred in Chicago in 1897, which resulted in the death of four firefighters and injuries to nine others. Numerous dust explosions have occurred before and since the incident in Chicago, and they continue to this day. In fact, the Chemical Safety Board (CSB), a U.S. government agency, is currently working on an initiative to deal with the continuing problem of dust explosions.

Petroleum Industry is Born

On August 27th, 1859, near Titusville, Pennsylvania, the U.S. petroleum industry was born, which forever changed the economy, standard of living, and culture. Before that time, flammable liquids were in use but on a much smaller scale. Following the discovery of oil, economies soared, production increased, technologies were developed, and response to fires involving these newly found flammable materials increased. Today over 50% of all hazardous materials incidents involve petroleum products. In 1892, a combination of heavy rain, flooding, and lightning, ironically near Titusville, Meadville and Oil City in Pennsylvania, resulted in a spill and fire involving benzene, a petroleum product made from crude oil, that killed over 50 people and injured hundreds. In addition to explosives causing damage, death, and fires fire, incidents involved large amounts of flammable liquids as well.

Hydrocarbon fuels began to replace coal and wood as heating and cooking fuels. Invention of the internal combustion engine increased the demand for petroleum products. Technology continues to advance to the present day. So far, even though I have talked about hazardous materials, they have not been named yet. In reality, in all the examples of fire incidents, fire departments did their best to extinguish fires and firefighters died, not knowing the dangers they were facing.

Reading through the incidents highlighted in this volume will show a pattern through the years that we have already touched upon. These are not all the incidents that have occurred to date. My focus has been on the worst of the worst. Many of these events resulted in firemen making the supreme sacrifice while doing the best they knew how. Also included are incidents that caused massive injury and death to the general public. For me there is a definite pattern between changes in the types of hazardous materials and the impact on responders who faced them. Safety concerns increased first through responders, then unions, and finally government regulation. Unfortunately, business people who are causing these hazards do not always give much thought to the safety of workers and emergency responders. Deaths have to occur before regulators and business owners attempt to make safety corrections. In dealing with hazardous materials or dangerous situations at the workplace, reactive rather than proactive actions have been the way of life throughout the early years. Unions became interested in addressing hazardous workplaces in the early 1900s, but the government did not begin getting involved in workplace safety until the mid- to late 1900s.

Explosive incidents involving materials used in the mid- to early 1800s become rare after World War II (WWII) except for agricultural ammonium nitrate, and the dangers of explosives were addressed through education, safety procedures, and regulations. We know about the dangers of ammonium nitrate or at least we should if it is in our communities. Yet, it still kills emergency responders in the 2000s. Over the years, explosives incidents have reduced, but other forms of hazardous materials incidents have replaced them. My research of these incidents has also revealed an interesting statistic. Over the years, deadly hazardous materials incidents have not occurred with "exotic" materials but rather with common chemicals such as liquefied petroleum gases, chlorine, anhydrous ammonia, natural gas, ammonium nitrate, and hydrocarbon fuels. However, often training has focused on what I call "exotic" hazardous materials based on potential rather than reality. Responders I have talked with who have experienced major loss of life in their organizations and/or significant property damage during hazardous materials incidents have told me that

Chronicles of Incidents and Response

they were not trained for the incidents they experienced. They were over-whelmed by the magnitude of what they faced. It has also become clear to me that we have not always learned lessons from previous incidents where others gave their lives. There seems to be a trend of complacency toward common materials that we are more likely to face in a hazardous materials incident than "exotic materials." The incidents presented in this volume are intended to help responders learn about mistakes that have happened in the past to help us avoid repeating them in the future, which unfortunately has already happened too many times.

> *Author's Note: As you read through these historical incidents, you will notice that in the beginning, even though not politically correct today, the terms "firemen" and "fireman" were used up until the mid-1900s. Personally, I do not remember when the changeover occurred. When I started out as a volunteer fireman in the late 1960s that was still what today's firefighters were called. When I taught at the Louisiana State University in the 1990s, the facility was still called the Firemen's Training School. However, I have kept the terms "fireman" and "firemen" in place where they are historically correct.*

Nebraska Firefighters Museum Hazardous Materials Display

Shortly following my retirement and the subsequent move to my home state of Nebraska, I discovered the Nebraska Firefighters Museum. What I noticed about this museum and other similar ones that I have visited over the years is the primary focus on firefighting. However, the field of fire-fighting has evolved into many other areas, including EMS, rescue, and hazmat response. So, I approached the museum director and enquired about creating a hazardous materials history display. She went to the board of directors and they decided it would be a good idea. I offered to spearhead the effort and visited hazmat teams across Nebraska to enlist their help in providing old outdated equipment to include in the display. Some personal items of history, like the complete set of DOT Emergency Response Guide Books that I had collected over the years, were also included in the display. What I have completed is a start, and is a work in progress. Many other ideas come to mind, but space is limited for now (Figure 1.1). The Nebraska Firefighters Museum and Memorial is located right off Interstate 80 East of Kearney next to the Great Platte River Road Archway. If you are looking, you can't miss either one. The archway goes right over the top of Interstate 80 and the fire museum is just west of the archway.

Figure 1.1 What I have completed is a start, a work in progress. Many other ideas come to mind, but space is limited for now.

I am not aware of any other hazmat display in museums in the United States.

Nebraska Firefighters Museum & Education Center
Nebraska Firefighters & EMS Memorial
2834 E. 1st Street
Kearney, Nebraska 68847
Phone: (308) 338-3473
E-mail: mail@neffm.org
https://nebraskafirefightersmuseum.org

Hastings, Nebraska "Gasoline & Oil Clean-Up Squad"

One of the earliest efforts I have been able to uncover, while not called a hazmat team, could be considered a precursor or even the beginning of an organized response to hazardous materials in 1940. The "City Service Truck," a 1940 Dodge Panel Truck (Figure 1.1A), was put into service carrying sawdust and equipment to handle gasoline or oil spills. Hastings Fire & Rescue began their actual hazardous materials team over 40 years later in the 1980s. Their first hazardous materials unit was Squad 7, purchased in 1977 as a rescue truck. Squad 7 was replaced in 2002 by Rescue 1 Hastings Fire Department.

Chronicles of Incidents and Response

Figure 1.1A The "City Service Truck," a 1940 Dodge Panel Truck, was placed into service carrying sawdust and equipment to handle gasoline or oil spills. (Courtesy: Hastings Fire Department.)

Milwaukee, Wisconsin's Petroleum Dispersal Unit 1

Milwaukee Fire Department established a response unit similar to the one in Hastings, Nebraska. This was also an organized unit that responded to petroleum spills before the advent of organized hazmat response as we know it today. During 1966, the Milwaukee Fire Department Repair Shop converted a spare pumper into the Petroleum Dispersal Unit 1 (Figure 1.2).

Figure 1.2 Petroleum Dispersal Unit 1 carried a 300-ft oil slick bar and another 100 ft in quarters for use in the Port of Milwaukee and local rivers. (Courtesy: Milwaukee Fire Historical Society.)

This unit carried a 300-ft oil slick bar and another 100 ft in quarters for use in the Port of Milwaukee and local rivers.

An electric oil skimmer was basically a floating pump which would suck accumulated oil from the surface of the water and carry it to a waiting tank or tanker using a plastic hose. In addition, 24, 25 gallon containers of Jansolv dispersant, a 2½ in. educator, and a 2,500 watt generator were also carried. Additional equipment included an explosion-proof electric exhaust fan, a 100 gpm portable pump, and 6, 5 gallon containers of light water. Petroleum Unit 1 went into service on March 8, 1972 and was located in the quarters of Engine 12 and available on special calls. Unit 1 was dispatched to water areas where gas or oil accumulation was reported, as well as to large spills on land. This unit remained in continuous operation until 1989. The functions of the unit were taken over by the hazardous materials team (Milwaukee Fire Historical Society).

Evolution of Present Day Hazmat Teams

During the 1970s the idea of organized hazardous materials response by fire departments and other organizations began to develop. Several significant high-profile incidents involving railroad tank cars in Crescent City, Illinois, Waverly, Tennessee, and Kingman, Arizona resulted in changes in tank car safety, hazmat transportation markings, and regulation at the federal level. Organization of hazardous materials response teams soon followed with many established by fire departments. Starting a hazardous materials team in the beginning was difficult because of a lack of guidance available. There were no guidelines for training from National Fire Protection Association (NFPA) or Occupational Safety and Health Agency (OSHA) at the time and there was little information on standard operating procedures (SOP's) and proper equipment. For expertise, equipment, and training, there was a heavy reliance on industry and military.

Teams developed as brainchildren of pioneer leaders, including Jacksonville Chief Russell Yarbrough and Ron Gore; Houston Chief V.E. Rogers, William "Bill" T. Hand and Chief Max McRae; and Memphis Training Chief Adelman, Captain James Covington, and Lieutenant Charles Smith. There efforts led to the formation of teams in Jacksonville, Florida, Houston, Texas, Memphis, Tennessee, among other locations. These team formations preceded the industrial accident in 1984 in Bhopal, India, which was the high-water mark in obtaining regulatory attention of the United States Congress. Legislation followed by regulations from OSHA, the Environmental Protection Agency (EPA), and the Department of Transportation (DOT) revolutionized hazardous materials response.

Jacksonville, FL's First Emergency Services Hazmat Team in the United States

Chief Yarbrough's Vision

The Jacksonville Fire Department, Florida, is the home of the first hazardous materials response team created in the United States. During the late 1970s, Chief Yarborough of the Jacksonville Fire Department envisioned the need to deal with hazardous materials response in a trained and organized manner. He realized that Jacksonville was the "Great Shipping Center of the South." Railroads had major operations in the city and there was significant military activity in the area involving nuclear materials. No one had ever taken an organized approach to hazardous materials response before this, except for the petroleum response units in Hastings, NE and Milwaukee, WI. Being the first at anything is a difficult undertaking because there is little, if any, guidance available. Chief Yarborough and Captain Gore believed a specialized team was the best approach and they needed 15–20 volunteers from firefighting ranks to form the nucleus of the new hazardous materials response team.

This came directly from the man, because many of us credit Captain Gore as the "Godfather of Hazmat." "It wasn't my idea," said retired Captain Ron Gore during my visit to Fire Station 7. In town for a reunion of Jacksonville Fire and Rescue Department's (JFRD) original hazmat team members Gore shared how the specialty team's concept originated with Fire Chief Russell Yarbrough in the 1970s. Tanker car derailments, shipments of nuclear weapons to local military installations, dangerous and volatile cargo passing through the city's main thoroughfares, and Jacksonville's growing chemical industry attracted the chief's attention. If something went awry, Yarbrough wanted JFRD to be as prepared as possible.

> **Author's Note:** *While Ron Gore doesn't want to take credit, even though it was Chief Yarbrough's idea, it was Captain Ron Gore who made it happen. Not just in Jacksonville but all across the United States. Ron Gore is one of the most influential people in the world of hazmat and has touched thousands of people during his training sessions. Ron, you are the undisputed "Godfather of Hazmat" in the American Fire Service. I for one do not think you get enough credit for your contributions to all of us in hazmat response.*

The "Godfather of Hazmat"

Engine 18's Gore was about to become hazmat's number one son. As a new lieutenant in the early 1970s, Gore was assigned to the Training Academy for a year. This was common practice back then, but Gore's experience would be far from common after he met Training Chief Simon Joseph King, Jr. Gore was still getting used to pinning bugles on his collar when King assigned him the task of teaching hazardous materials as part of a fire science course at Florida Junior College (FJC), now Florida State College at Jacksonville. There was just one problem: "I didn't know the subject matter," said Gore. Neither did anyone else on the job. Initially, Gore and his students, many of them firefighters, spent time with chemistry teachers. However, lectures on molecular structures and chemical reactions didn't translate well to emergency response.

A frustrated Gore soon found relief in practicality. He convinced FJC's administration to allow him to train his students in the field. In Ron's words, "We trained at the rail yard, ship yard, the tank farm," adding that private industry was very cooperative in providing access and institutional knowledge. Gore's class became an on-site assessment of industrial hazards, chemicals, production methods, and containers. Through that familiarity, Gore and his students began to understand the risks and devised response strategies.

America's First Emergency Services Hazmat Team

Together, the members focused on making America's first hazmat team as capable as it could be, often at their own expense and while off duty. They looked to military surplus outlets for protective gear, they collaborated with private industry and JFRD's machine shop to design effective tools, and they traveled the country to understand the volatility of new chemical products. Back then, Jacksonville provided very limited funding to the team, according to Gore. However, that didn't stop them; they willingly bore the costs of travel, training, and even some specialized equipment. Their first hazmat units are shown in (Figure 1.3).

The initial request for volunteers resulted in more than 50 firefighters wishing to join the team. There was no additional pay for being on the hazmat team. For selecting team personnel, Captain Gore looked for firefighters who had some previous experience with chemicals. These were firefighters who had worked for gas companies, had some military experience, and had experience with various chemicals. As with anything new, there was a lot of distrust and resistance to the new hazardous materials team from officers and firefighters alike.

According to Phil Eddies, JFRD retiree and an original hazmat team member, "We had the attitude that when we were working together, there was no event we couldn't handle." Theirs was an extensive and unrelenting discovery process. According to Gore, there were no precise OSHA standards for hazardous materials response to guide them. Although they lacked technical knowledge and defined procedures, the team had confidence. "When we formed that team, it was the best group of people

Chronicles of Incidents and Response

Figure 1.3 Jacksonville's first hazmat units were Engine 9 and Engine 9A. (Courtesy: Jacksonville Fire Department.)

you could have hoped for," according to JFRD retiree Phil Eddins original hazmat team member. And handle they did; propane tank fires, derailed train cars leaking hydrogen chloride or muriatic acid, and petroleum tank farm fires. There were exposures, but no serious injuries. Each incident was a response as well as a lesson for the team (Figure 1.4).

There were skeptics of the team from early on, and some who poked fun by labeling the hazmat team members as the "Clorox team" or "bleach drinkers." Eddies said he eventually took the monikers as a compliment, and JFRD retiree Jim Croft also embraced the identity. On occasion, when he got transferred to another station, Croft would show up with his gear

Figure 1.4 Jacksonville's original hazardous materials team members. (Courtesy: Jacksonville Fire Department.)

in one hand and a genuine bottle of Clorox in the other; "I'd come inside the station and set it down on the table". As the team continued to prove itself, the nicknames subsided, and interest from departments across the country increased.

Personnel from the Jacksonville Fire Department pretty much had to make things up as they moved forward. There were no response procedures to follow so they had to be developed as the team gained experience. Attempts at tactics were developed as they responded to incidents. Sometimes they worked and sometimes they did not. They developed procedures and equipment following incidents where they saw the need for something that would have helped them during an incident.

As there wasn't much hazmat equipment available commercially, much of what they used was created in their maintenance shop by the Chief of Maintenance. They literally invented things as they went along. Incidents that occurred were discussed on all three shifts, which is how many of the early team members learned to deal with hazardous materials, basically through trial and error. It would be difficult to train every firefighter to deal with hazardous materials. When the team was organized, little was known about personnel protective clothing for chemicals outside of the military. Early on they would use plastic trash bags over their firefighter turnouts for chemical protection.

They would soon find out that suits obtained from the military were good for some chemicals but not for all. During a tank car leak incident at Union Camp in July 1978, a technician's suit was breached during operations on top of a hydrogen chloride tank car dome (Figure 1.5). Little was known in the beginning about suit compatibility and the suits being worn were not compatible with hydrogen chloride. Decontamination didn't

Figure 1.5 During a tank car leak incident at Union Camp in July 1978, a technician's suit was breached during operations on top of a hydrogen chloride tank car dome. (Courtesy: Jacksonville Fire Department.)

Chronicles of Incidents and Response

exist at the time, and there were no hot or warm zones, like we use today. Most personnel in protective clothing or not were right up in what would be the hot zone. Much was learned about personal protective equipment (PPE) and suit failure from this incident. This incident, despite its issues, helped establish the credibility of the new hazardous materials response team.

Many of the procedures and equipment mentioned above would seem highly unusual in today's world of hazmat response. However, keep in mind that it was the early efforts of the Jacksonville Fire Department's trial and error, flying-by-the-seat-of-their-pants operations that have evolved into today's organized and structured hazardous materials response. Hazmat response was new to city government as well. The city did not provide money for equipment or training. Much of the training and travel expense was paid by team members. Despite the uncertainty and lack of procedural guidance available at the time the team was organized, there was no loss of life or serious injury among team members.

Gore's travel included trips with Chief Yarbrough, who was the president of the International Association of Fire Chiefs (IAFC). This gave Gore numerous opportunities to present to large audiences who were hungry for information that only he and JFRD's team could offer. Gore has said that he doesn't recall any other fire department challenging Jacksonville's position as the first to forge a municipal hazardous materials response team. "It makes me proud," he said.

JFRD retiree Andy Graham joined the team in 1980 and remained a member until retirement in 2004, longer than anyone else. During that time, he interacted with several visiting fire departments. According to Graham, "We had people from Canada, the Virgin Islands, New York." He also recalled, "They wanted to ride with us. A lot of times, we didn't have anything noteworthy, but they would ask questions. We would just talk about hazmat. They knew that Jacksonville was the leader in this, and they wanted to pick our brains." Graham said he joined the hazmat team to try something different. "After a couple of weeks, I thought Why didn't I do this sooner? he recalled. They were so eager to learn. That's what impressed me."

Graham also recalled how JFRD members initially had their doubts about the team, but after a few years, the field took notice of their effectiveness and cautious work habits. According to Graham, "We'd show up decked out in air packs." "People began to understand if we had our stuff on, they better have theirs on." Although the team faced plenty of unknowns, Graham said that Gore trained him and everybody so well that he "can't ever remember fearing" for his life. Graham: "Capt. Gore loved the job, and he taught me how to love it." "He gave me the push to love what I did, and I love it. When I retired, it took me about a year to get over leaving."

My Visits to Jacksonville

Not only was Jacksonville the first hazardous materials team in the country they also know how to make a fellow firefighter and hazmat responder feel very welcome. My trip to Jacksonville was the first time I was met at the airport and escorted to my destination, many thanks to District Chief Randy Wyse and Lieutenant Todd Smith for all their assistance while I was in Jacksonville and during my first visit to Jacksonville. During the trip to Hazmat Station 7, we took a short detour to visit old Station 9 (now a museum) at 24th and Perry Streets, where the first hazmat team in the country was originally housed (Figure 1.6). Station 9 was chosen as the new hazmat station because of its strategic location near I-95 and the 20th Street expressway, which allowed them quick access to all parts of the city including industry to the east and rail yards to the west. The trip to Station 7 was followed by a barbecue attended by former hazmat team members, including Ron Gore, Bob Masculine, Neil McCormack, Richard Morphew, Phil Eddins, Jim Croft, and Davis Love, who together represent over 60 years of hazardous materials experience (Figure 1.7). Also attending was Jacksonville Fire Chief Richard Barrett, current hazmat team members, and other Jacksonville fire department personnel. Over the next several hours, I experienced some of the most enjoyable conversations about hazardous materials of my career.

Figure 1.6 Station 9 was chosen as the new hazmat station because of its strategic location near I-95 and the 20th Street expressway. (Courtesy: Jacksonville Fire Department.)

Chronicles of Incidents and Response

Figure 1.7 The thirtieth anniversary was attended by former hazmat team members, including Ron Gore, Bob Masculine, Neil McCormack, Richard Morphew, Phil Eddins, Jim Croft, and Davis Love, who together represent over 60 years of hazardous materials experience.

Captain Ron Gore Retires

Although Gore retired from JFRD in 1989, he continues to teach the subject through his hazardous materials company, Safety Systems, which he founded in the late 1970s. Gore offers training and emergency response services. His clients include the military, private industry, and other fire and rescue departments. Gore estimates that he and his team of instructors have trained "hundreds of thousands of people" nationwide (Figure 1.8).

Figure 1.8 Although Gore retired from JFRD in 1989, he continues to teach the subject through his hazardous materials company, Safety Systems, which he founded in the late 1970s. (Courtesy: Jacksonville Fire Department.)

Gore is now the hazmat subject matter expert that he was desperately seeking in the early 1970s, when he was tasked with arguably one of the most challenging assignments ever entrusted to a firefighter in Jacksonville or elsewhere. He may not accept the "Godfather of Hazmat" label, but just Google "Ron Gore" and you'll learn how his place in the hazardous materials circle is renowned. At 80, Gore is still eager to teach what he's learned over the last 40 years and what he's learned from the intrepid firefighters who formed America's first hazmat team. He has no plans to retire either. After being forced to leave his comfort zone that first year, he became a JFRD lieutenant, Gore believes his career path was forged by Chiefs Yarbrough and King and, most importantly, God. "I was created to do this," he said.

In addition to Ron Gore's efforts in Jacksonville, he traveled throughout the United States and conducted hazardous materials training classes through Safety Systems Incorporated, a business he founded. During my talks with members of hazmat teams across the country, many have referenced Jacksonville and particularly Ron Gore as having influenced the formation of their teams (*Firehouse Magazine*).

> **Author's Note:** *My own early interest in hazardous materials was fueled by a Safety Systems training course conducted by Ron Gore at the Fire Museum of Texas in Arlington in 1981. Following that class in Texas, my Chief Bob Cox at the Verdigris Fire Protection District in Claremore, Oklahoma and I arranged for Ron Gore to bring his training class to our department. During this class I had the opportunity to get to know Ron and realized that he was likely the most knowledgeable person in the field of hazardous materials response in his time. Ron's impact on the formations of countless other hazmat teams in the United States is incalculable. During my travels across the United States visiting fire departments and hazmat teams, I've heard countless stories of how Ron Gore influenced other hazmat teams. His legacy, Captain Ron Gore is "The Godfather of Hazmat" response in the United States.*

Martin County Florida: First Volunteer Hazmat Team in United States

Fire Department History

District-2 (Station 18) is the original Martin County Fire Department. During the 1970s there were 12 volunteer fire companies, and one city department that was career and part volunteer. The Grumman Company Fire Department was located at the airport. The county career fire department began in 1976 with the hiring of the first career firefighter assigned to Station 21 in Palm City. Additionally, there were four volunteer BLS ambulances and one career ambulance operated by the City of Stuart Police Department. Chief Wolfberg was hired in 1981 to organize and place in service a career community-wide ALS ambulance service. One year later, the ALS service was operational. During this transition the county took over ambulance response in the city of Stuart. One of the volunteer ambulances closed, and many members began a new career with the county EMS service.

Around 1981 and 1982 a career fire marshal was added to the department. Within the next 2 years, the job transitioned into Chief of Operations/ Fire Marshal for the entire county. His responsibilities included overseeing all volunteer fire departments in Martin County. In 1994, the Martin County Fire Department and Martin County EMS merged and became today's Martin County Fire Rescue.

Hazmat Team History

Chief Edward B. Smith became chief of the Martin County Fire Department in 1976. Trade magazines were routinely publishing stories of hazmat incidents occurring across the United States. Outside his office window, trains traveled through Martin County with tank cars carrying hazardous materials. Hazardous materials response was a new and growing addition to the fire service. Recognizing the potential in his community for chemical incidents, Chief Smith began preparing himself and his firefighters for hazmat emergencies conducting classes and learning as much as they could.

In 1976, young captain Ron Gore of Jacksonville, Florida was also tasked with starting the first emergency services hazmat team in the United States. He also formed his own company, Safety Systems, Inc., to train emergency responders not only in Florida but across the country. His company was one of the first, if not the first, to provide hazmat training for emergency response personnel. Chief Smith attended a Safety Systems class and became friends with Captain Gore, which resulted in a partnership between Safety Systems and the District-2 Fire Department. District-2 would routinely host Safety Systems classes and District-2 members could also attend these classes.

On December 6, 1978 District-2 Fire Department placed their hazardous materials team in service specifically for responses to incidents involving hazardous materials. Their response vehicle had been donated by Chief John Caserta of the Grumman Fire Department (Figure 1.9) located at Witham Field in Stuart, who was also a volunteer firefighter with District-2 and formally the city of Stuart (*Firehouse Magazine*).

Figure 1.9 Martin Counties' first response vehicle had been donated by Chief John Caserta of the Grumman Fire Department. (Courtesy: Martin County Fire Department.)

Houston, Texas Pioneer Hazmat Team

Beginnings of Hazmat Response in Houston

Houston Chief V.E. Rogers had been injured severely during the Mykaw Road fire and explosion in 1971. While attending an IAFC conference in the late 1970s in Florida, where Captain Gore was a speaker. After hearing Captain Gore's presentation Chief Rogers went back to Houston and asked District Chief Max McRae to get a hazmat team together for the Houston Fire Department. According to retired hazmat team member Bill Hand, "District Chief McRae was happy at Station 28. However, Chief McRae was the type of guy that would do what asked and do it to the best he could."

> **Author's Note:** *During preparation for this volume, I had the pleasure of talking with Bill Hands, another pioneer in the field of hazmat, on the phone. He provided me with valuable insight into the beginnings of hazmat response in Houston. I am looking forward to meeting him.*

Training for hazardous materials response at the time when the Houston team was formed, as with Jacksonville, was limited at best. Bill and Jesse Ybarra attended Ron Gore's Safety Systems course in Florida, which was funded by the Shell Oil Company in Houston. Industrial hazmat teams were common in the Houston chemical industry along the shipping channel. They were, however, only formed and used for the facilities they

Chronicles of Incidents and Response

worked for. These teams were helpful with informal training, but nothing much was available in the Houston area. Bill Hand was apprehensive about joining the hazmat team, he drove the district chief and like Chief McRae, he was happy where he was. Bill Hand set about recruiting firefighters for the team from department ranks, but found the task very difficult. The first hazardous materials vehicle was an old rescue truck retrieved from the salvage yard (Figure 1.10).

Bill became the training officer for the team. He taught hazmat operations to recruit classes and looked for people among recruits with specialized backgrounds who would benefit the hazmat team. If someone would be of interest to him based on their background, he would recruit them for the team. From the rank and file, team members were handpicked based on their reputations in the department. Bill also told me that Alan Brunacini, former chief of the Phoenix Fire Department, once said to him, "People who do training are the most powerful people in the department."

Following several months of training and equipping the response vehicle and facing an uncertain future, the team was activated at 06:30 hours on October 5, 1979. The new hazardous materials response team in Houston was assigned to Fire Station 1 and would also run rescue calls with a new rescue unit as one of the eight rescue teams in the city. On their first day of service they were called upon to mitigate a small chlorine leak. An additional 29 hazmat alarms were answered that first year. Although more hazmat alarms occurred during the first year and through the years to come, the dispatchers did not always remember they had a hazmat team and other companies were dispatched sometimes instead.

Figure 1.10 Houston's first hazardous materials vehicle was an old rescue truck retrieved from the salvage yard. (Courtesy: Houston Fire Department.)

Despite early organizational difficulties and growing pains, Houston's Hazardous Materials Response Team became the busiest and one of the best teams in the country. They were used as a model for other fire departments starting their own teams during the 1980s and 1990s. Hazardous materials personnel from around the world have rode with the team in its Ride-Along Program.

> ***Author's Note:*** *During October 2018 I had the honor of riding along with Houston's Hazmat Team. It was a pleasure getting to know team members and learning how they function on a daily basis.*

Many of the visitors would experience more hazmat response in a week in Houston than they would in months in their own departments. Houston's Hazardous Materials Team is currently a standalone unit (since 1983) dedicated to hazmat response and other hazmat duties.

Team quarters moved from Station 1 (now a restaurant) to Station 17, to Engine 22, Ladder 22, and Medic 22, and then relocated 20 miles to its current quarters. Engine 78 is located at the former Engine 22 Station, and Engine 22 and Ladder 22 were taken out of service. Hazmat 22 currently operates out of relocated Station 22 at 7825 Harrisburg Street. Station 22 is close to the city's petrochemical area, and hazardous materials units have a 20–25 min response time anywhere in the city (Figure 1.11).

Figure 1.11 Today's HM 22 newest rig located at Station 22 which is close to the city's petrochemical area, and hazardous materials units have a 20–25 min response time anywhere in the city.

Houston Distribution Warehouse Complex

One of the largest fires to ever challenge the Houston Fire Department occurred on June 24, 1995 (Figure 1.12). Seven alarms were transmitted for the Houston Distribution Warehouse complex located at 8550 Market St., with the first at 8:33 a.m. Two-thirds of the city's on-duty force was called on to fight the stubborn fire over the next 31 h. Thirty-one engine companies and fourteen ladder companies fought the fire. Warehouses in the complex were loaded with organic and inorganic chemicals, including corrosives, motor oils, plastics, solvents, cleaning compounds, lubricants, organophosphorus pesticides, flame retardants, and metallic compounds.

Hazmat units responded on the second alarm. The fire overcame the building's sprinkler system and firefighter hose streams, and by 11 a.m., the building was completely involved. Exposures included five tractor-trailers loaded with organic peroxides and a liquefied petroleum gas rail car. Lines were placed in service to protect exposure, and protective booms were placed to keep runoff water from entering storm drains. Mutual aid was requested from Channel Industries Mutual Aid (CIMA) at 7:30 p.m., which brought ten foam tankers, four 2,000-gpm monitors mounted on trailers, and a 6,700-gallon foam tanker to the scene. After pouring foam on the fire all night, firefighters brought it under control at around 3 a.m. the following morning. Fire units remained on scene for

Figure 1.12 One of the largest fires to ever challenge the Houston Fire Department occurred on June 24, 1995. Seven alarms were transmitted for the Houston Distribution Warehouse complex. (Courtesy: Houston Fire Department.)

several days putting out flare-ups. Despite the magnitude of the fire and the chemicals involved, only three firefighters sustained minor injuries.

On October 5, 2019 the Houston Fire Department Hazardous Materials Team, Company 22, celebrated 40 years of organized hazardous materials response. A celebration was held at the Houston Fire Department Pension Center with retired team members, current team members, and other fire department personnel. (*Firehouse Magazine*)

Author's Note: Despite their reluctance and apprehension in the beginning, Chief Max McRae and Bill Hand went on to do an outstanding job for the Houston hazmat team and earned the respect and admiration of hazmat personnel throughout the United States and beyond.

Memphis, Tennessee Pioneer Hazmat Team

Hazmat Team History

Following the Waverly Tennessee derailment and explosion on February 24, 1978, for which the Memphis Fire Department provided mutual aid, Memphis began exploring ideas for the formation of a hazardous materials response team in Memphis. Captain James Covington, Charles Smith, and Chief Adelman (Chief of Training) spearheaded the formation of Memphis hazmat team in the Spring of 1978. Following the firefighter strikes in the summer of 1978, the hazmat team started. By late 1978, the entire Memphis Fire Department had been trained in hazardous materials awareness. Chief Adelman and several others went to Texas A&M for hazmat training. When they returned, they trained firefighters to use Aqueous Film-Forming Foam (AFFF) foam and prepared them for fighting fires involving propane and hydrocarbons. The hazmat team was phased in over 1978 and 1979 while beginning qualifications and training for team members.

Author's Note: Captain Jim Covington retired and went on to the National Fire Academy to teach in the Hazmat Curriculum. He was my instructor for Hazardous Materials Operating Site Practices in 1982.

Hazardous Materials Exposure

Highway routes for hazardous materials include I-40, I-55, and U.S. Highway 51. Railroads that operate in the Memphis area are the Burlington, Northern Santa Fe, and Canadian North. Barge traffic on the Mississippi often carry hazardous materials. A pipeline is located at President's Island on the eastern shore of the Mississippi river. Fixed facility hazardous materials exposures include DuPont Chemical, Praxair, Drexel Chemical, USZINC, Valero Refinery, Cargil, and numerous other chemical facilities within the city limits. Chemical exposures include solvents, fertilizers, hydrogen peroxide, anhydrous ammonia, cyanide, sulfuric acid fuming, sodium hydroxide, propane, chlorine, and pesticides. The greatest concentration of hazardous materials exposures in the city are located on President's Island. The island is also considered the largest commercial and manufacturing area in the South. President's Island is part of the International Port of Memphis.

This sea port is the second largest inland port on the shallow draft portion of the Mississippi river, and the fourth largest inland port in the United States. The International Port of Memphis covers the Tennessee and Arkansas sides of the Mississippi river from river mile 725 to mile 740. Within this 15-mile reach, there are 68 water-fronted facilities, 37 of which are terminal facilities moving products such as petroleum, tar, asphalt, cement, steel, coal, salt, fertilizers, rock and gravel, and of course grains. The International Port of Memphis is 400 river miles from St. Louis and 600 river miles from New Orleans and is ice-free throughout the year.

Incidents

Drexel Chemical Company Fire & Explosion

In the heat of summer on July 5, 1979 at 9:25 a.m., about a year after the formation of the hazmat team, Memphis Fire and Hazmat faced a third alarm fire at the Drexel Chemical Company, 2387 Pennsylvania Avenue, near the intersection of Pennsylvania and Mallory Streets (Figure 1.13).

Figure 1.13 In the heat of summer 09:25 hours on July 5, 1979, about a year after the formation of the hazmat team, Memphis Fire and Hazmat faced a third alarm fire at the Drexel Chemical Company. (Courtesy: Memphis Fire Department.)

An 8,000-gallon tank of parathion, a pesticide, caught fire. In an interview with the Commercial Appeal, maintenance mechanic David Trumble said the following of the initial explosion, "I heard one hell of an explosion and ran like hell." Another employee Robert Belden in a nearby warehouse "heard an explosion. It was like a dull boom and I saw the doors of the warehouse coming off. Then it knocked me down, and the flash burned my hands in a couple of places and I went outside."

Richard Arwood, who 25 years later became the Director of Fire Services (Fire Chief), was assigned to Engine 10 the day of the fire. Earlier that day he had driven by while performing a familiarization of the area and noting locations of hydrants. Back at the station, he was looking out the window when the explosion occurred. He told me during an interview that "it was a straight shot to Drexel, the wind was blowing North to South so he could approach the fire from upwind, the wind was blowing in the best direction it could have been" because of the toxic nature of the chemical involved and the toxic smoke blowing downwind.

> **Author's Note:** *Richard is a friend and former instructor of mine in a course I took at the National Fire Academy in the early 1980s. He is a very genuine person and a wise man. I had the opportunity to conduct a phone interview about the Drexel fire and the development of the Memphis Hazmat Team in preparation for this volume.*

The Drexel Chemical Fire was in a concrete block and a flat roof building 100 ft × 200 in. long. Firefighters were told by dispatch that organic phosphate pesticide was involved in the fire. When firefighters arrived on scene, they all put on Self Contained Breathing Apparatus (SCBA). Master streams devices were deployed, and at the height of the fire 4,000 gpm of water was being put on the burning chemicals. Numerous explosions occurred, sending 55-gallon drums flying through the air. Captain Jim Covington, who helped form the hazmat team, figured out early on that they would just let the fire burn. Following dangerous explosions, rocketing 55-gallon drums, and massive fire, Drexel Chemical Company was allowed to burn to the ground (Figure 1.14).

Water from the firefighting efforts spread the fire through a cache of methylparathion, which poured into the Mississippi River because the fire could not be contained. Sulfuric acid also dumped into the river as a toxic cloud spread beyond the plant, endangering the lives of nearby residents. Over 3,000 residents were evacuated including 400 who were taken by transit authority buses to an evacuation center set up at Whitehaven High School. Over 200 people visited hospitals complaining of bleary eyes, headaches, and vomiting. Treatment was hampered by lack of medical personnel familiar with the treatment of chemical exposures. There were no firefighter injures as a result of the fire.

Figure 1.14 Following dangerous explosions, rocketing 55-gallon drums, and massive fire, Drexel Chemical Company was allowed to burn to the ground. (Courtesy: Dick Adelman Collection.)

In 1984, the Memphis local government took steps to help prevent another disaster like the Drexel Chemical fire. A hazardous materials advisory committee was formed with representatives from emergency services, government, and industry, who secured a $100,000 federal grant to create a hazardous materials plan. As a result of the plan and extensive training, the Chemical Manufacturers Association declared in 1986 that Memphis was the best-prepared American city to cope with a hazardous materials emergency.

Pro-Serve Fire (Brooks Road)

In August 2006, the Memphis Fire Department faced an extraordinarily complex hazardous materials incident that was complicated by fire. Three chemical manufacturing facilities were involved that produced and stored a wide range of hazardous chemicals.

On August 2, 2006 at 13:47 hours, Memphis firefighters were dispatched to a fire next to Valley Products, 384 East Brooks Road (Figure 1.15). The first arriving companies radioed there was smoke showing at Pro-Serve, 400 E. Brooks Road. Size-up determined a small fire involving a pallet of sodium chlorate and a forklift, which had been extinguished by the sprinkler system before firefighters arrived. Employees at Pro-Serve who had been exposed to smoke were decontaminated and given medical evaluations by fire personnel on scene. Clean-up was handled by Pro-Serve. Command became aware that employees at Bucyrus International at 3057 Tranquility had also possibly been exposed to the smoke. Because of limited exposure, Bucyrus employees refused decontamination, medical treatment, and transport. Both incidents were declared stabilized at 15:28 hours.

On August 4, 2006 at 16:07 hours, the Memphis 911 center received multiple reports of an explosion and heavy smoke in the area of Brooks Road and Third Street. It was initially dispatched as a hazardous materials

Figure 1.15 On August 2, 2006 at 13:47 hours Memphis firefighters were dispatched to a fire next to Valley Products, 384 East Brooks Road. (Courtesy: Memphis Fire Department.)

incident. First-in companies reported a fire and heavy smoke at the Pro-Serve building, 400 E. Brooks Road. The first arriving engine company advanced a 2½-in. attack line into the building. Division 1 arrived on scene a short time later and assumed Brooks Road Command. A second alarm was requested at that time. Command decided to cease all offensive operations until more information could be gathered about the chemicals present inside. A hazardous materials branch was established. Runoff was reported to have entered and contaminated Nonconnah Creek. Hazmat operations decided to recommend extinguishment of the fire, and it had been knocked down at 20:29 hours approximately 4½ h after companies were dispatched. The incident was declared stabilized at 21:07 hours.

As the scene was downgraded, a fire ruins detail schedule was established. Brooks Road command was transferred several times. It was discovered that IBC Manufacturing at 416 E. Brooks Road had also been involved in the incident. Due to multiple unknown chemicals stored at that location and the unavailability of IBC personnel, Brooks Road Command gave orders not to enter the structure.

On August 5, 2006 at 02:55 hours, Brooks Road Command requested a hazardous materials assignment and Division 1. At 02:56 hours, a second alarm was requested. Brooks Road Command reported fire coming through the roof as well as explosions occurring in what was eventually determined to be the IBC warehouse. Due to the unknown nature of the products, the amount of fire involvement, and environmental concerns from runoff, Brooks Road Command made the decision to let the building free-burn. Efforts were concentrated on isolating the area, evacuating the citizens, and monitoring air.

At daylight, aerial surveillance was conducted via helicopter, MSDS sheets were obtained, and hazmat teams conducted recon. As the intensity of the fire subsided, the decision was made to attack the remaining fire that was in the IBC tank farm using limited water and AFFF. At 15:54 hours the fire was reported knocked down. The incident was downgraded and a ruins detail was scheduled including hazmat team personnel.

On August 9, 2006 at 16:42 hours fire companies were dispatched to Pro-Serve, 400 E. Brooks Road. Initial arriving companies reported light smoke coming from the tower area. United States Environmental Services (USES) and the Center for toxicology and Environmental Health (CTEH) had been on scene since the earlier incident on August 5th conducting clean-up and air monitoring operations. They reported that the material on fire was not water-reactive. The company officer made the decision to attack the fire with a 2½ in. hose line. Hazmat was dispatched and was conducting air monitoring. Brooks Road Command was set up at Brooks and South Center Road and a second alarm requested at 17:04 hours. Command ordered companies out of the building until more product information was obtained. Hazmat personnel placed unmanned monitors

Figure 1.16 The fire was reported knocked down at 20:04 hours and a ruins detail was set up. A fire department presence was maintained until September 1, 2006. (Courtesy: Memphis Fire Department.)

in place to extinguish the remaining fire. The fire was reported knocked down at 20:04 hours and a ruins detail was set up. A fire department presence was maintained until September 1, 2006 (Figure 1.16).

The Memphis Fire Department has a rich history and their hazmat team has been one of the early pioneers in hazmat response development. Their extensive and diverse hazmat exposures provide daily challenges for the current hazmat team.

Selected Historical Hazmat Incidents

The 1800s

Hundreds of incidents involving what we would today call hazardous materials can be found in old newspaper clippings and online. Several of these date back to the mid-1800s. As hazmat teams did not exist then, the entire community turned out to help when these events occurred. In some cases, there were organized fire brigades, which were no match for chemical fires and explosions. As no resources were available to determine the dangers posed by these chemicals, they just had to do their best to deal with the emergency. As a result, many lost their lives.

Details were often sketchy and sometimes inaccurate. In those days there was no television, radio, cable news network, or other modern methods of communication other than telegraph. Newspapers covered these stories on a local basis, but it took longer to get the news out. Thus, it is clear that hazardous materials and incidents have been around for quite some time and there have been many disasters over the years. Although we have significantly progressed since those days, unfortunately, we have not learned lessons well and end up making some of the same miscalculations others have made before us.

Early examples do not include boiler explosions, home explosions, criminal acts of arson, military incidents involving missile silo, or mining explosions. Many of those early explosions involved explosives related to mining industries and war efforts during times of war. Early newspaper coverage was included as written because writers of the time were very descriptive and I find their accounts of the incidents quite interesting. Not all of the incidents have photos, or the quality of available photos is not good enough to include. Many of the incidents in the 1800s and early 1900s are wire service versions, sent to locations far from an actual incident location, that someone thought might be interesting to their readers.

First Incident Found During Research

Syracuse, NY August 30, 1841 The Great Gunpowder Explosion

Friday night a fire broke out at Syracuse in Goings Carpenters Shop near the Oswego Canal (Figure 1.17). It spread with great rapidity and the building was soon enveloped in flames. The village's volunteer fire department companies rushed to the scene. Unknown to all, a gunpowder consignment totaling 650 pounds in 25 kegs, was secretly stored in the Goings Shop. Crowds of citizens flocked to the scene, and soon after a great number had collected. "Powder! Powder! There is powder in the building!" But this cry had but a momentary effect on the crowd. The mass moved back a step, then stopped, and in a few moments as firemen were starting their hoses, the twenty-five kegs of powder blew up with one explosion, scattering fragments of the buildings and limbs of human bodies in every direction. As near as could be ascertained, upwards of thirty persons were killed outright, and no less than fifty wounded, some very seriously, and perhaps fatally. From ten to fifteen were so mangled and cut to pieces that it was impossible to recognize them. Every exertion was immediately made to relieve the sufferers.

Everything was done that could have been done under the circumstances. An extra train of cars was run to Auburn for physicians, and our

Figure 1.17 Friday night a fire broke out at Syracuse in Goings Carpenters Shop near the Oswego Canal The village's volunteer fire department companies rushed to the scene. Unknown to all, a gunpowder consignment totaling 650 pounds in 25 kegs, was secretly stored in the Goings Shop.

hotel keepers threw open their doors for the reception of the wounded. As to the origin of the fire it is unknown; but it is supposed to have been the work of an incendiary! The fire appeared to have commenced in the top of the building.
The Adams Sentinel Gettysburg Pennsylvania 1841-08-30

> *Author's Note:* Records of this incident are scarce, and I could not find any lists of casualties. However, based on the description of the tragedy, I am certain there must have been firemen killed in the explosion. Syracuse fire department was not established until 1871, so they would not have any record of firefighter deaths.

Minneapolis, MN May 2, 1878
Washburn "A" Flour Mill Explosion

On the evening of May 2, 1878, the Washburn Flour Mill exploded in a fireball, hurling debris hundreds of feet into the air. In a matter of seconds, a series of thunderous explosions heard ten miles away in St. Paul destroyed what had been Minneapolis' largest industrial building, and the largest mill in the world, along with several adjacent flour mills (Figure 1.18).

Figure 1.18 On the evening of May 2, 1878, the Washburn Flour Mill exploded in a fireball, hurling debris hundreds of feet into the air.

It was the worst disaster of its type in the city's history, prompting major safety upgrades in future mill developments. The massive A Mill, only four years old at the time, had been built by Cadwallader Washburn, a businessman from LaCrosse, Wisconsin, in the heart of Minneapolis on the Mississippi River, near St. Anthony Falls. The seven-story flour mill was powered with river water, which was diverted through a canal that ran inside its lower level. With two hundred workers, it was one of the city's largest employers.

At six o'clock that evening, the mill's large day crew completed its shift. Fourteen men who made up the night crew arrived. An hour later, three deafening explosions echoed out, reverberating in waves all over town. All fourteen workers were killed. Within minutes, the fire had spread to the adjacent Diamond and Humboldt mills. They also exploded, killing four more workers. The city's fire department worked all night to contain the fire. Their efforts were futile. The intense heat prevented firefighters from setting up rigs and hoses close enough to the site to have much impact. Within minutes the fire had spread to the adjacent Diamond and Humboldt mills. They also exploded, killing four more workers.

The next day, on May 3, the *Minneapolis Tribune* told its readers, "Minneapolis has met with a calamity, the suddenness and horror of which it is difficult for the mind to comprehend." At the inquest into the deaths of the eighteen workers, John A. Christian, the A Mill's manager, explained that rapidly burning flour dust had caused the disaster. His explanation was later confirmed by two University of Minnesota professors, S.F. Peckham and Louis W. Peck, who reviewed a series of controlled experiments that caused flour dust to explode. Peckham and Peck concluded that two of the millstones, running dry, had rubbed against each other, causing a spark that ignited the dust.

As the furor over the explosions began to subside, the *Tribune* speculated about the impact of the event on the local milling industry, now that one-third to one-half of the city's production capacity had been destroyed (Figure 1.19). The newspaper noted that Minneapolis had recently surpassed St. Louis and Buffalo in milling production and was now the country's leading flour producer. Cadwallader Washburn, who had rushed to town from his Wisconsin home after hearing news of the calamity, announced that he would rebuild, even as the embers from his destroyed plant were still glowing.

Washburn kept his word. By 1880, his new A Mill was up and running. It was safer and more technologically advanced than its predecessor, with a greater production capacity. The milling magnate opened his new plant in time to take advantage of the economic boom that Minneapolis would experience during the last two decades of the nineteenth century and into the twentieth. The city's milling production continued to increase, peaking during World War I.

"Then each floor above the basement became brilliantly illuminated, the light appearing simultaneously at the windows as the stories ignited one above the other," he recalled. "Then the windows bust out, the walls cracked between the windows and fell, and the roof was projected into the air to great height, followed by a cloud of black smoke, through which brilliant flashes resembling lightening passing to and fro." The A Mill continued in operation even as the local milling industry underwent a decline.

Figure 1.19 As the furor over the explosions began to subside, the *Tribune* speculated about the impact of the event on the local milling industry, now that one-third to one-half of the city's production capacity had been destroyed.

Chronicles of Incidents and Response 35

But age eventually caught up with the riverfront landmark, and the mill was finally closed in 1965. It remained, abandoned, until the 1991 fire.

Minneapolis Tribune, Minneapolis, Minnesota May 3, 1878.

Chester, PA February 16, 1882 Jackson Fireworks Plant Explosion

A terribly fatal, and as yet mysterious, explosion occurred this morning at a fireworks factory. The factory occupied part of an old stone building. There was a two-story frame extension in the rear, and this was used as a part of the factory. Only four persons had been employed in the factory up to this morning. But the demand for railroad torpedoes, which received a sudden stimulus from the Spuyten Duyvil disaster, had become so great that two or three more young women were engaged, and a room was fitted up for them in the second story of the main building.

A fire was lighted in the stove this morning for the first time, and Superintendent Van Horn says that about 7:20 o'clock he went to look at the stove, found the fire was somewhat low, and went down stairs for some wood and coal to rekindle it. As he went down he heard an explosion in the room he had just left, which blew off part of the roof and set the building on fire. It was caused, he thought, by a puff of coal gas from the stove, which set fire to some loose powder scattered on the floor, and this in turn exploded some red fire and rockets which were stored in the room.

An alarm of fire was given, and the three volunteer companies Moyamensing Hose, Franklin Hose (Figure 1.20), and Hanley Hook and Ladder (Figure 1.21) which comprise the Fire Department of the city responded at once, though their members were tired by their heavy work at the burning of the Military Academy last night. Up third street came the Franklin Company, the engine horse drawn, and the hose carriage drawn by the members a foot. Over Third Street Bridge it went. When his wagon was half-way between Concord Avenue and Penn Street three men made an effort to "jump it"; one of them Tony Barber, was a member of the Hanley Company; another John Turnpenny, was a member of the Franklin Company; the third Jake Lamplugh, was too young for membership in any company. Tony and Jake made it, but John owed to his portly figure his fortuitous failure to get a "life," for he alone of the three survived that fatal day.

It wasn't long before the populace knew it wasn't a rekindle of the "Academy" fire from the night before. The discharge of explosives with ominous frequency and thunderous report soon frightened the neighbors into a realization that the fire was in the one place where it's happening was always dreaded; The Jackson Fireworks Plant. It was stated by the Superintendent and believed by the firemen, that there was little or no

Figure 1.20 Franklin Hose Company Chester, Pennsylvania.

Figure 1.21 Hanley Hook and Ladder Company Chester, Pennsylvania.

powder in the building, and only a small quantity of comparatively harmless pyrotechnics.

Consequently, the men worked without fear, and some of them mounted on the roof of the frame extension to get a good position from

Chronicles of Incidents and Response

which to direct their hose streams, while a great crowd of spectators pressed close about the firemen. Van Horn warned them away, but they would not go. He and a number of others entered the burning building and busied themselves in removing the machinery. While this work was going on there was another explosion, and this, though slight and unimportant, caused the crowd to withdraw further from the building, thus no doubt preventing a more serious loss of life than actually occurred a few minutes later.

The work of fighting the flames had been going on for about half an hour, and the building was surrounded by a throng of firemen and others, when suddenly a quantity of some terribly explosive substance was reached by the fire, and an entire wing of the stone building was blown into the air, fragments of stone, timber, and human bodies being hurled long distances and scattered over a wide area. Men 150 yards away, who thought they were watching the fire from an entirely safe distance, were struck and killed. Others were dashed against trees and houses with fatal violence. Others again were torn limb from limb, and one man, a laborer named janitor of the National Hall and an assistant fireman, who was standing on the roof of the wing in which the explosion took place, was thrown to the top of the main building, where he lodged among the burning rafters, and lay shrieking and howling, roasting to death, for half an hour.

He was rescued by a brave carpenter who volunteered to bring him down. He was alive when reached, but died a few minutes later. Anthony Barbour and John Vandegrift, firemen were standing on a ladder throwing water on the roof of the main building. Barbour was struck by a heavy stone and hurled 150 feet away, where he fell dead and horribly mangled. Vandegrift was blown off the ladder, but was only slightly hurt. Two other firemen were standing on the roof of the frame building when the explosion came, and their dead bodies were picked up nearly 200 yards away. These, however, were almost the only men who were instantly killed among those close to the building; the other victims were struck by flying pieces of the wreck.

Firemen dropped their hose and devoted themselves to the work of gathering up the dead and rescuing the wounded. They paid no more attention to danger than before, though two or three other slight explosions warned them that there was still a possibility of it. After seeing that all the dead and wounded who could be reached were cared for, the firemen turned their attention again to the flames and succeeded in putting out the fire about 10 o'clock, though by that time there was little left to burn. Search was then made in the ruins for a number of missing men, and this is still going on. This afternoon trains brought throngs of people from as far away as Philadelphia in one direction and Baltimore in the other, but they could only stand around the smoking ruins and watch the firemen as they searched for the bodies of those supposed to be buried there.

The number of the dead, according to the latest accounts, is 17. Of those killed 9 were firemen. The number of persons injured is not yet known, but it is believed that over 59 were wounded.

Firemen That Made The Supreme Sacrifice:

Franklin Volunteer Fire Company:

Thomas Anderson

William "Curly" McNeal

Joseph Kestner

Alex Phillips

William H. Franklin

Hanley Volunteer Fire Company:

Zach Vandergrift

Tony Barber

John Pollack

Robert Stinson

New York Times New York 1882-02-18

Cavan Point, NJ May 11, 1883 Standard Oil Works Fire Started by Lightning

The Standard Oil works of Jersey City were struck by lightning early this morning. Seven oil tanks have already exploded Figure 1.22). The fire is still raging with terrible force. More disaster is feared. The estimated loss is $500,000. The fire at the Standard Oil Company's works, at Jersey City, is still raging and is now near the tanks of the Jersey Central track, where 150,000 barrels of oil are stored. Thus far 500,000 barrels of oil have been destroyed, including 250,000 barrels of refined stored in the store house. It is now thought the loss will reach $750,000.

The burning oil poured out like an angry river and flowed down the avenues, between rows of enormous tanks, towards the river. In a short time the flames communicated with those huge tanks and they exploded simultaneously with a terrific force. Fragments of iron were hurled a distance of half a mile and the burning oil was scattered in all directions. A fireman of Engine No. 10, in Halliday-street, Jersey City, about a mile

Chronicles of Incidents and Response 39

Figure 1.22 The Standard Oil works of Jersey City were struck by lightning early this morning. Seven oil tanks have already exploded. (Courtesy Jersey City Free Public Library.)

away, saw the flames when they burst from the first tank, and at once gave the alarm. The engine was immediately started for the scene and was dragged with great difficulty across the marshy land intervening.

The firemen began work by trying to quench the fire with water, but soon found that their efforts were useless. The water only carried the blazing oil to sputter and crackle more. They then confined their work to

Hazmatology: The Science of Hazardous Materials

saturating with water those buildings which were still safe. Engines Nos. 8 and 9 and Truck No. 5 arrived soon after No. 10 and joined in the work. They were apparently successful in their efforts until nearly 8 o'clock, when tank No. 7, containing 20,000 barrels of oil, suddenly exploded with a terrific report and a concussion that was felt for many rods around. Shock waves shook houses 3 blocks away and he sound was heard four or five miles away. The heavy brick walls on which the tank rested, over a foot thick, were blown in every direction, and pieces of the heavy riveted iron were carried long distances. Several victims were blown into

As the burning streams of oil poured through the yard they fired the tank on the east side, and licked up the buildings. The store house, a structure where barreled oil was kept, was attacked by the flames and succumbed. The engine house went next, and the oil and water pump houses, sunken tanks and machine shop quickly followed, and the whole east side of the works is a seething mass of flames. Despite the tremendous heat and blinding smoke, the firemen are working valiantly in hope of checking the progress of the flames, but it appears at this hour that the vast works will be destroyed, and it is feared the flames will reach out and consume the properties in the neighborhood. Eleven tanks were destroyed containing 110,000 barrels of oil valued at $500,000.

Chief Terrier and a number of men were within ten yards of this tank when it exploded and were saved almost by a miracle. Dropping the hose, which was soon burned to ashes, they fled for their lives. There could be no doubt as to the fate of the six men, as it was utterly impossible for them to escape. Firefighters lost were from the Standard Oil Fire Brigade.

Firemen Who Made The Supreme Sacrifice:

Chief Terrier

John Herbert, 41, Assistant Plant Manager

Joseph Jenkins, 27

George J. Davis, 39, Engineer

Henry S. Kegler, 30

Richard "Dick" Conklin

Civilian:

William J. Curry, boy.

(Son of employee that father brought to work the day of the fire)

The Evening Journal Jersey City, NJ Thursday May 18, 1883

Oil City, Titusville, and Meadville, PA
June 1892 Flood and Explosion

The oil regions of Pennsylvania were visited today by a disaster of fire and water that is only eclipsed in the history of this country by the memorable flood at Johnstown just three years ago (Figure 1.23). It is impossible at this hour, midnight, to give anything like an accurate idea of the loss of life and property, as chaos reigns throughout the devastated region and a terrible conflagration still rages in Oil City. It is safe to say that not less than 150 lives have been lost. Nearly 100 bodies have already been recovered, and many people are missing. The number may far exceed 150, but his is regarded as a conservative estimate. The property loss will reach far into the millions.

At Titusville the loss is estimated at $1,500,000; Oil City, $1,500,000; Corry, $60,000; Meadville, $150,000, and surrounding country probably a million more. For nearly a month it has been raining throughout Western and Northern Pennsylvania almost incessantly, and for the past three or four days the downpour in the devastated regions has been very heavy. The constant rains had converted all the small streams into raging torrents, so that when the cloud-burst came this morning the streams were soon beyond their boundaries and the great body of water came sweeping down Oil Creek to Titusville, which is eighteen miles south of its source, and eighteen miles from this city. When it reached here, the most appalling calamity in the history of Oil City fell upon it today, resulting in the destruction of life and property which as yet can only be approximated.

Figure 1.23 The oil regions of Pennsylvania were visited today by a disaster of fire and water that is only eclipsed in the history of this country by the memorable flood at Johnstown just three years ago.

42

Hazmatology: The Science of Hazardous Materials

Thus far forty lives are known to be lost, hundreds of people are missing, and it is believed the loss of live will reach seventy-five or more.

At 11:30 o'clock this forenoon a large proportion of the population of the city was distributed along the banks and bridges of the Allegheny River and Oil Creek watching the rise of the flood in both streams, the chief cause of the rise of the latter being due to a cloudburst above Titusville last night, which resulted in the loss of many lives in that city. At the time mentioned this forenoon an ominous covering of oil made its appearance on the crest of the flood pouring down the Oil Creek Valley, and the dangerous foreboding waves of gas from distillate and benzene could be seen above the surface of the stream, which, at the bridge, is about 100 yards wide. People began slowly to fall back from the bridge and the creek. Hardly had they begun to do so when an explosion was heard up the stream, which was rapidly followed by two others, and quick as a flash of lightning the creek for a distance of two miles was filled with an awful mass of roaring flames and billows of smoke that rolled high above the creek and river hills.

Oil City is bounded on all sides by steep hills. Oil Creek comes down the valley from the north, and just before its junction here with the Allegheny is crossed by a bridge to the portion of the city embraced in the Third Ward, which lies along the west bank of the creek and the north bank of the river. Almost all that portion of the town was on fire within three minutes from the time of the explosion, and no one knows as yet how many of the inhabitants are lying dead in the ruins of their homes.

Almost as quickly as the words can be written, fully 5,000 people of that portion of the town were on the streets, wild with terror, rushing to the hills. Scores of men, women, and children were knocked down and trampled upon, both by horses and people, in the mad flight for places of safety. Just as this frantic mass of humanity had started up Centre Street the second explosion occurred, knocking many people down, shattering the windows in the main part of town, and almost transforming the day into night with the immense cloud of smoke preceding the second burst of flame.

The heat was intense and the spectacle presented to the panic-stricken people was that of a cloudburst of fire, bordered and over capped by a great canopy of dense black smoke, falling upon the city. The flood in the Oil Creek valley had inundated the upper portion of the town, flooding from fifty to seventy fine houses along north Seneca Street. Most of their inmates reached places of safety by means of boats or by swimming or wading, but a number of them were yet in the upper stories or in the water when the fire came, and their fate was quickly sealed. Some of them were seen to jump into the water to escape death in the flames. The distillate and benzene on the creek came from a tank lifted by the flood and

Chronicles of Incidents and Response 43

is supposed to have been ignited by a spark from an engine on the Lake Shore Railroad just above the tunnel at the northern part of the city. The fire shot up the creek as well as down, and several tanks are on fire at a number of the refineries up the creek. But it is believed that their burning contents can be kept within the tanks and that a second baptism by fire may be averted. The damage to property by fire alone cannot even be fairly approximated.

The Bellevue Hotel, the Petroleum House, the Oil City Barrel Factory, the new building of the Oil City Tube Works, the big furniture and undertaking establishment of George Paul & Sons, and probably one hundred dwelling houses have been destroyed. The Fire Department, by excellent work, kept the fire from crossing to the central portion of the town, except in one instance, when Trinity Church caught fire, but the department succeeded in saving the building as well as two bridges.

Paul & Sons' loss is estimated at $100,000; not fully insured. R. D. Naylor's loss is $3,000; no insurance. William Loots proprietor of the Petroleum House, totally destroyed, losses $5,000; loss partially insured. Notwithstanding the assurance that no danger will result from the fire at the tanks up the creek, few people will sleep a wink in the city to-night. Mayor W. G. Hunt has sworn in the members of Company D of the Sixteenth Regiment, National Guard, to serve as special police until further notice.

The New York Times, New York, NY 6 June 1892

Bradford, PA, May 14, 1894 Refinery Fire

The barrel house at Emery's refinery in this city took fire yesterday afternoon, presumably by spontaneous combustion, and was destroyed. The loading racks and five oil-tank cars standing on a side track of the Buffalo, Chester and Pittsburg road were also burned. The fire was a fierce one, and attracted immense crowds of people from all over this locality.

While the firemen were making a final stand and thousands of people were watching them, there was a tremendous explosion (Figure 1.24). A tank car holding 4,100 gallons of benzene had let go with a mighty roar. Fortunately the burning benzene, which was flung into the air in sheets of liquid, fire had consumed itself before settling down over the crowd.

The explosion was followed by a panic that cannot be described in words. Blind, unreasoning, pitiless instinct of self-preservation showed itself, and the weak went down before the strong in multitudes of cases in the frantic rush to escape what seemed to threaten a horrible death. In the stampede, men as well as women and children were thrown down and trodden over by the flying mass that surged up from behind. Thirty-five of the firemen were burned so that the skin peeled off their faces and

Figure 1.24 While the firemen were making a final stand and thousands of people were watching them, there was a tremendous explosion. (Courtesy Bradford Landmark Society)

hands and the hair was signed off their heads and faces. Of the many others who are slightly burned there is no record, and the total number of those burned and injured in the stampede there is no record and the number will probably reach 100 persons.

With all the suffering this fire will cause, the property loss will not exceed $5000.

Lowell Sun Massachusetts 1894-05-14

Butte, MT January 15, 1895 Warehouse District Dynamite Explosion and Fire
"The Great Disaster of 1895"

Regular Firemen as they are called in Butte live at the Central Fire Station. They are on duty 24 hours a day, 7 days a week, 365 days a year. They live together, share things that most workers don't share. It is a "bastion of male privilege". Calling themselves "Brothers" they ate, slept, trained, played, and worked together. Most were not married. Butte is a mining city with many of the underground work places. Being a fireman is a great job to have above ground. Central Station Firemen are divided into companies, Hook & Ladder and Hose Company #1. The hook and ladder company also operates a Chemical Engine with a 50 gallon tank. "Minutemen (volunteers) are housed in the Quartz Station (Figure 1.25). They are back-ups for the Regulars at Central Station. Both stations can respond at one time to the same incident is needed.

Figure 1.25 "Minute-men (volunteers) are housed in the Quartz Station they are back-ups for the Regulars at Central Station 1890's. (WesternMining History.com)

Hook & Ladder:

- Driver George D. Fifer
- Ladderman Peter J. Norling

Hose Company #1:

- Chief Angus D. Cameron
- Assistant Chief John Findlay "Jack" Sloan
- Driver Dave McGee
- Edward Findlay Sloan, Pipeman
- Samuel Ash, Pipeman
- David Moses, Pipeman
- Jack Flannery, Plugman

Horses:

  Hose Company #1 Jim & Danny (Figure 1.26).
  Hook & Ladder Youngest team of horses.

Police officer James Steinborn was making his rounds on foot through the warehouse district of Butte, when he heard a switch engine sounding a series of warning whistles. He investigated and found the engineer had spotted smoke from the warehouse district. Officer Steinborn headed

Figure 1.26 Jim the sole surviving horse of the "Great Disaster of 1895" recovered from his wounds and returned to active service. (Courtesy Butte Silver Bow Fire Department.)

Chronicles of Incidents and Response

directly to the nearest fire alarm box to sound the alarm. There are 30 Gamewell fire alarm boxes located throughout the downtown area of Butte. He came upon Box 72, which was to become known as "fateful" Box 72. When activated the fire alarm box transmitted the box location to 12 locations within the City of Butte. When the signal is received tape is punched with seven holes followed by a pause, then two holes. This indicates that the emergency is located near Box 72.

The fire bell at Central Station located in City Hall began to ring, alerting firefighters to respond. Horses automatically step from their stalls, each to a designated spot on the floor. Suspended above them are the rubber harnesses that drop from the ceiling and they are ready to respond. Firemen scramble to their helmets, boots and coats and slide the pole to the apparatus bay. Within one minute from the sounding of the bell, firemen are on their wagons responding to the fire. Hose Wagon #1 stopped at Arizona and Iron Street to catch a plug. Fireman Flannery pulled off a 2 1/2" hose line and hooked it up to the plug. It was his job to stay at the hydrant. When the horses heard the connection to the plug, they instinctively headed to the fire playing out the hose as they went. The fire was located at the Connell & Kenyon Warehouse. Unknown to the firemen and illegally placed by the owners, cases of dynamite had been stored in the warehouse. To protect the dynamite from stray bullets when the miners got off shift, drinking, fighting and in some cases gun fighting. They placed "rabble heads", tools used in the foundry business between the dynamite and any stray bullets. In addition they placed metal sheets and wire. Firemen were forcing entry into the warehouse and setting up fire fighting operations within feet of the building. Driver McGee was putting blankets on Jim and Danny. Firemen Copeland had brought along his family pet, a Newfoundland dog. Inside the warehouse fire had already reached the dynamite.

At 10:08 p.m. M.S.T. the first of three explosions occurred with massive force and flying shrapnel. Sounds of the explosion and the orange glow over the mountains was witnessed over 60 miles away. Still at his post at the plug, Firemen Jack Flannery watched helplessly as his Brothers were blown to bits. "Just as I was turning on the hydrant I heard an explosion, which shook the earth for miles around spreading ruin and death upon every hand. Chunks of debris of every kind and size were falling around me. Watching as Butte's finest team of Firemen vanished, " though Jack Flannery.

People from throughout the area rushed from their houses, from saloons, and businesses to see a "volcano of fire" rising hundreds of feet in the air. Every building in Butte shook. Percussion from the blast shattered street lights, windows and knocked people off their feet. One pregnant woman had a premature birth. A miner working 600 feet below ground felt the shock, checked his watch and it was 10:08. Rabble heads thrown from the blast site did much of the damage to people and structures. Shrapnel killed and injured many people. Jack Flannery at the hydrant and Dave McGee caring for the horses were the only survivors from Central Fire Station. McGee was sandwiched between the two horses, Jim and Danny. Danny who had taken the full force of the blast shielding McGee, saved his life. Danny died from injuries suffered from the blast wave. Jim was injured, but alive. Fireman Scotty Orr was blown 250' in the air and survived. Survivors regained their wits only to find themselves on fire.

Mining people know about accidents, loss of limbs and other injuries. They know how to triage and how critical it is to get in there and help. Many miners rushed in to help those who survived, fire was everywhere. Minute Men from Quartz Station responded with Hose Wagon #2. They were warned by spectators turned rescuers not to go into the area. A rumor was circulating that there was a plot to destroy the city. What came next only reinforces that notion. Butte Hardware Warehouse was damaged by the 1st explosion and caught fire. No one expected that an even larger dynamite supply was now burning inside. Without warning the 2nd explosion occurred involving the Butte Hardware Warehouse.

This was a much bigger explosion than the first one. Heavy equipment used for mining inside the warehouse was thrown into the air and fell back onto buildings causing damage. One massive wheel from a railcar was thrown eight blocks. A circular saw 4' wide was launched and landed in the town of Rocker five miles away. Every remaining window in the city as far north as Walkersville four miles to the north. There wasn't enough glass in the state to repair all of the windows, a gas pipe was launched into the air and fell on the ground forming a triangle with two others. This area came to be known as the triangle of death as 7 people died there (Figure 1.27).

Death and injury from the 2nd explosion was much greater than the 1st. A Hackman (taxi driver) in the process of removing three bodies was killed along with his two horses. John Sloan Sr. the father of Firemen John & Edward Sloan while searching for his sons, was critically injured and would die within the week. Dave McGee was helped by two civilians who freed him from the horses. By a miracle, he survived the 2nd blast. Rescue workers were leery of another explosion, but many continued to help victims. Shortly, a 3rd explosion would occur. Hot oil was blown into the air and it fell back down scalding and burning those it fell on, including the police chief. Although the blast was as strong as the 1st, there wasn't much left to be destroyed by the blast. Fire was now consuming the debris from the three previous explosions.

Most of the fire departments equipment was already destroyed and many firemen dead so there were nothing to suppress the flames. The

Figure 1.27 "Fatal Triangle" A gas pipe was launched into the air and fell on the ground forming a triangle with two others. This area came to be known as the triangle of death as 7 people died there. (WesternMiningHistory.com)

next day remains of Chief Cameron were found near the flour mill, he was identified by his belt buckle. Fireman Jack Flannery along with the remaining "Minute Men" from the Quartz Street Station returned to the scent and extinguished the remaining fires by 2:00 p.m. They were warned about potentially more dynamite and hundreds of downed power lines, but went ahead into harm's way anyway. Nothing but a heap of smoking rubble remained as people searched for bodies. At 11:00 the next morning the remaining firemen returned to their station.

The scene was chaotic at the morgue as people searched for loved ones. At hospitals fatigued doctors and nurses worked feverously to treat the injured that lined the hallways of the building on makeshift stretchers. Official death toll was 57 killed instantly or died within a week. Injuries were in the hundreds. The death count was a conservative figure. Population was fluid in Butte. People were coming in on the train to work at the mine, immigrants, and those just traveling to the west for a better life would come through the train station that was right next to the warehouse district. Some of those people could have been disintegrated by the explosions. County coroner and the funeral home began the impossible task of identifying the dead. Fireman Copeland was identified by his one remaining boot and his faithful Newfoundland dog, who was by his side. Positive identification was nearly impossible, fragments were too small and scattered over a large area. Some errors followed victims all the way to the cemetery. Among the ruins sticks of frozen, unexploded dynamite were found. Investigators wondered what would have happened if the temperatures were warmer when the dynamite exploded.

Firemen were buried with full military honors (Figure 1.28). Over 3,000 firemen came from all over Montana to pay their respects on the day of the funeral. Fireman Dave McGee though injured road on Hose Wagon #2 in the precession. Since so little remained of Chief Cameron, Fireman Ash, Fireman Norling and Fireman Moses, their remains were placed in one casket. With the sentiment that they "They were together in life, fighting fires side by side, they are no unified in death". There were nine caskets for the other fallen heroes. Fireman William Copeland's Dog was there, he survived all three explosions. He followed Copeland to the morgue, and now to his final place of rest. The dog never left the cemetery despite repeated attempts to relocate him. In six weeks the dog was found dead, lying on his masters grave.

Surviving Firemen Dave McGee and Jack Flannery along with Minute Men got together on January 15, every year for dinner. That tradition continued until age & time eventually reunited the entire fire department.

Figure 1.28 Firemen were buried with full military honors, over 3,000 firemen came from all over Montana to pay their respects on the day of the funeral. (WesternMiningHistory.com)

Chronicles of Incidents and Response 51

Jim the sole surviving horse of the "Great Disaster of 1895" recovered from his wounds and returned to active service. His first job was to help firemen train replacement horses. New horses spending time with a veteran horse like Jim would quickly learn to behave at the hitch and on the fire ground. Each Memorial Day firemen would decorate their new hose wagon with flowers and Jim would take them to the cemetery to decorate the graves of the lost comrades. Jim spent the next 16 years pulling ladder and hose wagons until all the horses were replaced my motorized apparatus. Jim spent his final years wondering freely through Butte untethered rewarded with sugar cubes and apples from school children who knew of his celebrated history. Perhaps the most spoiled horse in Montana.

Since half of the firemen were never married, it only took one generation for their memory to disappear as did their unmarked graves at the head stones deteriorated over the years and were no longer readable. Following the turn of the Century in 2000, members of the modern Butte Fire Department were concerned their pioneer Brothers and set about to restore their memory for future generations. Old records and surveys had pinpointed the lost grave sites. Along with the Montana State Firefighters Association in 2001 firefighters gathered from across Montana to Butte's Mt. Mariah Cemetery to honor five new grave markers.

Firemen and Police Officer That Made The Supreme Sacrifice:

Angus D. Cameron, Chief	John Fred Bowman-Minute Man
John Findlay "Jack" Sloan, Assistant Chief	W. A. Brokaw-Minute Man
Samuel Ash, Pipeman	Thomas Burns-Minute Man
Edward Findlay Sloan, Pipeman	William Copeland-Minute Man
David Moses, Pipeman	Steve Deloughery-Minute Man
Peter. J. Norling, Ladderman	W. H. Nolan-Minute Man
George D. Fifer, Driver Hook and Ladder	Fred Krambeck, Police Officer

Aspen Weekly Times, Aspen, CO 19 Jan 1895
Butte-Silver Bow Public Archives and Montana PBS "Hidden Fire".

Chicago, IL August 5, 1897 Grain Elevator Fire

Seven firemen lost their lives and 40 firemen injured in an explosion which took place this during a fire in the Northwestern Grain Elevator at Cook and West Water Streets (Figure 1.29). From the force with which the explosion swept the spot on which they were standing it is certain

Figure 1.29 Seven firemen lost their lives and 40 firemen injured in an explosion which took place this during a fire in the Northwestern Grain Elevator at Cook and West Water Streets.

that they must have been instantly killed. Explosion of grain dust caused the awful havoc. Dynamite is a child's plaything compared to grain dust held in suspension in a confined atmosphere. Distribute 10 percent finely powdered grain or fibrous matter, set free a spark and there will be an explosion that will wreck the building and kill or maim everybody within reach. The stronger the structure and the more confined the air, the more disastrous and destructive will be the explosion.

It was shortly after 5 o'clock when Ed Anderson, an employee of the elevator, ran into the engine room and declared breathlessly he smelled smoke in the elevator and there must be a fire somewhere. Engineer W.W. Grubb wasted no time in talking and turned on the fire pressure leading into the big standpipes which go to the top of the building. Then Grubb accomplished a feat which justly entitles him to be classed among brave men. Fully appreciating that an elevator fire is the most dangerous of all fires of the explosive tendency of the grain dust. Grubb started out on a journey of exploration. The smoke noxious and fraught with possibilities of instant death hung heavily in the dryer. Around the dryer was an iron spiral staircase leading to the bin floor a hundred feet above. Up the staircase Grubb stumbled and ran. He could see no fire, but the smoke told of its presence, and he hurried on.

There was no blaze in the grain bins. This Grubb satisfied himself of. These were below the bin floors. Still Grubb toiled on. He reached the landing on the bin floors. The atmosphere had begun to get keen with heat. Grubb stepped out where he could get a good survey of the bin floor. He saw flame and smoke bursting through the walls. One look was enough. Another moment and he might be hurled into eternity.

Fire Marshal Anderson of the Fifth Battalion, stationed at Milwaukee an Chicago Avenues, was the first assistant chief to arrive. Anderson

Chronicles of Incidents and Response 53

saw the volume of smoke blocks away. His buggy ran across the viaduct. Anderson did not get out of it until he stopped at the box, when he sounded a 4-11 alarm. He had seen all he wanted from the bridge. He knew there was hard work ahead for him and his men, and that it would require half the department before that torrent of flame still shrouded in the pall of smoke should have worked itself out. The 4-11 brought Swenie, Musham, and a half dozen other Fire Marshals. The fire boat Yosemite was due on the first alarm. The second had summoned the Fire Queen.

Chief Swenie and Marshal Musham arrived almost simultaneously They found Anderson doing all he could. The fire was still confined to the upper part of the building. Tongues of flame and fire swept 50 feel upward. Swenie congratulated himself possibilities of a grain explosion had been minimized. It looked as if the roof had been burned away and that the dust had been exhausting itself gradually instead of going all at once. Swenie prepared himself to take the brunt of the battle. By his side stood Anderson busily executing the orders of the chief. These orders were brief: "Get all of the engines possible along the riverside." Within ten minutes there were eleven of them in a row. Nine streams were ranged alongside the building on the West side.

Chief Swenie, the incident commander who survived the explosion said "I never in my life was so surprised when that elevator let go. Never in all my connection with the fire department, have I known of a grain elevator exploding. And the peculiar part of it was that the explosion did not occur before the fire, but nearly 40 minutes after it began. If I had known, or even dreamed, that there was the possibility of that elevator going up I would have never have risked the lives of my men or my own life."

Without warning, the explosion tearing the building asunder, burying 100 firemen fighting the flames burying the men in a hail of white hot bricks and metal and scattering destruction for many blocks. Lieutenant Noon, the lone survivor of Company 8 escaped death by a fraction. It was one of the chances of conflict. He was there with poor Coogan, Schnur, and Straman. With set faces they had turned their line on the fire. They were only a few feet from the building. For a time it seemed to Noon that he had been struck in the head by a canon ball. He might as well have been, for the flat side of a ragged section of sheeting had knocked him unconscious.

"Firemen are used to those types of things, if you can use that expression." said Noon, "but that one was terrible. It was a boom and it was all over. When I came to my senses I found myself under a mass of debris, bricks, stone, pieces of wood and iron. I felt as if I was burning up. I managed to struggle to my feet and pushed the stuff off of me. Everything was a whirl of fire. I could not distinguish anything a foot away. I could see nothing of the men who had been with me. I groped around blindly until I got to the side of the freight building, Then someone gave me a helping hand and pulled me inside. I lay down on the floor and got a good breath. I was weak, sick. I am glad I am alive."

Chief Swenie and Marshal Anderson were not far from one another to the windward side of the fire. When the explosion came they with the rest of the firemen were knocked flat to the earth. The chief declares that it was his closest call so far. He was felled and turned over and over, his mouth, nose and ears filled with cinders, his eyes were filled with sparks and every inhalation of his lungs gave him torment. Staggering and groping the chief crept away from the fierce heat. His clothing was burning, his right foot and left arm felt as if they had been partially wrenched off, his rubber coat began melting and running down in rivulets. The chief finally reached the freight house. He was gasping for breath, and worked his way along to where a stream was still playing. He fell into it and soaked himself thoroughly. He made up his mind he was still able to command and stuck to his post until the last ember was extinguished.

It was only a moment until the disorganized firemen rushed to the rescue of their brothers. Firemen from other companies rushed to the spot and faced the wall of fire. Those streams that had escaped uninjured were speedily manned and for a time the fire was neglected. The water was turned on the rescuers and held there until the last unfortunate was dragged away.

Firemen That Made The Supreme Sacrifice

Char les H. Conway, Driver
William A. Hanley, Pipeman
Jacob J. Schnur
John J. Coogan
Jacob F. Straman
Thomas Monohan, 35, 15 year veteran
John Hamps
Titusville Morning Herald Pennsylvania 1897-08-06

Philadelphia, PA August 5, 1897
Naphtha Explosion

A naphtha explosion in the chemical works of the Barrett Manufacturing Company, in Frankford, yesterday, caused a dangerous and disastrous fire. Thirteen firemen and two employees were badly burned and scalded. Two men at work in the building first observed the danger from

Chronicles of Incidents and Response 55

six large naphtha tanks in the room. One of the men pointed out to his companion that the oil in two of them was boiling to the bubbling point and likely to explode. The men started for the door. Hardly had they reached it when the two tanks in rapid succession blew up, and the entire building containing them, a brick structure 100 feet long by 48 feet deep, was a mass of flames. Both were frightfully burned and scalded.

The flames spread with fearful rapidity, but notwithstanding the danger that threatened the entire block of buildings, the firemen believed that by heroic exertions they might succeed in confining the more destructive blaze to the structure in which the explosion took place. Fifteen fire companies were on the ground, and the firemen were distributed around the plant in such a way that their combined efforts would do efficient service. In the yard attached to the refining house were 180 barrels of naphtha, each containing 100 gallons, and it was their immediate effort to prevent the flames from reaching them.

Most of the firemen managed to escape, but many who had been in the storage yard among the barrels, and who were retreating from the pending explosions, were caught in the darting flames and struck by the flying iron and stone and brick work that followed each explosion. Their companions carried them to places of safety and had them conveyed to the hospital. After the explosions had ceased and the danger there from abated, the firemen succeeded in confining the fire to the immediate vicinity of the refining house.

The names of the burned and injured firemen are as follows:

William Mc Dade
Samuel L Cook
Samuel White
Fred Hampshire
Aaron Knight
James Nevling
John Duffield
James Ridgeway
Charles D. Myers
Charles Miller
John Mair
Thomas O'Dair
Jacob Leonard

None are expected to die, but several may be disfigured by the burning oil. The loss is estimated at $20,000.
The News Frederick Maryland 1897-08-05

The 1900s

The 1900s witnessed more technological development then at any period of time before, with the advent of automobiles and airplanes to computers and other electronics. Firefighting changed from bucket brigades, hand pulled, horse pulled to mechanized apparatus. However, tactics remained much the same as they had been until the 1950s when tactics and protective equipment began to evolve. Organized response to hazardous materials incidents began in the 1970s. However, the types of materials involved in the incidents and the types of incidents did not change much for the first 50 years. Explosives, crude oil, and products continued to be the primary killer of firefighters and the public until the 1950s.

New York, NY November 2, 1900 Tarrant & Co Drug House Explosion

(Greenwich Street Volcano FDNY)

A terrific explosion of chemicals in the wholesale drug store of Tarrant & Co., on the northwest corner of Warren and Greenwich streets, totally wrecked about twenty buildings and damaged three score others more of less; made more than thirty firms homeless, entailed a property loss of about $1,500,000, injured over 200 persons, some of them seriously, and killed over a score of others. The loss of life may be finally determined when the ruins have been searched. Flames were seen breaking from the third floor of the drug manufacturing house of Tarrant & Co. some minutes before the crash came. At ten minutes after 12 o'clock a terrific explosion in the Tarrant building threw the entire front of the structure into the street (Figure 1.30).

One fire company had just arrived on the scene when the awful explosion occurred. It threw the whole engine crew down the stairway. Captain Devanney, of the company, ordered his crew back into the building again. They were dragging the line to the doorway for the second time when a second explosion, more terrific that the first, come, and the whole crew was hurled across Greenwich street. In the meantime the other engines that had responded to the alarm had collected, and the firemen were busy rescuing people from surrounding buildings. Firemen had already taken many girls down the only fire escape upon the building, and more persons had been carried down the escapes of the Home-Made Restaurant, next door, and the buildings adjoining upon Warren Street.

Reliable persons who witnessed the second explosion declare they saw a column of smoke and flame shoot 300 feet into the air from the top of the drug shop, and nearly all of them assert that they clearly saw bodies in the fiery fountain. The great explosion was followed by half a dozen

Figure 1.30 At ten minutes after 12 o'clock a terrific explosion in the Tarrant building threw the entire front of the structure into the street. (Photograph by Albert Dreyfous, from the Collection of the Connecticut Firemen's Historical Society.)

more, scarcely less intense, and by a countless number of smaller ones. By this time the fire apparatus was arriving from every direction. Deputy Chief Ahern came about two minutes after the second series of explosions, and he at once ordered a fifth alarm sent out, followed by a general call for ambulances. The explosion and the fire together had now assumed the proportion of a great catastrophe. Throngs of people were rushing about in the near-by streets, many of them panic-stricken, fleeing from the fire. Half an hour after the explosion the streets for blocks around the fire were crowded with fire apparatus with a score of ambulances, while hundreds of police were being rushed from all the lower precincts of the city to form lines, and many priests from near-by parishes were going here and there in the smoke-obscured thoroughfares, seeking for injured who might need their aid.

The second explosion carried destruction in every direction. Just after the outbreak of fire from the windows of the building a downtown bound train stopped at Warren street station of the Ninth Avenue Elevated road (Figure 1.31). It passed on in time to escape the explosion. The explosion completely carried away the station, and the mass of masonry that fell with it broke through the flooring and almost demolished the structure just below the building. Captain McClusky, of the detective Bureau, was appealed to protect the funds of the bank, he being told that they were in the vault, the door of which was supposed to be unlocked. When the Captain and his men went in however, they found about $10,000 scattered in confusion over counters and floors. This was hastily thrown into the vault and the door locked.

The explosion was supposedly caused by naphtha. In the great confusion that existed it was impossible to find any person who could give the slightest detail regarding the work that was being done at the time the calamity occurred and the possible cause other than the escape of naphtha gas. The whole lower part of the city felt the shock and streets for blocks leading to the scene were strewn with glass from windows and doors, whose empty frames told of the force of the quakes.

The Cranbury Press New Jersey 1900-11-02

Figure 1.31 The second explosion carried destruction in every direction. Just after the outbreak of fire from the windows of the building a downtown bound train stopped at Warren street station of the Ninth Avenue Elevated road. (Photograph by Albert Dreyfous, from the Collection of the Connecticut Firemen's Historical Society.)

Sheridan, PA Naphtha Car Explosions May 12, 1902

Twenty-one persons are dead and not less than three hundred others injured as the result of the terrible catastrophe at Sheridan last evening. Of the three hundred injured the physicians say that at least fifty will die. At 9:30 o'clock this morning nine bodies were at the Pittsburg morgue and five at the Carnegie morgue. The last body was brought in at 5 o'clock this morning. Many of the dead have not been reported to the coroner, and an accurate list cannot be obtained at this time.

Officials of the Pan Handle Road fear a worse explosion than the three which wrought so much damage yesterday. A danger line has been established 500 yards on all sides of the burning wreckage and the railroad police are keeping the curious crowd back. A few feet below the burning wreckage lies the big 36-inch main of the Philadelphia Company, which comes from the gas fields in the southwestern portion of the state and which supplies the McKee's Rocks and lower Allegheny districts with natural gas.

It is feared that the concussion was so great yesterday that some of the joints or even the pipe itself might have been damaged, and if such is the case the gas, which is under great pressure, will soon force its way through and another terrific explosion would follow. About twenty cars are piled up between the Sheridan station and Cork's Run, in the Sheridan yards. This is still a mass of flames. It covers an area of 40 by 150 feet. In there are all kinds of merchandise.

The volunteer fire department of Sheridan and No. 10 Engine Company from the West End are playing streams on the burning debris, but little headway is being made owing to the fact that the entire wreckage is saturated with naphtha and kerosene, and every now and then a fresh volume of flame shoots out from various portions of the smoldering ruins.

The accident happened in the railroad yards at Sheridan, where the Panhandle Railroad makes a turn near Cork's Run. Banked in by two high hills, hundreds of people were caught. At this point, which is about one-fourth of a mile from the city line, there are thirty-three tracks. Upon these tracks there were several hundred cars. About the middle of these tracks at 4:45 o'clock, a heavy freight train was being made up for the West.

On this train were ten tank cars containing refined petroleum and naphtha. In the shifting that was necessary to prepare the train for her journey five tank cars, two of them filled with refined petroleum and two with naphtha, were switched with too much force and one of the cars of naphtha was broken. Instantly the inflammable by-product poured out in a stream. The trainmen, seeing that one of the cars was damaged, started to pull them all out of the way. Already the man in charge of the switch lights had made his rounds and the lights were burning. As the tank car passed over one of the lights the dripping naphtha caught on the little flame and almost instantly an explosion followed. Two brakemen were on the car. One of them jumped and escaped injury. The other was seriously injured. He was taken to the hospital.

The first explosion sent showers of burning naphtha over the freight station near at hand and also enveloped a number of carloads of coke and lumber that were close by, and in a moment all were blazing.

Knowing that there was imminent danger of another explosion, there was no attempt to save the freight station or the cars for the time, and all sought safety in flight. This fact saved many lives, for at 5 o'clock, fifteen minutes after the first explosion, the two cars of refined petroleum that had been damaged by the bursting of the tank of naphtha and were leaking blew up with a terrific report. No one was injured by this, however.

The noise of the two explosions, the smoke of the fire and the crackling of the flames told the people of Sheridan that something unusual was going on in their usually peaceful valley, and they crowded down the streets of the borough to watch the scene. School had been dismissed for some time and troops of children ran from their play to see the fire. The yards at Sheridan lie in a narrow valley, the hills on each side running up from the tracks. On the south side of the yards, cut out of the hillside, is a road on which are the tracks of the Sheridan Branch of the Pittsburg Railroad. Above these tracks is a rounded hill, bare of trees. Across the cut, on the other side of the yards, a hill fully two hundred feet high rises sheer above the round house, so steep that the thick trees and the score or more of homes there seem to cling to their hold with difficulty.

In less than fifteen minutes after the first explosion both these hills were black with men, women and children, who were eagerly watching the flames in the cut below, unconscious of the fact that two tanks filled with naphtha and with power enough to bring death to many and suffering to scores, were lying in the heart of the flames. Great clouds of smoke arose in masses from the burning debris on the tracks, the flames cracked and hissed as they took hold of the masses of coke that had been waiting for shipment, and great tongues of flames leaped from the burning skeleton of the freight station.

Suddenly a third explosion was heard, muffled this time as though from a distance, and eventually coming from the mouth of the valley on the Ohio River. Between the time when the first tank of naphtha was damaged in switching and sprung a leak, and the moment when it exploded enough of the dangerous liquid had escaped to work its way into the sewer that empties into the Ohio near the mouth of Cork's Run. Burning oil, too, found its way into the sewer, and as soon as it reached the open air at the mouth, an explosion followed. This third explosion, separated as it was by a mile from the scene of the burning cars at Sheraden, served

Chronicles of Incidents and Response

as a warning to but a few of the spectators there. Hillsides, Packed With Spectators, Enveloped in Flames.

Without a moment's warning there was an awful roar loud enough to be heard in the heart of Pittsburg, five miles away, and a sheet of solid flame shot up from the wreckage and enveloped both hillsides even to their tops. There was a moment's lull as though every living thing in the vicinity had been annihilated, and then the cries, the screaming for help and the blanks in the crowd that told of the explosion's dread results. Both of the two remaining tank cars had blown up. Aid was sent to the suffering communities from both Pittsburg and Allegheny. Surgeons hurried to the scene of the catastrophe and worked hard for hours to aid the injured and alleviate suffering. All the story cannot be told, for no words can picture the horror of the scene, the terror it inspired, the havoc it wrought. Grim fate invited the populace to a magnificent spectacle so that they might be made victims. The telescoping of the cars was one of those things that not infrequently happen without serious consequences. Its subsequent ignition by an open switch lamp is a matter which will later be given full investigation.

The cause of the explosion at the point where it once again broke out into the open is still a matter of conjecture. It may be that the flame followed fast the naphtha that had gone before and resented the cramped quarters of the culvert's opening, sending out a shot of naphtha and flame that ruined half a dozen houses in the neighborhood, wrecking them, setting them on fire, injuring not less than a hundred people incidentally. The shock of the shot from this cannon culvert was felt across the river in Allegheny where windows were broken. Speedily all of the local hospitals were notified of the second and greater catastrophe.

It was soon seen that the ambulances of all the hospitals in the city would not be able to adequately give attention to the injured. Doctors were sent for in all directions, and they were hurried to the spot on special engines provided by the railway company. The southern division of the Pittsburg Railway Company moreover sent out a number of cars especially for the transportation of the injured, and these were lined up both on the Sheraden road and the McKee's Rocks road, and the unfortunate ones were tenderly carried to them.

Not the least exciting of the many scenes attendant on the catastrophe was the scene of the arrival of the victims in Pittsburg. People lined the roads all the way in and eagerly looked at the vehicles innumerable that carried the dead and dying to the morgue and the hospitals.

Despite the fact that the rain came pouring down all morning thousands went to Sheraden today and the cars and trains were crowded with curious people from all parts who went for the sake of seeing what had happened.

Four distinct explosions wrought ruin in the Chartiers Valley this evening. It is estimated 25 persons are dead and between 100 and 150 are burned, many of whom will die. The first explosion occurred in the Sheridan yards of the Pan Handle Railroad, about five miles from the Union Station in this city. Switchmen were shifting a train of cars, to which were attached two cars of refined oil and one of Naphtha. The Pan Handle Railroad Company has thirty-six tracks through Sheridan and has succeeded in keeping communication open. The property loss will amount to at least $600,000.

Kasey, Samuel, Captain No. 6 Volunteer Hose Company, McKee's Rocks; badly burned about face.

Omaha Daily Bee May 14, 1902.

Milwaukee, WI February 4, 1903 Schwab Stamp & Seal Acid Spill

February 4, 1903 an acid spill occurred at the Schwab Stamp & Seal Company. About 2:00 p.m. an acid carboy broke open on the 2nd floor of the facility. Employees scattered after calling the fire department. A leaking carboy of nitric acid leaked creating corrosive and toxic vapors. Cleaning up the mess was only a salvage job for the trucks and Insurance Patrol. Other companies were sent home. Chief of Department Foley and Trucks 1&2 located the leaking carboy and carried it outside. Then companies spent time scattering saw dust on the floor, raking it up and tossing it outside. After venting the building, Captain White of Truck 1 became ill. During the next few hours man after man complained of choking up. At 9:00 p.m. was at his bedside. Captain White died followed by Ed Hogan.

A little earlier Tom Droney of Chemical 1 had responded to another alarm. While there he collapsed in the snow. Revived he went back to the firehouse where within a short time he also passed. By then Chief Foley was sick. Shortly after White's death he remarked I'll bet $100, that I'll be dead by morning. He was right, he died at 4:15 a.m. The fumes had seared the men's lungs dooming them to slow suffocation.

Firemen That Made The Supreme Sacrifice

James Foley, Chief of Department
Andrew White, Captain Truck 1
Edward Hogan, Pipeman, Engine 1
Thomas Droney, Pipeman, Engine 1

Injured:

>Daniel McCarthy, Truckman, Engine 1
>Thomas Clancy, Assistant Chief
>Peter Lancaster, Captain Truck 2
>William Meloy, Truckman
>George Hanrahan, Truckman
>William Kennedy, Truckman
>John Linehan, Truckman
>Joseph Nunwash, Truckman
>George Ryan, Truckman
>Jack J. Henneasey, Lieutenant
>Milwaukee Fire Historical Society

Avon, CT September 15, 1905 Fuse Co. Explosion

A town without any fire protection and citizens who are powerless to do anything following the explosion and fire. An explosion of a fuse was followed by fire in the building of the Climax Fuse Company (Figure 1.32). Caused a panic among the twenty employees in the building and resulted in the death of fifteen and injury that doubtless will prove fatal to several

Figure 1.32 An explosion of a fuse was followed by fire in the building of the Climax Fuse Company. (Courtesy: Avon CT Public Library.)

64   *Hazmatology: The Science of Hazardous Materials*

others. There was no way of coping with the flames which soon spread rapidly and in less than an hour after the explosion occurred those who were unable to escape were in the clutches of the fire that eventually burned their bodies to ashes. As the day wore on the great crowd that collected in the hamlet saw the bodies of men and women roasting in the fire and were powerless to check the flames.

The exact cause of the accident which caused the loss of fifteen lives may never be known, but it is the accepted theory here that in the effort to burn out the stoppage in one of the machines a workman caused an explosion with the hot iron he held in his hands. Inflammable material was set on fire and in a few minutes the room was a mass of flames. In an instant there was a mad rush for doors and windows and during the scramble many were pushed back into the building while others were severely burned. There were only sixty hands at work in the entire plant. It was stated tonight that some of the dead may have been overcome by smoke before they had a chance to flee and that it is improbable that any one of the seven victims lost his life by the force of the explosion. The scene at the fire was heartrending.

Friends and relatives of the missing were almost frantic with grief when it became known that the bodies of those close to them were being burned to a crisp in the ruins of the factory. As there is no fire department in the town it was utterly useless to cope with the flames with buckets. As one of the walls of the second largest building fell several bodies could be seen entangled in the mass of masonry in the basement and in a short time they, too, were reduced to ashes. Early in the evening the body of a woman was seen near the edge of the fire and an effort was made to pull it away from the flames, but on account of the great heat this was impossible. The loss on the four buildings burned is estimated to be $100,000.

This was not the first fuse factory in Avon nor the first to be destroyed by an explosion. Franklin M. Alford in 1870 converted an old cotton mill into a fuse factory. Three years later an explosion destroyed that facility and fuse making died out in Avon for almost a decade. Another fuse factory was opened by Albert Andrews with his nephew John in 1880. This factory burned down three years later, but was quickly replaced by two new buildings and production resumed. This business was the parent company for the Climax Fuse Company that was formed in 1892.

Grand Forks Herald, Grand Forks, ND 16 Sept 1905. Avon CT Public Library

Portland, OR June 27, 1911 Oil Company Fire

Chief David Campbell of the Portland fire department was instantly killed today and three firemen received serious injuries in a blaze at the plant of the oil company on the East Side. Campbell had led a small squad of firemen into the blazing warehouse when an explosion occurred. Two of them were hurled back through the door and Campbell was not seen again. His body was recovered.

Fireman That Made The Supreme Sacrifice:

Chief David Campbell
Indianapolis Star Indiana 1911-06-27

Eddystone, PA April 10, 1917
Ammunition Works Disaster

A trio of explosions of unknown origin in the shrapnel plant of the Eddystone Ammunition plant at 10:10 o'clock this morning killed fifty, seriously wounded 200, while several hundred other employees are slightly injured (Figure 1.33). A fire which is being battled by the entire fire companies of this city and nearby boroughs in the yards of the plant, at the noon hour was still under control. Residents of this city immediately after the appeal for "help" was flashed here, rushed to the plant in motor trucks and automobiles and donated their services in carrying the wounded to the hospitals. Guards from the Remington Arms, Baldwin Locomotive Works and other nearby big industries have sent their guards

Figure 1.33 A trio of explosions of unknown origin in the shrapnel plant of the Eddystone Ammunition plant at 10:10 o'clock this morning killed fifty, seriously wounded 200, while several hundred other employees are slightly injured.

to watch over the damaged shrapnel plant and help to battle the flames that are threatening the entire plant.

Heads of the ammunition works telephoned Chief of Police Vance a few minutes after ten o'clock appealing for assistance and earnestly requesting that the fire companies and all the doctors possible be rushed to the scene. A few minutes later every fire fighter in the city responded to the call and hurried to the borough. Doctors immediately upon hearing the news hurried to the plant and joined in rendering medical assistance to the injured men scattered over the plant. The force of the explosion was terrific and sent terror into the hearts of the workmen who ran for all exits. However, many hundreds of them were baffled, for directly after the explosion two other outbursts rippled through the air and men in their flying effort to escape from death were stricken down while hundreds of others fell by the wayside seriously injured with little hopes of recovery.

Motor trucks were put into service and injured men were rushed to the Chester and Crozer Hospitals while the police patrol and private owned automobiles rushed dead men to the morgue. Fire companies from all boroughs between Eddystone and Philadelphia joined forces with the local fire fighters near the noon hour in battling the flames. Hundreds of residents of this city are giving valuable service to the ground in caring for the wounded.

All physicians of the city and nearby towns were hurried to the scene of the explosion. Hurried calls were sent to the city fire department and to those of the adjoining boroughs and townships. These responded in short order and did what they could to allay the flames and get the dead and injured out of the burning structure. Their work was badly handicapped by reason of the dense smoke. All the nurses that could be pressed into service from the Chester, Crozer, Taylor and Media Hospitals were sent to the scene, as well as the private nurses of the city and nearby places. The excitement which was at high pitch in this city extended all over the country. From distant points telephone and telegram messages poured into the Times office for information in regard to the extent of the explosion.

The hospitals were unable to care for all of the injured and the Armory at Eighth and Sproul Streets was pressed into service shortly after 11 o'clock as the trucks and other vehicles brought the maimed into the city. At the hospital the Red Cross rendered assistance while some of the more trained were sent to the Armory to care for the injured. One of the injured men, at the Chester Hospital said that the first explosion was slight and occurred in the big shrapnel shop, where about 40,000 loaded shells were stored. Following this in quick succession there were intermittent smaller explosions. Then came the big explosion in the smaller building close by the shrapnel shop, which was filled with black powder. "We had but a minute to reach the door, but many of us never got that far. Some were killed and others were injured by the flying bullets."

Despite the cool-headed efforts of the Chester Hospital physicians, the nurses, Red Cross workers and volunteers, scenes that dismayed the strongest hearts were enacted during the afternoon at the entrance doors. The official list was posted after diligent work by Dr. George Cross and a corps of nurses, and frantic parents and relatives pressed into the corridor to learn if their loved ones were hospital patients. Several had visited the Crozer Hospital and the Taylor Hospital without learning any news, and with the alternative of visiting the morgue if they failed at the Chester Hospital, they pitifully begged physicians and others if some mistake had not been made in omitting a name of a loved one who might only be injured.

Chester Times Pennsylvania 1917-04-10

Morgan, NJ October 4, 1918 T.A. Gillespie & Company Explosions

On Oct. 4, 1918, explosions rocked the T.A. Gillespie Company Shell Loading Plant in New Jersey, killing scores of people and destroying the largest munitions factory in the world (Figure 1.34). In 1918 the United States was standing at the latter years of World War I, a conflict that tore the globe apart at the seams and let out incalculable waves of questions and fear. The munitions that brought such horrors had backgrounds as diverse as the soldiers themselves; in the United States production plants sprung up across the country in order to meet the product demand.

Figure 1.34 On Oct. 4, 1918, explosions rocked the T.A. Gillespie Company Shell Loading Plant in New Jersey, killing scores of people and destroying the largest munitions factory in the world.

One small town transformed by the need for munitions was Morgan, New Jersey, and in October of 1918 the town would feel the horrors being delivered overseas when a massive explosion rocked the central coastline, plunging it into a raging inferno.

The T.A Gillespie Company Shell Loading Plant was one of many constructed in New Jersey as Middlesex County was a prime location situated on the coast. It employed 5000-6,000 people many of them women. Those who sought the new jobs in munitions manufacturing were mostly immigrants who could not serve in the war and could successfully perform their jobs despite a lack of mastery in the English language. Despite the seaside advantage, some were uneasy with the placement of the Morgan facility due to its proximity to towns, something normally avoided with the construction of a munitions factory. When production began in June of 1918, the massive complex covered approximately 2,200 acres including 700 buildings used for the complicated process of manufacturing explosives. Composed of galvanized steel, concrete, and brick, some buildings were constructed to be protected against accidents and explosions.

Brick firewalls were installed in buildings with loading rooms, melting kettles, and ammonium nitrate/TNT storage, and service magazines (each capable of storing 150,000 pounds of TNT) surrounded by earthfilled bulkheads served as a buffer should an explosion occur. Other structures meant for the storage of completed shells were constructed of wood and concrete, a use of materials that would be criticized in the days following the disaster.

It was 7:40 pm on October 4, 1918 that Morgan, New Jersey, and its surrounding towns felt the earth shake. The Gillespie Plant's building 6-1-1 exploded into a raging fireball, cutting power and severing water mains. There were 12 explosions in all. Ambulances and Doctors responded to the scene from all over. Over 100 employees were killed in the explosions. Firefighters arriving on scene faced severely depleted water pressure and were left helpless while the fire ignited other magazines, which caused further destruction. Additional firefighters were unable to access the plant's hydrants situated in between and near buildings that were igniting. Explosions threw firemen into the air as they raced to other buildings in the attempt to save them and any occupants. Plant employees flew into a panic and those that could ran leaving behind all belongings in their attempts to escape.

Shells whistled through the air and massive fires and explosions shot into the sky providing both source of terror and a source of light for those desperately looking for an escape. At one point a group of firemen on site went into a nearby firehouse for coffee when two massive explosions rocked the earth and blew in the end of the firehouse. They were able to flee the site and take shelter under a hill before they were told they needed to evacuate. Employees ran into swamps, some arriving at their nearby

Chronicles of Incidents and Response 69

homes just in time to see them be ripped apart by shells. People climbed over barbed wire fences, plunged into surrounding bodies of water, and scrambled in the punctuated darkness looking for escape while trying to avoid glass, shrapnel, and explosive shells falling like heavy rain that had no end in sight.

While employees of the Gillespie plant were trying to escape the inferno, citizens of nearby South Amboy, New Jersey were trying to escape their town crumbling around them. The stream of people evacuating upon the initial explosions grew to a mass exodus with blasts shaking houses and shattering glass. People who went to bed as residents woke up as refugees forced to flee and walk the streets seeking shelter in neighboring towns while those who stayed were forced to camp outside out of fear that their homes would collapse on top of them. While many were moving as quickly as possible away from the scene, others headed *toward* the burning complex looking to help in any way possible.

Chief O'Connell of the South Amboy police reported seeing four young local priests racing toward the flames in an attempt to help anyone who may be trapped inside with one later suffering injuries from a shell explosion. The Coast Guard rushed into action guarding still standing parts of the plant, helping with home evacuations, tending to the wounded, transporting people to hospitals, and assisting in the moving of loaded barges and trainloads of TNT away from the site.

The accounts from survivors of that night paint a picture where words like "horror" simply do not seem to suffice. One member of the Coast Guard was speaking to a plant employee at a gate when the man was decapitated in front of him. Others simply stayed put with one female telephone operator at the plant and the operator of the plant drawbridge staying at their posts throughout the entire ordeal. Witnesses of the refugees from local town recall seeing people walking the streets in a daze, some holding bread, others holding items like lamps that served no use but were simply grabbed in the panic. Women fainted in the street, children looked for their parents, and masses of people simply began leaving by car and on foot with no destination in mind.

One man working in the first building to explode claims his life was saved due to the initial force of the explosion throwing him out of harm's way:

> *"Suddenly from a point a few yards away from me, there came a blinding flash. The floor seemed to rise beneath me. As conscious left me I was conscious of a deafening crash and a deep red glow that seemed everywhere. When I regained my senses I was laying on the top of a box car which had been standing near building 6-1-1. The tremendous force of the explosion had simply lifted me from my feet, then rocketed me bodily through the air, out of the building, and on the roof of the car."*

70 *Hazmatology: The Science of Hazardous Materials*

The most unique view of the disaster as it unfolded may be claimed by airmail aviator Robert Shank who flew over the site on his way to Long Island:

"The whole eight square miles that the plant covered was one vast volcanic crater of dull red flames, bursting shells and bright flashes as one magazine after another went sky high. Big cavernous holes told where a magazine or shop once stood"

Nearby towns like South Amboy were hit especially hard with windows shattered, trolley lines torn, and buildings knocked down or facing severe damage. Though the material loss was astounding the human toll was devastating. While the nature of the disaster made an exact death toll difficult to confirm (there were many accounts of family members disappearing that night without a body ever being found) it is estimated that approximately 100 people died on the grounds of the plant. The total death toll was raised not only by people involved in the immediate explosion but also by those left unprotected against the elements. Of the estimated 62,000 people that were displaced by the explosions nearly 300 people would die as the result of a flu epidemic that was sweeping the area at the time of the incident.

It would be days before the explosions and fires would cease and the extent of the devastation could be assessed. Of the 700 buildings at the T. A. Gillespie Loading Company, 325 were destroyed. The property was heavily pocked with craters, with one measuring 30 feet deep, 140 feet wide, and 150 feet long where an ammonium nitrate storage containing 1,000,000 pounds of the substance had once stood. Of the 30 million pounds of explosives located in the magazine storage and freight cars, 12 million pounds were destroyed, and of the over one million loaded shells on site that night, over 300,000 were detonated or destroyed.

Theories about the cause of the accident spread as quickly as the fires, with some wondering if this was an act of sabotage. Days later T.A. Gillespie himself spoke to newspapers, describing his belief that a kettle of amatol a mixture of TNT and ammonium nitrate had overflowed, resulting in the first blast despite employee efforts to keep the material under control. A later investigation by the army determined that the blast was not preceded by a fire and had occurred without warning. Further investigation confirmed that the initial explosion occurred in building 6-1-1, and subsequent explosions occurred in areas including shell storage, shipping cars, and finishing rooms causing massive fires and the launching of shells. An official cause of the explosion was never confirmed.

The blast that shook central New Jersey is now a faded memory, with many current residents unaware of what happened in their town.

Of the 700 plant buildings, only two remain, hardly recognizable in their transformation into local businesses and as part of a marina. The Ernst Memorial Cemetery located in Parlin, New Jersey is home to a memorial honoring those lost in the explosion. Erected in 1929, a stone sits above a mass grave reading: "In memory of the unidentified dead who gave their lives while in the service of the United States of America, at the Morgan Shell Loading Plant in the explosion of October 4–5, 1918."

Evening Public Ledger (Philadelphia, Pa.), October 5. 1918.

Rediscovering the Ruins of a Catastrophic WWI Explosion Everyone Forgot.

Sarah Blake April 2, 2014

Boston, MA January 15, 1919
'Molasses Flood' Tank Explosion

Molasses was used in the manufacture of liquor and munitions. World War I had just ended and the Eighteenth Amendment (Prohibition) set to take effect January of 1920. Owners of the molasses tank were trying to maximize their resources. Thus the leak-prone tank was filled to near capacity, setting the stage for disaster.

Twenty persons were killed and 50 injured by the explosion of a 50 foot high tank of molasses, with 2,000,000 gallons inside, on the waterfront off Commercial Street. Significant damage occurred to several buildings including a firehouse (Figure 1.35). Tonight the only bodies identified were those of a fireman, George Layhe, and two residents of tenements in

Figure 1.35 Damage occurred to several buildings including a firehouse.

the vicinity. A large number of the injured were taken to the Relief hospital. The tank was owned by the Purity Distilling company a subsidiary of the U. S. Industrial Alcohol company of Cambridge. Two million gallons of molasses rushed in a mighty stream over the streets and converted into a sticky mass the wreckage of several small buildings which had been smashed by the force of the explosion. The greatest mortality apparently occurred in a city building where a score of municipal employees were eating their lunch. The structure was demolished. Another city building also was torn from its foundations and two women occupants were severely injured. A section of the tank wall fell on a fire house crushing it. Three firemen including Leahy who was killed, were buried in the ruins. The rest of the tank wall crashed against the elevated structure of the Boston elevated railway in Commercial Street damaging three spans, suspending all traffic on the line which connects the north and south stations.

A trolley freight car on the street was blown from the tracks. Several persons passing were knocked down. Wagons, carts and motor trucks were overturned. A number of horses were killed. The first rescue party was a squad detailed from the State Nautical school ship Nantucket. Scores of ambulances army, navy, police, hospital and Red Cross, were quickly on the scene. The bodies recovered were taken to the Northern Mortuary and the injured were hurried to Relief hospital. Many firemen and city employees began the task of removing the wreckage.

The work of all the men was greatly hampered by the oozing flood of molasses (Figure 1.36). It covered the street and the surrounding district to a depth of several inches and very slowly drained down into the harbor. To hasten this process the firemen turned on several streams of water. By nightfall all of the injured had been cared for, and nine bodies had been taken to the mortuary. Throughout the night the search for additional bodies in the wreckage was kept up. The district was closely patrolled tonight. The men killed were teamsters and employees of the city who were at work in the city street department yard adjoining the electric freight yard where the explosion occurred. The molasses spread over the street to a depth of two or three inches. Many of those killed or injured were covered with molasses and could not be readily identified.

The explosion knocked over the fireboat house of Engine 47. One of the firemen was blown into the harbor. Two others were pinned in the ruins and a fourth was not accounted for. A nearby tenement house fell in. Two women and a man were taken from the ruins, all injured. Thirty-five persons were removed to hospitals and many others received medical attention and were sent to their homes. Eighteen city employees, eating their noon luncheon in an office building in the public service yard were caught in the building when it collapsed. Virtually every man in the structure was either killed or hurt. The police tonight still were searching the district for possible additional victims.

Figure 1.36 The work of all the men was greatly hampered by the oozing flood of molasses. It covered the street and the surrounding district to a depth of several inches and very slowly drained down into the harbor.

Fireman That Made The Supreme Sacrifice

George A. Layhe, 37, Third Engineer, Fireboat 31 (Figure 1.37)

Bridgeport Standard Telegram Connecticut 1919-01-16

Long Island City, NY September 14-15, 1919 Oil Tank Blaze

One fireman killed and 24 injured. With more than 50 persons injured and the damage already done estimated at from $5,000,000 to $10,000,000, weary fire fighters still were fighting tonight a threat of further explosions of oil tanks at the scene of the fire which practically wiped out the Stone and Fleming Oil Co's plant in Long Island City yesterday (Figure 1.38). Five tanks of crude oil were burning late today. Should there be a sudden shift in the wind from north to northwest, many additional tanks in plants nearby will be threatened as well as thousands of tons of coal. The firemen were working in short shifts. So exhausted had they become, that when relieved for a brief rest, they lay in the streets near the fire zone and went fast asleep. The twenty acres of fire swept territory looked like a zone in war-devastated France or Belgium. Tanks were crumpled up, huge steel girders lay in a tangled mass, few walls were left standing, and burning oil continued to flow along the surface of Newtown creek.

Hoseman
George Layhe
Engine Co. 31
January 15, 1919

Figure 1.37 George A. Layhe, 37, Third Engineer, Fireboat 31 died in the explosion.

Figure 1.38 Weary fire fighters still were fighting a threat of further explosions of oil tanks at the scene of the fire which practically wiped out the Stone and Fleming Oil Co's plant in Long Island City yesterday. (Courtesy: New York City Fire Museum.)

Chronicles of Incidents and Response

There were many spectacular deeds of heroism. Early today Lieutenant Louis Semansky threw a rubber coat over his head, rushed through the flames and turned off three valves, preventing the flow of burning oil from tank to tank. Another tank blew up a few minutes afterward and had it not been for Semansky's courage, three more undoubtedly would have gone. Thomas Whitcome of the fire boat New Yorker was fighting the fire in a tank this morning when the structure fell and burning gasoline was thrown into Newtown creek.

Whitcome tried to jump into the fire boat, but blinded by dense smoke fell into the water, the surface of which was covered with blazing oil. Hearings his screams, Fireman Benjamine Moore jumped overboard to rescue him. Fireman Framl Lannon also seized a rubber coat, and jumped but struck the two men in his dive. This submerged the men and put out the flames which were enveloping them. Lennon then threw the coat over the men and they were hoisted aboard the boat to safety.

Firemen That Made The Supreme Sacrifice:

Unknown

> **Author's Note**: *From the details of this incident I am sure there were firefighter fatalities and likely injuries as well. However, I have been unable to locate names of those who may have been killed or injured.*

Bridgeport Standard Telegram Connecticut 1919-09-15. Boston Historical Society

New York, NY July 18, 1922 Chemical Explosion

One fireman was killed and many policemen, firemen and spectators injured, and scores of others had remarkable escapes this morning in a fire and series of explosions in the six-story warehouse of the Manufacturers Transit Company, at No. 10-14 Jane Street, near Greenwich Avenue. The fire was discovered shortly after 8 o'clock. Assistant District Attorney Morgan Jones said he had learned that the building contained large quantities of magnesium, sulphur and potash. A combination of these makes the flashlight powder used by photographers. There was also a quantity of celluloid and a shipment of German toys belonging to the New York Merchandise Company, and there was much rubber.

Frederick Francis, Treasurer of the company that operates the warehouse, first refused to say what chemicals were stored there, saying he would have to see his lawyer first. The owner of the building is Edgar Bluxton, who lives at the San Remo Hotel. A rough estimate of the property loss is $1,000,000, including the value of the building as about

$300,000, print paper of the Star Publishing Company at about $300,000, and paper belonging to the Tribune Company, at $30,000. Mayor Hylan arrived early at the scene and wading through puddles of water, in some spots half way to his knees, crossed over to the west side of Fourth Street and plunged into the heavy pall of smoke in search of "Smokey Joe" Martin. The Assistant Chief came out of the building and shaking hands with the Mayor, said: "Mr. Mayor, this is the toughest and worst fires I've ever witnessed in my career in the department."

An early report that an engine company had been lost entirely to an explosion that tore off part of the roof proved to be erroneous but the real story of the incident is remarkable. The men concerned in it were six members of Rescue Squad No. 1, under Lieut. Kilbride. They were playing a hose down into the flames. Kilbride and Fireman Charles Rogencamp were at the nozzle. When the explosion came those two and their four helpers were blown back across the roof to the edge of the coping, where they lost their hose and scrambled to their feet.

A part of the coping dropped to the roof of a five-story tenement at No. 16 Jane Street, and there was another explosion at once which wrecked the two upper floors - from which, fortunately, all tenants had been removed a few minutes before. The members of the Rescue Squad found themselves stunned by this second explosion, but recovered and made their way down the fire-escape. Acting Chief "SMOKY JOE" MARTIN, who was blown through a door and slightly injured, said the building apparently contained quantities of magnesium and phosphorus, which added to the difficulties. Water poured on burning phosphorus accomplished nothing.

Fireman Made the Ultimate Sacrifice:

Schoppemeyer, Lieut. John. J. Fire Department, Engine 13

Twenty-four firefighters were injured:
Alexander, Phillip J. Engine Company No. 14; cuts.
Brown, Harry, fireman; Hook and Ladder Truck No. 5, to St. Vincent's Hospital.
Burke, Anthony E.; Hook and Ladder No. 3; lacerations to left leg.
Butts, William, fireman, Engine Company No. 33; cuts.
Calamari, Michael, Fire Patrol No. 2; both hands badly burned.
Corkery, Edward, fireman, Engine No. 33; cut and bruised.
Donlin, John, Hook and Ladder No. 3; blown through a roof scuttle by explosion.

Chronicles of Incidents and Response 77

Doughty, Thomas, Hook and Ladder No. 3; both legs crushed.

Havilane, Rudolph, Engine Company No. 72; eyes affected by smoke and chemicals; taken to

Kahn, Morris, Engine Company No. 33; lacerations.

Lewis, Joshua, Engine Company No. 3; cuts.

Lynt, E. D., fireman, Engine No. 13; cut and bruised.

McCaffrey, Michael, Lieutenant Fire Department; to St. Vincent's Hospital.

McConville, Hook and Ladder No. 5; to St. Vincent's Hospital.

Martin, Joseph, Acting Chief, Fire Department; cut and bruised.

Mullaly, Edward, Captain of Engine Company No. 74; lacerations of left foot; taken to Bellevue Hospital.

O'Brien, James, Engine Company No. 21; overcome by smoke; taken to Bellevue Hospital.

O'Connor, Capt., Fire Department Engine No. 18; cuts.

Reilly, Patrick, fireman, Truck No. 5; to St. Vincent's Hospital.

Reynolds, Peter, fireman, Engine No. 14; overcome by smoke; Bellevue Hospital.

Rosenkamp, Charles, Rescue Company No. 1; injured hand.

Rotunno, Salvatore, Engine Company No. 24; knee and hand injured.

Williams, Stephen, fireman, Engine 33; cut and bruised.

Yonholtz, Charles, fireman, Hook and Ladder Truck No. 5; to St. Vincent's Hospital.

The Evening World New York New York 1922-07-18

The Last Alarm by Michael Boucher, Gary Urbanowicz and Frederick Melahn

Pekin, IL January 3, 1924 Corn Products Refining Corn Starch Plant Explosion

Forty-two persons were killed and 100 injured, according to estimates, in an explosion and fire in the starch building in the Pekin plant of the Corn Products Refining company, early today. At 10 a. m. seven bodies, six identified, had been recovered. At that hour, two buildings of the big plant were in ruins; building No. 33, the starch powder house, where the blast occurred, being reduced to a mass of smoldering debris, and building No. 27, the starch house, still blazing, its walls standing but giving off heat so intense that no efforts could be made to search for bodies. Although ordinarily 250 men were employed in the starch houses, only 72 men were in the starch powder house, where the explosion occurred, according to the best information available. The cause of the blast has not yet been determined, but it is believed to have been caused by a dust explosion.

The explosion was so terrific that several box cars alongside the plant were shattered and blown off the tracks. The force wrecked the starching department, table and retable houses and the kiln house, causing more than $500,000 damage to these departments. Fire companies from both Peoria and Pekin responded to the calls. The fire quickly followed the explosion, and the distance and severe cold handicapped the fire fighters and rescue workers. Water froze on the ruins and gave an icy mantle to the blackened walls of the starch house and debris of the starch powder house. Only the walls of building 27 were standing at 10 a. m.

Firemen working in five box cars about 30 feet from building 33 were missing and believed to have been killed as the cars were destroyed by the explosion. Building No. 29, known as the table house, also was partly wrecked by the concussion which shattered all windows, leaving only the steel skeletons of the window frames. There are no records of firefighter deaths in either Pekin or Peoria that day. Pekin Sesquicentennial 1974, Pekin Illinois Public Library.

Milwaukee, WI April 22, 1926 Sawdust Explosion

Three firemen were killed and 14 injured in a sawdust explosion at the Marsh Woodworks Company here this afternoon. Fire earlier in the day had been brought under control. Firemen detailed to clean up were at work when the explosion occurred. The Marsh Wood Products was the scene of several earlier fires. On the afternoon of 4/20/26 a small fire got started in the boiler room where a huge bin held tons of sawdust used for fuel. Engines 14, 3, 19, 20, Fireboat 15, Trucks 8-4 and District Chiefs 3 & 4 responded just after lunch. Only a few employees were in the plant. Although no fire was showing when companies arrived, there was a thin haze of smoke and several sprinkler heads were operating. Some companies began to pick up. Engine 14 and Truck 8 dug through the smoldering sawdust looking for sparks. Then a blinding flash occurred. A dozen men were afire from head to toe and ran and stumbled from the building. They came out screaming and writhing in pain. Engine 3's crew who had been outside picking up their line went to help their brothers. A third alarm was sent in. Engine 14 and Truck 8 were wiped out.

The first man to die of his injuries was Stanley Strzeminski of 14's. Lt.Tom Hanlon passed away just after 11 that night. Next was Al Schultz of T-8 who died in the next morning. Dead that night at Deaconess Hospital was Ambrose Skorzewski (E-14). Three days later, his brother Captain John Skorzewski (T-8) succumbed. Last of the six to die was George Liefert of E-14. At least 9 others were hospitalized.

Firemen That Made The Supreme Sacrifice

Captain John Skorzewski, Truck 8
Al Schultz, Truck 8
Lieutenant Tom Hanlon, Engine 14
Ambrose Skorzewski, Pipeman, Engine 14
Stanley Strezeminski, Engine 14
George Liefert, Engine 14

Nevada State Journal, Reno, NV 22 Apr 1926

Dover, NJ July 10, 1926 Picatinny Arsenal Munitions Explosion

One of components of the Picatinny Arsenal is the Lake Denmark Naval Ammunition Storage Depot (DNASD). July 10, 1926 an explosion occurred at DNASD that killed nearly two dozen people, shaking the U.S. Military to its core. The cause was not mechanical failure, human error, or sabotage; it was Mother Nature. A single lightning strike during a thunderstorm was the unlikely source. Shortly after 5:00 in the evening, the thunderstorm produced a bolt of lightning that struck the storage depot at Picatinny Arsenal (Figure 1.39). More than 600,000 tons of explosives stored inside the depot detonated, resulting in one of the most catastrophic explosions

Figure 1.39 Shortly after 5:00 in the evening, the thunderstorm produced a bolt of lightning that struck the storage depot at Picatinny Arsenal.

80 *Hazmatology: The Science of Hazardous Materials*

in the United States. The blast completely destroyed nearly 200 buildings in a half-mile radius, resulting in $47 million in damages (more than $631 million today) twenty-one deaths and dozens more injuries. The explosion was so powerful that people reported finding debris nearly 22 miles away.

The incident at the Picatinny Arsenal prompted the United States Government to get serious about explosives safety. Shortly after the explosion, Congress created Department of Defense Explosives Safety Board (DDESB), a board that exists to "provide oversight of the development, manufacture, testing, maintenance, demilitarization, handling, transportation and storage of "explosives" within the military". The board exists to this day.

Plattsburg Sentinel New York 1926-07-13.

Chicago, IL March 11, 1927 Chemical Plant Explosion

The ranks of engine company's numbers three and six were depleted late today when an explosion killed one fireman and seriously injured ten other, two probably fatally.

They had controlled a fire in the plant of the Daigger Chemical Company, just outside the downtown district, when a terrific explosion occurred and was followed by several of lesser degree. All the other injured firemen suffered serious body burns when their clothing caught fire.

The firemen killed and injured were trapped in the basement. Fumes filled the air and the men, bruised shocked and burned, were brought out only after comrades had donned gas masks.

Five firemen collapsed at a street door they sought to escape from the gas-filled basement and were dragged to safety by bystanders.

Max Woldenberg, president and several other officials of the chemical company, were taken to the police station for questioning. Twenty-five employees of the chemical company fled the building just before the explosion occurred. More than 200 workers in nearby buildings fled to the street when they were rocked by the explosion.

Carbide of sufficient quantity to damage the buildings in the entire block was found in the basement of the two-story brick plant, Samuel Kugelman, battalion chief said, explaining that the chemical explodes upon contact with fire or water. M. J. Corrigan, another battalion chief, in charge of an investigation, said the chemical company had been under surveillance for some time in connection with reports that chemicals were stored in the basement.

Firemen Who Made the Supreme Sacrifice:

> Edward Hirschhorn
> Thomas Bender

Seriously Injured Firefighter

Patrick Kelly inhaled flames and was burned about the hands, face and body.
   Independent Helena Montana 1927-03-12

San Diego, CA August 8, 1927
Acetylene Factory Explosion

Five persons are in hospitals as the result of an explosion of acetylene gas that wrecked the plant of the Pacific Acetylene company at Thirtieth and Main Streets and a collision between a fire engine and an automobile as the firemen were answering the alarm. The explosion, resulting from unknown causes, demolished the gas plant, doing damage estimated at $36,000. So great was the force of the blast that a nine inch brick wall was thrown out, pieces of the roof were hurled 100 feet into the air, and windows in nearby houses were shattered (Figure 1.40).

Responding to the alarm of fire that followed the blast, squad company No. 1 collided with a roadster at Twelfth and Market Streets. H. B. Haley, fire engineer, received concussion of the brain and possible internal

Figure 1.40 So great was the force of the blast that a nine inch brick wall was thrown out, pieces of the roof were hurled 100 feet into the air, and windows in nearby houses were shattered.

injuries, while H. T. Card, driver of the car sustained abrasions and lacerations, and Rosalie Melaney, who was riding with him, is suffering from concussion of the brain and possible broken back. Several others of the nine firemen riding on the truck received minor injuries.
Oakland Tribune California 1927-08-08
San Diego Library Central Library Special Collections

Heidelberg, PA December 25, 1928
Refinery Fire and Explosions

Fire broke out Christmas Day at the Carnegie Refining Co. in Heidelberg Pa. (Figure 1.41). Two firemen lost their lives, nine others were injured and damage estimated at $500,000 was done when fire broke out three times at the Heidelberg plant of the Carnegie Refining Company near here. The flames were brought under control today after first sweeping through the plant yesterday morning and again breaking out last night around 11:00 p.m. The fire broke out for a third time around 4:00 p.m. the afternoon of the 26th.

Ten members of the Stowe Township VFD (Hose Co. #2 and Hose Co. #4) were riding aboard a 1924 American LaFrance Pumper when it plunged over a hillside in the fog, along Windgap Road on their way to the Heidelberg fire. The dead, both members of a fire truck of the Stowe Township volunteer fire department which crashed en route to the fire, were Adolph Sonnett, 30, and Joseph DE Petro, 45, both of Stowe Township. Eight other members of the engine company No. 2 were injured while one

Figure 1.41 Fire broke out Christmas Day at the Carnegie Refining Co. in Heidelberg, PA.

Chronicles of Incidents and Response 83

fireman was injured in battling the blaze. Additional injures occurred in subsequent breakouts of at least 15 additional firemen.

Twelve nearby boroughs sent fire companies to battle the spectacular blaze, and an army of more than 150 volunteers fought the better part of twenty-four hours to bring the fire under control and prevent its spread over the entire 20 acre plant of the company. As the flames spread to the small tanks in the distilling and refining departments, calls were sent for volunteers from Crafton, Ingram, Glendale, Kirwin Heights, Bridgeville, Stowe Township, Bower Hill, Carnegie, East Carnegie, Dormont, Mt. Lebanon and Castle Shannon, as well as for two Pittsburgh fire companies.

Firemen Who Made The Supreme Sacrifice:

Adolph Sonnett, 30

Joseph DE Petro, 45

New Castle News Pennsylvania 1928-12-26

Lakehurst, NJ May 3, 1937 Hindenburg Disaster Hydrogen Explosion & Fire

The airship Hindenburg, the largest dirigible ever built and the pride of Nazi Germany, bursts into flames upon touching its mooring mast in Lakehurst, killing 36 passengers and crewmembers (Figure 1.42). Frenchman Henri Giffard constructed the first successful airship in 1852. His hydrogen-filled blimp carried a three-horsepower steam engine that turned a large propeller and flew at a speed of six miles per hour.

The rigid airship, often known as the "zeppelin" after the last name of its innovator, Count Ferdinand von Zeppelin, was developed by the Germans in the late 19th century. Unlike French airships, the German ships had a light framework of metal girders that protected a gas-filled interior. However, like Gifford's airship, they were lifted by highly flammable hydrogen gas and vulnerable to explosion. Large enough to carry substantial numbers of passengers, one of the most famous rigid airships was the Graf Zeppelin, a dirigible that traveled around the world in 1929. In the 1930s, the Graf Zeppelin pioneered the first transatlantic air service, leading to the construction of the Hindenburg, a larger passenger airship.

On May 3, 1937, the Hindenburg left Frankfurt, Germany, for a journey across the Atlantic to Lakehurst's Navy Air Base. Stretching 804 feet from stern to bow, it carried 36 passengers and crew of 61. While attempting to moor at Lakehurst, the airship suddenly burst into flames, probably after a spark ignited its hydrogen core. Rapidly falling 200 feet to the ground, the hull of the airship incinerated within seconds. Thirteen passengers,

Figure 1.42 The airship Hindenburg, the largest dirigible ever built and the pride of Nazi Germany, bursts into flames upon touching its mooring mast in Lakehurst, NJ, killing 36 passengers and crewmembers.

21 crewmen, and 1 civilian member of the ground crew lost their lives, and most of the survivors suffered substantial injuries.

Radio announcer Herb Morrison, who came to Lakehurst to record a routine voice-over for an NBC newsreel, immortalized the Hindenburg disaster in a famous on-the-scene description in which he emotionally declared, "Oh, the humanity!" The recording of Morrison's commentary was immediately flown to New York, where it was aired as part of America's first coast-to-coast radio news broadcast. Lighter-than-air passenger travel rapidly fell out of favor after the Hindenburg disaster, and no rigid airships survived World War II.

Wikipedia, the free encyclopedia.

Atlantic City, NJ July 16, 1937 Gasoline Explosion

Shortly after 1:00 p.m. more than 200 persons were injured, some seriously today, when 40,000 gallons of gasoline exploded and shot skyward from storage tanks at the Pure Oil companies plant here at Virginia and Drexel Avenues. Thirty Firemen, including 8 officers along with three civilians were admitted to Atlantic City Hospital with burns and other injuries. Two firefighters were placed in oxygen tents. The flaming liquid dropped over a full city block, showering firemen, spectators and employees of the company and setting fire to a coal yard and an oil plant adjoining. A general alarm was sounded bringing every piece of city fire apparatus to the

Chronicles of Incidents and Response

scene. No accurate estimate of damage was immediately obtainable. But firemen said it would run into the hundreds of thousands of dollars.

The first blast occurred shortly before 3 p.m., in a 10,000 gallon gasoline storage tank. The roar shook an area for blocks around, and the streets surrounding the tank were converted into rivers of flaming fuel. Within a 10-minute period six other explosions occurred blasting 10,000 and 5,000 gallon gasoline storage tanks. Fire Chief Leeds indicated there were 40 hose line playing water on the fire at the height of the fire. By that time, thousands of resort visitors and additional thousands of residents were at the scene. Police had difficulty keeping the fire lines intact. Shortly before 4 p.m. after at least 35 firemen were put out of action by burns and injuries, Fire Chief Joseph Leeds announced no further attempt would be made to save the Pure Oil plant. He directed his remaining men to concentrate on surrounding buildings that were fired by the blasts.

When gasoline entered the streets, the public director assembled 300 men and 52 trucks to dam up the running gasoline with sand. Burning gasoline entered the sewer system and blew manhole covers 20 feet in the air. Pure Oil Companies plant is located in the inlet section of Atlantic City and a stiff breeze carried acid fumes, smoke and burning embers to all parts of the city. The Atlantic City hospital with its limited facilities was not able to care for the stream of injured brought by patrol wagon, ambulance and private automobile. Some of the injured were given emergency treatment at a contagious disease hospital while others were aided on the scene by doctors and nurses. The cause of the first explosion could not be determined.

Charleston Gazette West Virginia 1937-07-17

Hastings, NE September 15, 1944
Explosion Naval Ammunition Depot

Hastings was buzzing with excitement on June 10, 1942, six months after Pearl Harbor, when Senator George Norris and Congressman Carl Curtis announced that the Navy had authorized the establishment of a $45,000,000 Naval Ammunition Depot southeast of Hastings. The Hastings area was chosen because of its abundance of electrical power from the Tri-County project, its location equidistant from both coasts and the availability of railroads. The government immediately began the process of taking 48,753 acres of farmland, located mostly in Clay County, from 232 owners.

Construction began July 14, 1942 with Maxon Construction Company as the prime contractor. The initial construction lasted 18 months and employed over 5,000. To relieve a critical housing shortage, the Navy established a 20-acre trailer court at 14th and Burlington, site of the Adams County fairgrounds, and constructed Spencer Park in southeast Hastings.

86　　　　　　　　　　　*Hazmatology: The Science of Hazardous Materials*

The first Marine guards arrived in December 1942 and a year later the Coast Guards K-9 patrol dogs arrived. The Navy began to hire men and women for the manufacture and storage of ammunition in January 1943. The NAD was commissioned on February 22, 1943, with Captain D.F. Patterson as commander. The first test run of projectile loading was made July 1, 1943 and three days later on July Fourth, the first loaded ammunition came off the production lines ready for the fleet. Employment and production reached their maximum in June and July 1945, when the Depot was manned by 125 officers, 1,800 enlisted men, and 6,692 civilians. An additional 2,000 civilians were still working for construction companies.

In 1945 a special train of ammunition left Hastings for the West Coast and eventual use by the Pacific fleet. Among the types of ammunition produced at the Hastings NAD during the war years were 40mm shells, 16 inch projectiles, rockets, bombs, depth charges, mines and torpedoes. One of the ingredients used on some loading lines was "Yellow-D" a powder which left a residue of yellow on workers' skin. Pay for line workers started at 74 cents an hour for a 60-hour week. A typical sales clerk in Hastings was being paid 25 cents an hour at the time. The need for workers was so great that almost everyone who applied was hired. The depot operated 24 hours a day seven days a week during the war years. The NAD maintained a good safety record, which is remarkable considering the vast amounts of ammunition manufactured and the speed with which it was produced.

There Were Occasional Explosions

The official Navy history of the NAD lists only two explosions, but four occurred, all in 1944. The first was on January 27, when a six-inch shell which was being gauged exploded in the black powder room. Three men of the Negro Ordnance Battalion were killed: Adolph Johnson, Jesse Wilson, and J. C. miles. Three streets in the depot were named for them. Four months later during the early morning hours of April 6, a second explosion occurred in the bomb and mine loading area. One hundred thousand pounds of explosives blew up in a dual blast that occurred first in a boxcar being loaded and then in a cooling shed filled with mines and depth charges. Bodies of three persons were identified - Chester Arthur Curtis, Norris Elmer Frey, and Lida Sarah Mitchell - and buried along with the ashes of five others - Lois Lillian Nevins Adams, Vera E. Conant, Mary E. McQuaid, Keith Clark Mathiasen, and LaVerne L. Tompkins - in a little cemetery on the depot land, south of Inland.

Their monument erected by fellow employees reads "They gave their lives that liberty might not perish." Streets in the depot were named after these people also. This blast was felt as far away as Omaha and south into Kansas. Eyewitnesses said a blinding sheet of flame lighted up the entire sky. Glenvil the town closest to the explosion was badly damaged. Every

Chronicles of Incidents and Response

house in the village seemed to rock on its foundation. Downtown store fronts were shattered. Other nearby towns - Harvard, Fairfield and Clay Center - suffered considerable damage.

A small explosion on June 10, 1944 killed one man when a detonator in the 60mm building went off, decapitating Walter Michaelsen, a civilian employee. The largest explosion occurred at 9:15 a.m. on September 15, 1944, when the south transfer depot of the railroad line blew up, leaving a crater 550 feet long, 220 feet wide, and 50 feet deep. Reportedly, nine servicemen were killed and fifty-three injured. Those killed were Coast Guard S1/C Bert E. Hugen, and Navy S1/C Leslie Williams, S1/C Freeman Lorenzo Tull, S1/C Willie Williams, S2/C Daniel Casey, S2/C Frank William David, S2/C Samuel Burns, S2/C Clarence Randolph, and S2/C Ulysses Cole, Jr. There is still speculation that the number of dead and injured was higher. The blast was felt as far away as Kansas and Iowa. There was damage in all the towns around. A portion of the roof at the Harvard school caved in, injuring ten children.

The earthen barricades in front of the storage igloos loaded with explosives held, preventing an even greater loss of life and property. Newspaper accounts of all the explosions are limited due to the wartime security issues involved. A complete study of Navy records is yet to be done. The U.S. Naval Ammunition Depot, known locally as "The NAD" was the largest of the Navy's WWII inland munitions plants, covering almost 49,000 acres of Adams and Clay County farmland.

Ordnance was loaded, assembled and stored from groundbreaking in July 1942 to final closing in June 1966. By VJ Day in 1945, the NAD employed 10,000 military and civilian workers. At one point in WWII the site produced nearly 40% of the Navy's ordnance, including 16 inch shells. Built at a cost of $71 million, the NAD had 207 miles of railroad track, 274 miles of roads and 2200 buildings, including 10 miles of distinctive "igloo" storage bunkers. The plant embittered farmers whose land was taken by the government but also produced an economic boom for the region as Hastings' population jumped from 15,200 to 23,000 in 1943. Eventually the NAD consisted of more than 2,200 buildings, warehouses and bunkers with a value of 71 million dollars.

The Hastings NAD was the largest of four inland Navy ammunition depots, the others being located in Oklahoma, Indiana and Nevada. In April 1945 the work week was reduced from 60 to 54 hours, and in August it was reduced to 40 hours and the number of employees was reduced to 3,000. By 1949 personnel numbered 1,189. The outbreak of the Korean War brought about reactivation of the depot in August 1950. Peak employment during the Korean War occurred in January 1954 when 2,946 civilians were employed. The depot was placed on maintenance status in April 1957 and in December 1958 disestablishment was ordered, to be completed no later than June 30, 1966.

In September 1964, 10,236 acres near Clay Center were transferred to the U.S. Department of Agriculture for the Meat Animal Research Center. The administrative headquarters, comprising 640 acres and 28 buildings were transferred to Central Community College in 1966. During 1966 and 1967 surplus land at the NAD was sold to various business and industrial firms who comprise what is now known as Hastings Industrial Park East. In its lifetime the NAD produced and stored vast quantities of ammunition. From its lines came a constant flow of ammunition from 40 mm shells to 16 inch projectiles, plus rockets, bombs, depth charges, mines and torpedoes. At one time during World War II, the depot supplied 40 percent of the Navy's ammunition needs. In August 1992 the Adams County Historical Society erected a historical marker at the entrance to Central Community College.

Air Force Presence

During the Vietnam War, a portion of the NAD was turned over to the U.S. Air Force. This became a radar bomb scoring detachment that helped train pilots in electronic bombing techniques that were used in Southeast Asia (Nebraska History Museum).

Cleveland, OH October 20, 1944
LNG Leak, Explosion, Fire

Two liquefied natural gas tanks explode in Cleveland, Ohio, killing 130 people, on this day in 1944. It took all of the city's firefighters to bring the resulting industrial fire under control. At 2:30 p.m., laboratory workers at the East Ohio Gas Company spotted white vapor leaking from the large natural gas tank at the company plant near Lake Erie. The circular tank had a diameter of 57 feet and could hold 90 million cubic feet of the highly flammable gas. Ten minutes later, a massive and violent explosion rocked the entire area. Flames went as high as 2,500 feet in the air. Everything in a half-mile vicinity of the explosion was completely destroyed.

Shortly afterwards, a smaller tank also exploded. The resulting out-of-control fire necessitated the evacuation of 10,000 people from the surrounding area. Every firefighting unit in Cleveland converged on the East Ohio Gas site. It still took nearly an entire day to bring the fire under control. When the flames went out, rescue workers found that 130 people had been killed by the blast and nearly half of the bodies were so badly burned that they could not be identified. Two hundred and fifteen people were injured and required hospitalization.

The explosion had destroyed two entire factories, 79 homes in the surrounding area and more than 200 vehicles. The total bill for damages

Chronicles of Incidents and Response 89

exceeded $10 million. The cause of the blast had to do with the contraction of the metal tanks: The gas was stored at temperatures below negative 250 degrees and the resulting contraction of the metal had caused a steel plate to rupture. Over 680 people were left homeless. Newer and safer techniques for storing gas and building tanks were developed in the wake of this disaster (Ohio History Connection). The Cleveland Memory Project.

Westport, CT May 3, 1946 Truck Explosion

An explosion which suddenly sprayed blazing liquid on firemen and others as they fought a blaze in an accident damaged truck today left a toll of one fireman killed and "ten or twelve" others badly burned when a large trailer truck which had left the road and overturned, caught fire and exploded. Francis P. Dunnigan, 53, a fireman and former chief of the Westport department, was almost instantly killed. Frank L. Dennert, 55, Chief of the Westport Fire Department died later at Norwalk Hospital. At Norwalk hospital a spokesman said that "ten or twelve" men apparently "very badly injured" had been brought to the hospital. Two other burn victims died including the truck driver.

No names were available, the hospital reported. At state police headquarters the desk officer said that a detail had been dispatched to the scene of the accident, which occurred shortly before 11 p.m. on the Boston post road a mile from the center of the town. An officer on duty at the Westport fire headquarters said he had only fragmentary information as the department still was at the scene, but he understood that the injuries occurred when "something in the truck's cargo" exploded.

Before the department arrived, Chief Dolan, said, he heard several muffled explosions from within the truck. When the department truck reached the scene the firemen approached the truck with the hose from the chemical tank just as the door blew out and the men were sprayed with a blazing liquid. The police chief said that spectators rushed to the aid of the firemen, helping several of them roll on the ground to extinguish their blazing garments. A call was sent for state police aid and the injured men were rushed to the hospital in state and local police cars and in private automobiles.

Five 50 gallon drums of "Black Vulcanizing Cement" (BVC) were in the back of the trailer. (BVC is an adhesive used to repair exposed cords such as Nylon, Rayon, Fiberglass, Polyester, and Polyaramid). Heat from the burning trailer floor caused the cement in the drums to expand until two of the drums exploded. Flaming material was sprayed over the firemen severely burning them. Three of the drums did not explode, but he ends were bulged out from the heat.

Firemen Who Made The Supreme Sacrifice

Frank L. Dennert, 55
Francis P. Dunnigan, 53

Kingsport News Tennessee 1946-05-03
The Kingston Daily Freeman New York 1946-05-03

Atlanta, GA December 30, 1946 Drug Fire

Authorities sought a clue today to the type and origin of fumes which overcame 75 firemen, sending 52 to hospitals, while they fought a stubborn basement blaze in a drug store for eight hours early Sunday (Figure 1.43).

Figure 1.43 Authorities sought a clue today to the type and origin of fumes which overcame 75 firemen, sending 52 to hospitals, while they fought a stubborn basement blaze in a drug store for eight hours early Sunday.

Chronicles of Incidents and Response   91

As the gases poured up from the smoldering sub-floor stockroom, the firemen dropped one by one. At least 27 received emergency treatment at the scene at one time, but there were no fatalities. Ambulances made repeated round trips to hospitals.

Alton Evening Telegraph Illinois 1946-12-30

Texas City, TX April 16, 1947
Ammonium Nitrate Ship Explosions
The Day Texas City Lost Their Fire Department

Texas City, Texas is located approximately 40 miles Southeast of Houston and 11 miles Northwest of Galveston, Texas on Galveston Bay just north of the entrance to the Gulf of Mexico. Currently the population is estimated at just over 44,000. Founded in 1911 with settlements dating to the early 1800's Texas Citie's development has always been influenced by its location in proximity to the Gulf of Mexico and the petrochemical industry. The Port of Texas City is the eighth largest port in the United States and the third largest in Texas currently exceeding 78 million net tons of cargo annually.

Texas City has had a fully paid fire department since 1956. Prior to that time the fire department staffing consisted totally of volunteer personnel until just after the Texas City Disaster of 1947. Following the disaster the department went through a period of a combination department with some career and some volunteer firefighters before becoming all career. David Zacherl is the present fire chief and leads a force of 60 personnel operating from three fire. A fourth fire station is in the planning stages. Texas City has three front line engine companies, one Quint with a 100' ladder, foam tank, water tank, hose bed, and 1500 g.p.m. pump; a 2005 Pierce heavy duty rescue hazmat truck; a 2006 American La France rescue truck and a 1987 Zodiac Boat with 2002 motor and trailer. They have added an additional Zodiac boat and a small rescue boat. Each engine and the quint also carry rescue tools that were formerly carried on a rescue unit. All Texas City firefighters are paramedics and engines are paramedic engines. They also have in service a Water Rescue Team, High Angle Rescue/Confined Space Response Unit and Hazardous Materials Response Team. Every firefighter is trained as a hazmat technician and in all other forms of rescue as well. Texas City responds to around 40 hazmat calls a year. The majority of them are routine such as gas leaks. They usually have about 10 a year that require more knowledge and skills, and they usually have at least one a year give or take that would be classified as major incidents that last several days or weeks. Texas City Fire Department responds to an average of 4,600 calls for assistance each year.

April 16, 1947 started out like any other spring day along the Gulf Coast in the town of Texas City, population 16,000 people. Longshoreman were busy loading ships at the Port of Texas city and other town's people were going about their daily activities. The SS Grandcamp was at the Port taking on a load of ammonium nitrate fertilizer to be shipped to Europe as part of the rebuilding process following World War II. Initially the Grandcamp was scheduled to load the ammonium nitrate cargo at the Port of Houston. However, ammonium nitrate is not permitted to be handled at the Houston Port.

Transported to Texas City by rail from Nebraska and Iowa the 32.5% ammonium nitrate fertilizer was placed into the hold of the ship along with a cargo of small arms ammunition. Approximately 17,000,000 pounds (7,700 tons) of ammonium nitrate was loaded onto the ship. Also in the harbor that fateful day was the SS High Flyer located approximately 600 feet from the Grandcamp on the same dock and loaded with 2,000,000 pounds (900 tons) of ammonium nitrate and 4,000,000 pounds (1,800 tons) of sulfur.

By comparison the bomb used in the bombing of the Oklahoma City Federal Building contained 5,000 pounds of ammonium nitrate and the bomb used at the World Trade Center in New York City contained 1,500 pounds of ammonium nitrate. Ammonium nitrate at a concentration of 32.5% is classified by the United States Department of Transportation as an oxidizer and used primarily as a fertilizer for agricultural and home use. Ammonium nitrate is also used as an oxidizer in the manufacturer of explosives.

Ammonium nitrate that was loaded on the two ships at the Port of Texas City had been coated with paraffin (a hydrocarbon product) and other chemicals to prevent caking of the material. By adding paraffin to the ammonium nitrate oxidizer, it is like combining fuel and oxygen fulfilling two requirements of the fire triangle. You now have two of the three materials necessary for a fire to occur. All that is missing is heat. Because ammonium nitrate is also an oxidizer under certain conditions it can become explosive.

For an explosion to occur involving ammonium nitrate confinement is necessary for the fuel and oxidizer before the heat is applied. As longshoremen were loading the ammonium nitrate on the ship they reported that the bags were warm. This could have been caused by an exothermic (heat releasing) chemical reaction going on within the bags of ammonium nitrate. A chemical process resulting in spontaneous combustion can occur with chemical oxidizers at elevated temperatures.

Around 8:10 a.m. fire was reported deep in cargo hold number 4 of the Grandcamp. Hold Number 4 was located towards the rear of the ship at the lower level (Figure 1.44). Two fire extinguishers and a gallon jug of drinking water were applied to the fire area by the ship's crew with little

Figure 1.44 Around 8:10 a.m. fire was reported deep in cargo hold number 4 of the Grandcamp. Hold Number 4 was located towards the rear of the ship at the lower level. (Courtesy: Moore Public Library.)

effect. The captain of the ship ordered the hold sealed and steam injected into the burning hold. A common method of shipboard firefighting at that time was to seal the cargo hold where the fire was occurring and inject steam into the hold to extinguish the fire. This confinement and injection of steam is likely to have resulted in the elevation of the temperature of the ammonium nitrate and the explosion that occurred. Texas City's volunteer fire department responded to the initial report of fire with their four fire engines led by Chief Henry J. Baumgartner.

It is unknown if the fire department had any knowledge of the dangers of fires involving ammonium nitrate or potential explosives (Figure 1.45). A fire boat from Galveston was requested at about the same time by the ship's captain. Twenty seven of the department's twenty eight volunteers answered the call along with the Republic Oil Refining Company firefighting team. They set up their hose streams along the dock and applied water into the burning hold of the Grand Camp. Smoke poured from the hold followed by flames at around 9:00 a.m.

The unusual color of the smoke caught many people's eye. Some called it a peach color, some called it reddish orange. At approximately 9:12 a.m. an explosion occurred within the hold of the Grandcamp. Within an instant all 27 members of the Texas City Volunteer Fire Department at the scene were killed, some bodies were disintegrated by the heat and blast pressure of the explosion (Figure 1.46). All that remained of their fire

Figure 1.45 It is unknown if the fire department had any knowledge of the dangers of fires involving ammonium nitrate or potential explosives. (Courtesy: Moore Public Library.)

Figure 1.46 Within an instant all 27 members of the Texas City Volunteer Fire Department at the scene were killed, some bodies were disintegrated by the heat and blast pressure of the explosion. (Courtesy: Moore Public Library.)

Chronicles of Incidents and Response

95

engines were piles of twisted metal. Texas City lost all but one of their firefighters and all of their apparatus in the explosion.

Until the September 11, 2001 terrorist attacks in New York City, this was the largest single loss of life incident experienced by any fire department in the United States. In addition to the loss of life there was extensive property damage. The force of the explosion knocked two small airplanes out of the sky and was heard over 150 miles away. Debris fell on homes and businesses setting many buildings on fire. People on the streets of Galveston, 11 miles away, were knocked to the ground by the force of the blast.

Chemical plants and petroleum storage facilities along the shores of the bay were set afire as well. The explosion set fire to the USS High Flyer which was at the dock near the Grand Camp, also loaded with ammonium nitrate. An anchor from the Grand Camp weighing over 3,000 pounds was propelled two miles away and landed in a 10 foot crater. A seismologist in Denver, Colorado recorded the shock waves from the explosion and thought an atomic bomb had been detonated in Texas. In Omaha, Nebraska the Strategic Air Command briefly elevated the United States defense condition (Defcon) believing there was a nuclear attack taking place.

Following the explosion firefighters responded to Texas City from communities across Southern Texas some up to 60 miles away. A lone Texas City Firefighter Fred Dowdy who had not responded to the initial call coordinated the mutual aid firefighters coming to Texas City to offer assistance. His late response likely saved his life as Fred Dowdy was the only member of the Texas City Fire Department that was not killed by the explosion. Eventually over 200 firefighters responded from as far away as Los Angeles, California.

At approximately 1:10 a.m. (April 17) that night the High Flyer exploded just like the Grand Camp had earlier in the day. By many accounts the High Flyer explosion was greater than the one that occurred in the Grand Camp. However, the dock area had been cleared and casualties were light in the second explosion. Additional property damage occurred from the force of the blast and more fires were set throughout Texas City.

A small tidal wave created from the water in the bay displaced by the explosion traveled over 150 feet inland. It took almost a week to control and extinguish all of the fires caused by the two explosions, which destroyed the Grand Camp and High Flyer. The last body was removed from the debris nearly a month later. The Grand Camp and High Flyer explosions resulted in the worst industrial accident, producing the largest number of casualties, in the history of the United States.

When it was all over, more than 405 identified people were dead, more than 3,500 injured. There were also 63 people who died and could not be identified and more than 100 were presumed dead, as their bodies were never found. Some believed that hundreds more were killed but unaccounted for including visiting seamen, non-census laborers and their families and untold numbers of travelers. There were people located as close as 70 feet from the ship when it exploded actually survived.

Refinery infrastructure and pipelines, including about fifty oil storage tanks, were extensively damaged or destroyed from the blast pressure and resulting fires which burned for days. Monsanto Chemical Companies plant was heavily damaged and 143 of the civilian deaths occurred there among employees. Over 500 homes were destroyed and hundreds of others damaged leaving over 2,000 people homeless. Property loss was listed at over 100 million dollars. That is close to one-half billion dollars by today's standards. Bulk cargo-handling operations never again resumed at the Port of Texas City. However, the Monsanto Chemical Plant was rebuilt and back in operation within a year. The cause of the fire has never been officially determined but it could have been caused by spontaneous combustion, a carelessly discarded cigarette or intentionally set.

Services for the dead who could not be unidentified were held Sunday morning, June 22, 1947 at 10:00 A.M. Nineteen volunteer firefighters were listed as missing and some may have been among the unidentified buried on that June morning. Even though there was very little advance notice of the funerals, cars were parked a mile and a half on both sides of the highway leading to the cemetery. The crowd was estimated at 5,000. Sixty-three caskets were transported from Camp Wallace by separate hearses from fifty-one funeral homes representing twenty-eight cities. It must have been a striking procession, likely one of the longest to have ever occurred in the United States. Caskets with each victim's remains were carried by volunteer pallbearers provided by the American Legion, V.F.W., Labor Organizations and Volunteer Firefighters. Each casket was decorated with flowers donated by the Florist's Association.

When the Texas City Disaster occurred in 1947 there were no fire department hazmat teams or hazardous materials training in existence in the United States, SCBA's were generally unheard of, no placard and labeling system existed and there was little government regulation of hazardous materials. Government today plays a more active role in regulation in preventing a Texas-City-Disaster-like event from happening again. Safe handling of ammonium nitrate is covered by National Fire Protection Associations, "NFPA 490 Storage and Handling of Ammonium Nitrate." Additionally, state and local agencies may have regulations for hazardous materials including ammonium nitrate.

Besides the regulations above, the National Fire Protection Association and other government agencies have established regulations and guidelines on fire fighting techniques including potential explosives fires such as ammonium nitrate. It is required that the employees who are expected to fight fires undergo specialized training and be properly equipped as

Chronicles of Incidents and Response

stated in OSHA 1910.156. Basically, the OSHA requirements for fighting this type of fire are to wear appropriate protective clothing to prevent contact with skin and eyes and self-contained breathing apparatus (SCBA) to prevent contact with thermal decomposition products. For small fires, water spray, dry chemical, carbon dioxide or chemical foam can be effective. Water should be sprayed on fire-exposed containers to cool down the temperature and avoid explosion. Flooding amounts of water are recommended for fighting a fire involving ammonium nitrate. Cargo holds of burning ammonium nitrate should never be sealed or injected with steam. Firefighters should exercise extreme caution when fighting fires involving explosives or explosive materials. If there is any danger of explosion firefighters should withdraw, evacuate the public and let the incident take its course.

Texas City Firefighters Who Made The Supreme Sacrifice on April 16, 1947

Henry J. Baumgartner, Fire Chief

Joseph Milton Braddy, Assistant Chief

Sebastian B. Nunez, Captain

William Carl Johnson, Captain

Marshall B. Stafford, Lieutenant

William D. Pentycuff, Lieutenant

Privates

Zolan Davis

William C. O'Sullivan

Roy Louis Durio

Marcel Pentycuff

Archie Boyce Emsoff

Harvey Alonzo Menge

Henry John Findeisen

Jimmy Reddicks

Virgil D. Fereday

Robert Dee Smith

Edward Henry Henricksen

Joel Clifton Stafford

William Fred Hughes

Maurice R. Neely

Lloyd George Cain
Marion D. Westmoreland
Frank P. Jolly
Clarence J. Wood
William Louis Kaiser
Clarence Rome Vestal
Jacob Otto Meadows
(*Firehouse Magazine*)

Minot, ND July 22, 1947 Oil Storage Tank Explosion

Residents of neighborhoods near Minot's factory district returned to their homes today after a day of terror in which flames fed by gasoline and oil swept through four blocks of the industrial section and killed four men. The fire started with explosions which destroyed three bulk gasoline and oil storage tanks of the Westland and Texaco Co. plants. Many persons living on the fringes of the industrial area moved their possessions from their homes as the flames approached. Firemen limited their efforts to attempting to stop the spread of the fire. They were driven back repeatedly by the intense heat (Figure 1.47).

Figure 1.47 Firemen limited their efforts to attempting to stop the spread of the fire. They were driven back repeatedly by the intense heat. (From National Transportation Safety Board Investigation.)

It was under control today but authorities isolated the district and assigned watchers to sound an alarm if the flames started up again. Fire Chief Gilbert Malek said he would attempt to determine the total property loss today. It was expected to reach more than $1,000,000. Three other persons were burned severely when the explosions sent sheets of flaming oil onto the surface of the Mouse river running through the center of the town. It flowed along the river for a quarter of a mile but burned itself out before the flames endangered railroad bridges and structures along the river bank. But in the four-block area shaken by the explosions, the flames destroyed two bulk storage companies, two creameries, a grain elevator, a tavern and a cafe.

Wisconsin State Journal Madison 1947-07-22

Reno, NV August 16, 1948 Massive Dynamite Explosion

Fire, a shattering explosion and panic among several hundred spectators left a toll of five dead, one missing and 137 injured in Reno's worst tragedy in its history. The fire broke out about 10 a.m. in a false-fronted frame and brick building in the old business district (Figure 1.48). It drew several hundred onlookers, massed around the scene just a few blocks east of the

Figure 1.48 The fire broke out about 10 a.m. in a false-fronted frame and brick building in the old business district. (Nevada Historical Society.)

100 *Hazmatology: The Science of Hazardous Materials*

city's gambling quarter. The sudden explosion and fire killed the fire chief of a nearby city and four men. A sixth, a Reno fireman, was missing. The blast sprayed injury among the spectators. Some were trampled in the rush to get away.

As firemen fought to prevent the blaze from spreading to adjoining buildings, a policeman warned the crowd to get back that "there's dynamite in there". "The rest pushed closer. There was a roar. Smoke, shot through with spurts of flame, blanketed a half block of Lake Street." Property damage was estimated, unofficially, at between $100,000 and $300,000.

The blast blew out the front of the building, which housed small shops, a vacant cafe, and a small rooming house.

A slot machine one of those under repair in one of the shops went booming across the street. Shoes were blown off some spectators. The crowd, showered with smoking debris and splinter glass, broke in panic. Some were trampled underfoot. Others were badly cut. All available ambulances were summoned, some racing to Reno from points as distant as Truckee, Calif., 35 miles west of here. While the injured were being rushed to three hospitals, rescue crews went into the still smoldering ruins of the building to recover the bodies.

Firemen Who Made The Supreme Sacrifice

Fire Chief Hobson, 55 of adjacent Sparks.

Capt. Glenn Davis 35, Reno

Fireman Earl B. Plattt, Reno

Fireman Harold Hodson, 26, Reno

Domingo Galli, 28 Volunteer Rescue Worker

Bill Byron, 50 Volunteer Rescue Workers

The Abilene Reporter-News Texas 1948-08-16
Nevada State Historical Society

Perth Amboy, NJ June 24, 1949
Asphalt Plant Explosion

Three men died two of them buried alive under flaming asphalt as a crackling series of explosions destroyed a $500,000 asphalt plant here yesterday. The shriveled, tar covered bodies of two volunteer firemen could not be recovered for several hours after they were blown into a ditch filled with boiling asphalt. A third victim, a workman, died of burns later. Eight others were injured, two critically. Black, greasy smoke rose hundreds of feet into the air over the ruined California Refinery Co. plant. It was visible as

Chronicles of Incidents and Response

far away as Manhattan, 25 miles to the north. The first explosion let go at 2 p.m., and fire spread rapidly to adjoining stills and storage tanks. Then a 10,000 gallon asphalt tank blew 50 feet into the air, spewing its blazing contents.

Firemen Who Made The Supreme Sacrifice.

Lawrence Dambach 50

Howard Adams 36

Lowell Sun Massachusetts 1949-06-24

Philadelphia, PA October 10, 1954
Chemical Plant Explosion

Three ranking fire department officers were killed and 24 other firemen and policemen injured Thursday in an odd explosion confined entirely to the rear yard of a north Philadelphia chemical manufacturing plant. The violent blast, without any fire, occurred minutes after a telephoned warning summoned the fire company to investigate the source of escaping fumes, thought to be ammonia. A 15-foot high steel tank mounted on a wooden platform in one corner of the yard exploded without warning as firemen searched for the source of the fumes.

So powerful was the explosion that it slammed the firemen and several policemen who stood nearby, with terrific force against two 25-foot-high brick walls enclosing the yard.

Rescuers found the dead and injured lying grotesquely on the ground. The steel tank was torn in pieces. Several windows in nearby buildings were shattered.

The firemen, led by news, were called to the two-story Charles W. at 5:56 a.m. (EST). Outwardly there appeared to be no sign of trouble. The blast came 10 minutes later. Deputy Fire Commissioner George E. Hink said the cause of the fumes and the blast are a mystery. "We searched the plant and we don't know where the fumes came from," Hink said. "There was no damage inside the place. We also have no explanation for what caused the tank to explode. There will be a thorough investigation."

Firefighters That Made The Supreme Sacrifice

Deputy Chief 2 Thomas A. Kline, 59.

Battalion Chief 3 John F. Magrann, 61.

Battalion Chief 6 John J. News, 61.

Injured

Firefighter Joseph J. Bandes, 54, Engine 2.
Firefighter James F. Tygh, 33, Engine 29.
Firefighter James E. Doyle, 32, Engine 29.
Firefighter Thomas W. Wilson, 36, Engine 29.
Fire Lieutenant Charles C. Holtzman, 30, Ladder 3.
Firefighter Joseph J. Vivian, 32, Ladder 3.
Firefighter Bernard Junod, 32, Engine 2.
Daily Chronicle Centralia Washington 1954-10-28

Dumas, TX Oil July 29, 1956 Shamrock Oil & Gas Corporation Tank Farm Explosion

The Shamrock Oil & Gas Corporation tank farm fire and ensuing explosions burned and shattered three other big storage tanks and blackened others nearby (Figure 1.49). The fire started when pentane was released from a relief valve on a tank and the heavier than air vapors ignited from a flame on an asphalt tank and flashed back to the pentane tank. Firemen were attempting to cool other tanks were caught in the first explosion at 6:53 a.m. Greater loss of life was averted because many of the men had

Figure 1.49 The Shamrock Oil & Gas Corporation tank farm ensuing explosions burned and shattered three other big storage tanks and blackened others nearby. (Courtesy: Dumas Fire Department.)

Chronicles of Incidents and Response 103

backed away from the tank because of the intense heat. The other tanks, ranging from 10,000 to 20,000 barrels in capacity, quickly caught fire from the initial blast. They contained crude oil or by products.

Pentane and Hexane tank fires and explosions killed 19 firemen on July 29th and subsided on July 30th to a single fire in one tank. Apparently there was no further danger. Pentane and hexane are liquids at normal temperatures and pressures. Fifteen men burned to death almost instantly when a hot wall of fire shot across the ground when the first of four tanks exploded and burned. Four others died later of horrible burns. Some 31 other persons were burned by the blast that shot an orange fireball thousands of feet and seared everything within a quarter-mile radius.

"The fiery blast snuffed out their lives and they crumpled in their tracks," Editor Bill Lask of the Moore County News said. Bob Hamilton, a News reporter, said his hair almost caught fire 200 yards from the explosion. A shirtless boy stumbled out with his naked back in flames. He said the dead were caught in a "pool of fire." Some of the burning bodies set fire to blankets used to wrap them. Others exploded from internal steam pressure. A railroad bridge burned a quarter mile away. A Shamrock public relations official said the men were fighting a small ground fire when the 20,000 barrel tank containing pentane, a butane-like fuel, caught fire.

Nine of the dead were from Shamrock's Industrial Fire Brigade and the others volunteer firemen from nearby Dumas and Sunray. Flames from the burning tanks were shooting at midnight as high as they were at noon. But winds that earlier had blown the flames away from the farm of 50 to 75 tanks subsided and fear that the other tanks would ignite subsided. The fire was expected to burn itself out. A skeleton fire crew stood by but a Shamrock official, S. G. Wait, said there wasn't "a thing any fireman could do to prevent the entire tank farm from going up" if the fire spread.

Big Spring Daily Herald Texas 1956-07-30

Firemen Who Made The Supreme Sacrifice:

Sunray FD Ray M. Biles, 43

Sunray FD Lewis Albert Broxon, 42

Sunray FD Gilford Ralph Corse, 41

Sunray FD Claude L. Emmett, 47

Sunray FD Alvin Walter Freeman, 35

Sunray FD Durwood Carl Lilley, 40

Sunray FD James Lester Rivers, 29

Sunray FD Virgil Wayne Thomas, 39

Sunray FD Gayle D. Wier, 31

Dumas FD Ollen Wynell "Shine" Cleveland, 38

Dumas FD Billy Joe Dunn, 25

Dumas FD Sam Anderson "Sam" Gibson Jr., 38

Dumas FD Charles Wad "Charlie" Lummus, 46

Dumas FD Albert Oliver Milligan Jr., 32

Dumas FD Paschel Lavert Pool Sr., 31

Dumas FD Meryl Wayne Slagle, 33

Dumas FD Donald Walter Thompson, 27

Dumas FD Ruebert Sam "Cotton" Wier, 37

Dumas FD Joseph Warlick West, 45

Author's Note: A museum about the explosion has been established at the refinery. Shamrock Oil and Gas is currently owned by Valero Oil Company. During the preparation of this volume, I made numerous calls and sent emails to the refinery fire department, Valero Oil Company to gather further information about the incident and victims. I did not get a single response.

Brownfield, TX December 23, 1958
Butane Truck Explosion

A butane gas transport truck blew up Monday night in the midst of 500 persons standing as close to it as 10 feet. Spectators said flames shot 1,000 feet into the air and the blast blew chunks of red hot metal two miles. The explosion killed three firefighters and the truck driver and injured 168. Most of the injured were treated for burns and sent home, but at least 63 remained in hospitals, some in critical condition.

The tragedy started with a wreck between the butane transport and a trailer truck. There was a small explosion, probably of gasoline, when the trucks collided, and a fire. The Brownfield volunteer fire department rushed to the scene and spectators gathered. Twenty minutes after the wreck and original explosion, the powerful butane gas on the truck blew up.

"About 20 of us firemen were on the south side of it, all within 20 feet," volunteer fireman Phillip Thompson said. "Suddenly there was an awful roar and it seemed like the whole sky was on fire. I was knocked down. Then I was up and running. I don't know where, just running."

"A boy passed me running hard. His clothes were on fire. I jerked some of his clothes off and put out the fire." Thompson was wearing a heavy fireman's coat and trousers over his regular clothing. His clothing

Chronicles of Incidents and Response

inside the firemen's coat and trousers got "scorching hot," but he escaped with minor burns.

The injured overflowed the local hospital and ambulances and automobiles rushed them to hospitals in Lubbock, Levelland and LaMesa, about 40 miles away. Residents of Lubbock, Muleshoe and Big Spring reported seeing the flash of the explosion. Big Spring is 80 miles from Brownfield. The explosion knocked out Brownfield's power supply and there was no power for more than an hour. It blew out windows all over town. The glass in a filling station two blocks away simply disappeared.

Butane is a gas, widely used for heating in rural areas that liquefies under pressure. It is carried about as a liquid and stored in tanks on farms and ranches, but the moment it touches air, it becomes a gas.

Firemen Who Made The Supreme Sacrifice:

Jim Cousineau, 45, fire marshal

Wayland Parker, about 43, fireman.

James Ray, 40, fireman

Yuma Daily Sun Arizona 1958-12-23

Schuylkill Haven, PA June 2, 1959
Propane Truck Explosion

A propane gas truck exploded on a heavily traveled eastern Pennsylvania highway today killing at least 11 persons. Some of the victims were as far as 100 yards from the blast.

"The tanker let loose and flames swept along the road like a ball of fire, killing people milling about a trooper directing traffic," State Police Sgt. Melvin Clouser reported. Police said a tractor trailer rammed the rear of the gas truck, setting it afire. It was half an hour after the crash, while firemen fought the flames that the truck exploded (Figure 1.50).

Motorists who climbed out of their cars to see what was holding up the early morning traffic, and just the idle curious, were watching at distances of from 100 to 300 feet up the highway. Some of them and a number of the firefighters were killed. At least 15 others were injured, 5 seriously. Frank Toohey Pottsville Republican reporter, said parts of bodies were strewn over the highway. His partner, photographer Vince Ney, said the sight was as sickening as anything he had seen in military service. "The flames shot 150 feet in the air from the gas truck and there was a terrific explosion," Shoener said. "Parts of a stone wall outside an historic church were blown forward. Groups of spectators about 200 to 250 feet away from the fire were mowed down. The tank of the truck landed in a field 220 yards away."

Figure 1.50 It was half an hour after the crash, while firemen fought the flames that the truck exploded.

Several school buses were on the highway near the scene just before the explosion at about 8:45 a.m. The highway was jammed with people headed for work. Klinger said a school bus was stopped to pick up children when the propane gas truck pulled up in back of it. A tractor trailer gasoline tanker then hit the rear of the gas truck. The school bus pulled away. Klinger said that when he heard the roar he ducked down behind his police car. He was showered with pieces of rock when a boulder smashed through the windshield and out through the rear window of the car.

The gas truck contained a trailer cylinder of 7,000 gallons of propane under pressure.

Ron Kramer, 39, a school bus driver with 40 children on board, said he saw the tanker and the propane gas truck collide. When he saw the flames shoot up, he served his bus across a field to safety. Another bus, with 45 kids aboard, had just passed the accident scene.

"Heads, legs and bodies are scattered everywhere. I saw six bodies lying in one field alongside the highway." Ney said he was told that a tractor trailer and a propane gas truck had collided. State police, however, said there was no collision. State Police Trooper Earl Kilinger, sitting in his car, said parts of bodies and debris crashed through his windshield. He avoided injury by hurling himself beneath the dashboard.

Sgt. Clouser Gave This Official Version of the Tragedy

"About 8:15 a.m. we got a report of a tanker on fire south of Schuylkill Haven at a place known as Red Church on Route 122." "We sent a trooper

Chronicles of Incidents and Response 107

who set himself up about 300 feet away to divert traffic from the scene. Two fire companies arrived to fight the flames." "A number of people left their cars to see what the hold-up was. They gathered around our trooper handling the traffic. About 8:45 a.m. the tanker let loose. There was a terrific explosion." "The flames swept up the road like a ball of fire killing eight or nine around the trooper. He was saved by ducking behind the police car." "As far as we know there was no collision. We don't know what caused the fire. Sometimes these tankers catch fire for some unexplainable reason."

Fireman That Made The Supreme Sacrifice:

Earl Hillibish, Orwigsburg, Friendship Fire Co.
Indiana Evening Gazette Pennsylvania 1959-06-02

Meldrim, GA June 28, 1959
Railroad Trestle Disaster

First Train Derailment Releasing Hazardous Materials Kills 23

Hundreds of people had been killed in train disasters throughout the 19th and early 20th centuries. Those accidents did not involve freight trains carrying hazardous materials, but rather they involved passenger trains. The Meldrim Trestle Disaster appears to be the first recorded incident where people were killed by hazardous materials released from a train derailment. Involved was an eastbound Seaboard Air Line mixed freight train that derailed over the Ogeechee River. Loaded butane tank cars from the train plunged into the river below and ruptured. The resulting explosion and fire killed 23 people (Figure 1.51). The derailment was caused by the movement of rails on the trestle, as they were compressed by the moving train. An Interstate Commerce Commission (ICC) investigation faulted the railroad for not installing guard rails along the trestle, which might have helped to keep the derailed equipment on the trestle deck, minimizing the risk of a hazardous materials release.

The blast occurred about 3:30 Sunday afternoon. Two railroad tanker cars spewed flaming death on some 175 fun-seekers in a recreation area beneath a 30-foot river trestle which the Seaboard Air Line train was crossing. No one could say whether all bodies had been recovered, but Coroner Harold M. Smith said reports that dozens were accounted for were greatly overdrawn. Smith said at the time he had accounted for 15 dead. Since then two children died of burns in a Savannah hospital. A complete casualty list was not available but 14 dead have been identified. Some bodies were so badly burned that identification was difficult.

The railroad said the explosion of one butane tanker set off a second loaded with 10,000 gallons of the cooking and heating fuel. Two trainmen

Figure 1.51 The Meldrim Trestle Disaster appears to be the first recorded incident where members of the public were killed by hazardous materials released from a train derailment, 23 citizens were killed.

were injured in the blast and the pileup of freight cars near the end of the long train. The blast turned the Ogeechee River bank into blackened ruins several hundred yards from the trestle. It caught some of the victims in the water, others on the bank. Children seared by the flames floundered in the river or ran screaming from the spot. A second but lesser explosion came after rescuers reached the scene near this East Georgia town about 18 miles northwest of Savannah.

Most of the 124 cars in the Seaboard Air Line Railroad freight had cleared the 30-foot high trestle before the crash at 3:30 p. m. None fell in the water, but several piled up on the bank and burned through the night. Some of the cars were telescoped. Sparks caused by the wreck may have ignited the gas, turning the area into an inferno, or the escaping gas may have been ignited by a riverbank campfire. "The explosion came over the water with a big boom, and after that you could hear children screaming and yelling," David Parker, one of the injured, said. "It was awful after it happened not to be able to save the children, but there was just nothing to do." The scene of the explosion is five miles or more from a paved highway, and about two miles from Meldrim. The area is reached by a bumpy, dusty one-lane road which is little more than a logging trail. The freight was en route from Montgomery, AL. to Savannah.

Witnesses said several cars started tumbling off the trestle and gas began shooting out of a tank car. Butane is transported under pressure as a liquid, but becomes a vapor when exposed to the atmosphere. Survivors said the gas spread like a ground fog over the river, then flashed into flame with a tremendous explosion. This ignited a second car loaded with 10,000 gallons of butane. The impact of the double explosion wrecked several cars on the banks. Within seconds the area became a screaming mass of humanity. Sobbing fathers searched for their families. Injured persons yelled with pain as they struggled to escape the fire in woods along the bank.

George Hodges, Jr., 21, of near Meldrim said "the train made a funny bump, bump sound when it came across the trestle and I stopped swimming and looked at it. Then some of the cars fell into the water and we began to get scared." "When I saw the gas coming out of a tank car I grabbed my wife and ran. We heard the explosion, but we didn't look back until we got safely back in the woods." "When we looked back we saw a great sheet of fire sweeping over the water and into the woods. It burned everything it touched.

When it first went off it sounded just like a bomb. We went back and tried to help some of the people who weren't killed. They were all burned, charred black. It was horrible." Just as we stopped, gas began to shoot out in every direction. It was like water coming out of a fire hydrant under tremendous pressure. One stream was shooting up the river and one was shooting down. I ran out on the platform of the caboose and started hollering to the people swimming to get out of there that the gas was going to explode any minute. Some of them started to swim away, but most just stood and looked at me and talked among themselves.

The Anniston Star Alabama 1959-06-29

A memorial plaque was placed in memory of the 23 victims of the Seaboard Air Line railroad disaster on the Ogeechee River trestle near Meldrim. The inscription on the plaque reads: "A fuel car derailed into the river and the ensuing explosion took the lives of many who had gathered on a nearby sandbar."

First Victims to Be Killed by a Hazardous Materials Release from a Train Derailment (ICC)

Interstate Commerce Commission Report			
Jimmy Anderson	Charles Carpenter	L. B. Lamb	Timothy Smith
Elizabeth Dixon Barnes	Billy Dent	Terry Lane	Wayne Smith
Ted Barnes	Joan Dent	Elbie Lane.	
Julian Beasley	Frank Dixon	Florence Lane.	
Linda Beasley	Edna Dixon	Leslie Lee	
Reba Lamb Beasley	Barbara Hales	James Smith	
Michael Bland	Claudia Johnson	Margie Hales Smith	

Roseburg, OR August 8, 1959
Explosives Truck Explosion

Truck Explosion in Roseburg, killed 14 people and injured more than 125. It is amazing that given the location of the truck and people in harm's way, that the death toll wasn't much higher. 300 buildings within a 30 block area smashed and a raging fire spread over the downtown area in this city of 12,200. This incident was eventually became known as simply: "The Blast" (Figure 1.52).

In the truck were two tons of dynamite and four and a half tons of ammonium nitrate, a fertilizer with explosive nature that caused the Texas City, Texas, disaster a number of years ago.

The blast came soon after fire was discovered in a building beside which the truck was parked, three blocks from the center of the business district. The force of the blast was so great it bowled over people walking several blocks away, crushed walls of buildings, and damaged numerous houses and small apartments. Hospitals said they had 52 patients, several of them in critical condition. Blood plasma and physicians were flown from nearby Grants Pass. A coroner's deputy said some of the dead would be identified only through dental records.

Figure 1.52 Truck Explosion in Roseburg, killed 14 people and injured more than 125. It is amazing that given the location of the truck and people in harm's way, that the death toll wasn't much higher. (Courtesy: City of Roseburg.)

George Rutherford the truck driver, came into town for a delivery at Gerretsen Building Supply and parked next to the building in downtown Roseburg. It was late and the business was closed for the night. Mr. Rutherford left the truck where it was and went a hotel for the night. That night the hardware store caught fire and the fire reached the truck loaded with explosives causing the worst human caused disaster in Oregon history (Figure 1.53). Approximately 1:00 a.m. a passerby called the fire department and they arrived in about 2 minutes. Heavy smoke from the hardware store obscured the "Explosive" placards on the truck. Firefighters not realizing the hidden danger were within several feet of the explosives truck then it exploded. Some of them were literally torn apart by the force of the blast. Several people nearby were killed by broken glass, a woman and her daughter slain by the picture window through which they had been watching the fire. The explosion created a crater 54 feet wide and 12 feet deep.

The fire, fought by men and equipment from as far away as Eugene, 75 miles north, was controlled some four hours after the blast came at 1:20 a.m. The truck blew up moments after the fire siren had sounded the alarm for a blaze at the Gerretsen Building Supply Co (Figure 1.54). The driver said he was walking back toward it and was knocked down

Figure 1.53 Mr. Rutherford left the truck where it was and went a hotel for the night. That night the hardware store caught fire and the fire reached the truck loaded with explosives causing the worst human caused disaster in Oregon history. (Courtesy: City of Roseburg.)

Figure 1.54 The truck blew up moments after the fire siren had sounded the alarm for a blaze at the Gerretsen Building Supply Co. (Courtesy: City of Roseburg.)

by the blast. Volunteer fireman Tony Shukle said he was knocked down too, blocks away. He got up, he said, and "the sky was red with embers." Then, he said, fire began to spread. "There was fire all over," he said, "the big one and probably four or five small ones." For a time there was fear that a propane gas storage tank might explode but firemen cooled it with water. As the flames spread they caught a building beside which a railroad tank car of propane gas was parked. After early fears that it might explode, firemen said they had it cooled down and danger was averted.

In the wake of "The Blast" Roseburg residents came together to help each other get through. Lumber-company trucks came downtown and handed out sheets of plywood so people could board up their blown-out windows. Shopkeepers labored for days picking up debris and sweeping up glass. Neighbors helped neighbors.

Firemen and Police Officer Who Made The Supreme Sacrifice

Roy McFarland, Asst. Fire Chief
Harry Carmichael 50, Firemen
William C. Unrath, 46, volunteer fireman
Donald De Sues, 32, Policemen
Nevada State Journal Reno Nevada 1959-08-08
City of Roseburg Website

Kansas City, KS August 18, 1959
Southwest Boulevard Fire

It was a beautiful but very hot summer day in the Kansas City metropolitan area, sunny with temperatures in the 90s and a south wind of 13 mph. Before the day would end, five Kansas City, MO, firefighters and one civilian would die in an inferno of burning gasoline referred to by KMBC-TV reporter Charles Gray as "when all hell broke loose." Gray, always a strong supporter of the Kansas City, Fire Department, called it "one of the darkest days in modern history of Kansas City firefighting."

It was the second-largest loss of life in Kansas City, MO Fire Department history (Figure 1.55). The fire changed the way flammable liquids were stored at automotive service stations and how flammable-liquid fires were fought involving horizontal storage tanks. National Fire Protection Association (NFPA) codes were changed to require flammable-liquid storage tanks at automotive service stations to be placed underground following the Southwest Boulevard fire. New procedures for fighting fires in horizontal flammable-liquid storage tanks involved approaching the tanks from the sides and not the ends. Fire officers stressed the importance of wearing full turnout gear during all fires.

Many of the firefighters burned at the Southwest Boulevard fire were not wearing full turnouts. Mutual aid played a major role in the Southwest Boulevard fire and became more common as a result. The monetary loss

Figure 1.55 It was the second-largest loss of life in Kansas City Fire Department history. (Courtesy: Kansas City Missouri Fire Museum.)

114   *Hazmatology: The Science of Hazardous Materials*

to the service station and tanks of approximately $30,000 paled in terms of the human loss and suffering caused by the fire. However, the lessons learned and code changes brought about by the Southwest Boulevard fire likely saved the lives of countless firefighters who would face similar fires over the years. At 8:20 A.M., on Aug. 18, the Kansas City, KS, Fire Department received a report of a fire at the Pyramid Oil Company (Conoco Station) at 2 Southwest Blvd.

The fire started on a loading rack at the combination bulk plant and service station in Kansas City, KS, near the Kansas-Missouri state line. On the initial alarm, Kansas City, KS, dispatched three pumpers, two ladder trucks and two district chiefs (a fire apparatus equipped with a pump is referred to as a pumper in the Kansas City area). At 8:35, two additional pumpers were dispatched from Kansas City, KS.

Additional equipment was called for at 8:45, including a specially built deluge truck and foam, although foam was not effective on the fire because there was no way to contain the leaking gasoline. The only foam available at the time was protein foam. Although protein foam a durable blanket over the surface of a flammable liquid, it spreads slowly and is not as effective as aqueous film-forming foam (AFFF) in fire suppression. In order for the foam to work effectively, the flammable liquid needs to be contained.

At 9:30, two additional pumpers and off-duty firefighters were summoned from the Kansas City, KS, Fire Department. Chief Edgar Grass of the Kansas City MO, Fire Department noticed the fire from his office window (the fire was visible for 15 miles in all directions). Knowing the Kansas City, KS, Fire Department was already there and the fire was near the state line, he sent a district chief to investigate. Upon arrival, the district chief immediately requested a first-alarm assignment from the Kansas City, MO Fire Department. The first alarm was dispatched at 8:33 and a second alarm was requested at 08:37, followed by a third at 8:45, a fourth at 8:54, a fifth at 8:59 and a sixth at 10:00 a.m., following the rupture of the tank.

Even before the fatal tank rupture, dozens of firefighters had been treated for heat exhaustion from the combination heat, humidity and radiant heat created by the burning flammable liquids. Ambulances dispatched from across the city stood by in line in case they were needed to transport injured firefighters to a hospital. At the time of the fire, ambulance service in the Kansas City metropolitan area was provided by 16 private companies. Firefighters had advanced first-aid training and responded to some medical calls, but did not provide transportation.

Despite the best efforts of firefighters the burning gasoline from the leaking fuel extended underneath four 11-by-30-foot cylindrical horizontal storage tanks resting on concrete cradles, each with 21,000 gallons of fuel capacity. Three contained gasoline and one kerosene. From left to

Chronicles of Incidents and Response

right at the fire scene, Tank 1 contained 6,628 gallons of gasoline, Tank 2 contained 15,857 gallons of kerosene, Tank 3 contained 3,000 gallons of gasoline and Tank 4 contained 15,655 gallons of premium gasoline. All of the tanks failed during the fire, but Tanks 1, 2, 3 did not leave the concrete cradles they rested in. This lack of movement of the first three tanks may have given firefighters a false sense of security while fighting the fire involving Tank 4.

The tanks began to fail at approximately 10 A.M., about 90 minutes after the fire started. Tank 4 was the last to fail and when it did, it moved 94 feet from its cradle into Southwest Boulevard through a 13 inch brick wall, spreading burning gasoline and flying bricks in its path. Firefighters with 2 1/2 inch hose lines were just 74 feet from the tank when it ruptured, so their positions were over-run by the tank and burning gasoline that completely crossed Southwest Boulevard. Chief officers had ordered personnel back from the fire lines when Tank 4 began to roar like a jet engine. It was during their retreat that the tank failure occurred. The rear of Tank 4 failed and the force coming out of the tank contributed to its forward movement. Two pumpers were destroyed and three damaged by the fire.

Following the failure of Tank 4, Kansas City, MO put out a call for six reserve companies and recalled one shift of firefighters. All available ambulances were requested from the metropolitan area, and station wagons were placed into service as make-shift ambulances. Area hospitals put their disaster plans into effect and prepared to receive injured firefighters, police officers and civilians. Twenty-two firefighters were admitted to hospitals, five in critical condition, along with civilian "firefighter" Francis J. "Rocky" Toomes.

All five critically injured firefighters and Toomes died, the first at 2:45 P.M. the day of the fire and the last on August 24th. An additional 35 firefighters were treated at hospitals and released. Approximately 40 firefighters were given first-aid at the scene. All suffered from burns caused by contact with the burning gasoline from Tank 4. The cause of the tank ruptures, including the fatal rupture of Tank 4, was determined to be over pressurization of the tanks on fire because of inadequate venting of the tanks. Uninjured firefighters picked themselves up following the rupture of Tank 4 and continued to fight the fire with more determination than before. By 11 A.M., the fire had been extinguished.

All five of the firefighters who were killed were from the Kansas City, MO, Fire Department and were from two companies, Pumper 19 and Pumper 25. Pumper 19 lost its entire crew-Captain George E. Bartles, Firefighter Neal K. Owen and Driver Virgil L Sams. Pumper 25 lost two of its three crew members-Captain Peter T. Sirna and Driver Delbert W. Stone. Driver Earl Dancil was scheduled to work that day, but was on sick leave. He heard about the fire from radio and TV news reports, and responded to the scene to relieve Stone, who joined his fellow crew

116   *Hazmatology: The Science of Hazardous Materials*

members on the fire behind Sirna. Toomes and firefighter Tony Valentini were helping on the hoseline behind Stone. Toomes was a civilian who was a friend of some of the firefighters. When he appeared on the scene, he approached Sirna and asked whether he could help. The captain told him yes. According to Gray, the KMBC-TV reporter, "Toomes had arrived on scene that day as a civilian, but was a firefighter by the time he left". Because of his actions in assisting firefighters he was honored by inclusion on the Southwest Boulevard Fire Memorial.

Valentini, a firefighter from Pumper 25, had been on the department for only eight months when the fire occurred. He was the only firefighter close to the tank who escaped the inferno. He suffered injuries from flying bricks, but was not burned. Valentini told me that members of the crew of Pumper 25 had discussed their escape route should they need to abandon the hoselines - they were to move to the left or right of the burning gasoline, staying away from the center. When the tank ruptured, Sirna and Stone went to the left and Toomes went to the right. Valentini went straight away, the wrong procedure, but it ultimately saved his life. Valentini was taken to a hospital because of his injuries and was held for observation overnight. He recalls that when he returned home, his wife and mother asked him when he was going to retire. "Tomorrow" he replied, but he stayed for 40 years, retiring in 1999.

The irony for Pumper 25 was that the crew should not have even been at the fire that day. Pumper 25's crew had been dispatched earlier to a fire and saw the smoke from the Southwest Boulevard fire as they were finishing up. They went back to Station 25 to replace their wet and dirty hose with fresh, dry hose, a common practice following fires. Radio calls from dispatch were piped through all stations and any dispatches were heard by all companies on the department. Pumper 2 had been dispatched to the fire scene to stage two blocks from the fire scene in case it was needed.

Enroute to the fire, the company was involve in an accident at 11th and Broadway and taken out of service. Pumper 25 was sent to replace Pumper 2 and stage two blocks north of the fire scene. Valentini said they arrived at the staging area and sat on the apparatus, drinking soda and watching the fire. Pumper 11 was deployed on Southwest Boulevard in front of the fire, supplying water to 2 1/2 inch hoselines, when its pump failed. Pumper 25 was sent to the scene from the staging area to replace Pumper 11. When they arrived, the firefighters placed a 2 1/2 inch hoseline in service from their pumper and began fighting the fire.

The magnitude of the fire brought reporters from Kansas City TV stations and newspapers to the scene. Gray and photographer Joe Adams were among those on scene. Adams was covering another event across town and Gray was at home when he got the call about the fire. Gray looked out his window at home and saw the smoke from the fire, which was quite a distance away. Gray and Adams arrived on the fire scene at about the same time. They had been at the fire approximately an hour before the rupture of Tank 4 occurred. TV reporting in the 1950s at Channel 9 centered on still photos from news scenes that were placed into a slide show on the air with narration. They did not use video and the station did not have video equipment that could be taken to a news event.

Gray was the acting KMBC news director the day of the fire and made a decision that both he and photographer thought would get them fired. He let Adams film the Southwest Boulevard fire with his personal movie camera. When the Tank 4 rupture occurred, Channel 5 had already run out of film and the Channel 4 photographer was changing the film in his camera. As a result, KMBC had the only footage of the rupture of Tank 4 and subsequent engulfing of firefighters and equipment by the burning gasoline. This film has been used in countless training sessions over the years to teach firefighters how to fight flammable liquid fires.

Dedication of Memorial

Several hundred firefighters and families gathered on August 18, 2009, at 10 A.M. for a memorial service to honor the five firefighters and one civilian who lost their lives 50 years earlier at the Southwest Boulevard fire. The author of this Volume was honored to be invited to the memorial service and moved by the numbers of firefighters and families and general public that attend the ceremony. A beautiful memorial that was built on the spot of the fire donated by the Union Pacific Railroad and dedicated in 1993 was refurbished and rededicated on this day. Honor Guards from the Kansas City MO and KS Fire Departments presented the colors while bagpiper John Tootle played. Twenty-eight retired firefighters from Kansas City, MO, and twelve from Kansas City, Kansas who were at the fire in 1959 were at the service along with families of the firefighters who died.

Also attending was Firefighter John, the grandson of Captain Peter Sirna, who was killed in the fire. John Sirna wears badge 440, which was his grandfathers badge number. He joined the Kansas City, MO, Fire Department over the objections of his grandmother. Gray was master of ceremonies, with an invocation by Deputy Chief (ret.) Arnett Williams, Fire Chief John Paul Jones of Kansas City, KS, and Fire Chief Richard "Smokey" Dyer from Kansas City, MO, gave keynote presentations, along with Robert Wing, president of International Association of Fire Fighters (IAFF) Local 64. Remarks were made by Deputy Chief Robert A. Rocha from the Kansas City, KS, Fire Department. (Editor's Note: Chief Rocha who has become a friend over the years is now the fire chief in Corpus Christi, TX.) Commemorative 50th anniversary medals were given to the families of the fallen firefighters and an award was given to Gray by the chiefs of the two fire departments. Thanks to Ray Elder who is the curator of the Kansas City, MO., Fire Museum who was a great help in

118   *Hazmatology: The Science of Hazardous Materials*

the assembling of information, contacting people and locating resources, without which this section would not have been possible. Thanks also to Chief Jones, Chief Dyer and my friend Chief Rocha for their assistance during the preparation of this section.

Firemen Who Made The Supreme Sacrifice

Pumper 19

Captain George E. Bartles, 33, Served KCMOFD 12 years 3 months.

Firefighter Neal K. Owen, 28, Served KCMOFD 2 years 3 months.

Driver Virgil L Sams, 27, Served KCMOFD 6 Years 7 months.

Pumper 25

Captain Peter T. Sirna, 46, Served KCMOFD 16 years.

Driver Delbert W. Stone, 29, Served KCMOFD 2 years.

Civilian Fireman Francis J. Toomes

(*Firehouse Magazine*)

Auburn, NY March 31, 1960
Service Station Explosion

An explosion blew apart a gasoline service station in downtown Auburn Wednesday night and killed three firemen, the station operator and an oil company employee. Two women residents of an apartment building next door, were injured and taken to a hospital in this central New York City. The dead were inside the 30X40-foot, cinderblock building when it disintegrated. Only a large pit filled with debris was visible. The firemen had been summoned to the station because a gasoline odor had been noticeable. Firemen said employees of the station had been using gasoline to wash a grease rack. The blast shattered windows in a four-block area. Every window in the First Universalist church, across the street, and in the three-story apartment building, was broken. Occupants of the apartments were evacuated. Cinderblocks were tossed 50 to 100 feet into the street.

Fire Chief Luke J. Bergan said today a spark from a compressor motor probably had ignited the fumes. He said gas apparently had leaked from a tank into a pit area of a building attached to the station. Fireman Sidney Burridge, the only one of a four-man fire detail to survive the explosion, said his group had found three to four inches of gasoline covering the cellar of the small, attached building. The firemen had been summoned to the downtown area because of the gas odor. The five bodies were recovered from the ruins.

Firemen Who Made the Supreme Sacrifice:

LT. Alfred Murphy, 52
John Searing, 27
Anthony Contrera, 36
The Oneonta Star New York 1960-03-31

Kingsport, TN October 5, 1960 Eastman Chemical Plant Explosion

There were reports of an unusual plume of yellowish-brown smoke pouring from the top of Tennessee Eastman's new aniline plant. At approximately 4:45 p.m. E.S.T a thunderous explosion rocked the sprawling Tennessee Eastman company chemical plant late today, killing at least 16 persons and injuring more than 60 (Figure 1.56). The nitrobenzene process in Eastman's Aniline Division blew up. Witnesses said it looked like the whole plant looked like it raised up in the air and just flew into pieces. The

Figure 1.56 At approximately 4:45 p.m. EST a thunderous explosion rocked the sprawling Tennessee Eastman company chemical plant late today, killing at least 16 persons and injuring more than 60. (Courtesy: Archives of the City of Kingsport.)

blast followed by several smaller explosions was heard for miles around in upper east Tennessee. It set off a fire which raged for more than three hours before it was controlled. Shock waves vibrated through Kingsport. Windows crashed in homes and businesses across a three-mile radius. Parts of the shattered building, still searingly hot, rained down on streets and houses.

Firemen, ambulances and rescue teams from throughout the area converged on the big plant whose 160 buildings are spread over 400 acres. The holocaust was centered in a city block area of aniline buildings where dyes are made. The blast rocked downtown Kingsport, about a mile and a half away, and shattered the windows in at least 14 businesses. It was heard 20 miles away at Johnson City. Most of the injured were taken to the Holston Valley Community hospital here. Convalescing patients were discharged to make room. Hundreds of blood donors answered radio appeals for blood. Company officials said some parts of the plant which operates around the clock were being shut down as a precautionary measure. However, they said most of the plant would be operating tomorrow.

The Bridgeport Telegram Connecticut 1960-10-05

Philadelphia, PA April 17, 1961
Gasoline Station Explosion

Three firemen were killed and 25 others were injured Saturday when an explosion blew apart a three-story building housing a gasoline station. The blast occurred after firemen arrived and found smoke filling the building from a fire in the basement of the service station. Firemen John Murphy, who was working inside with a hose line, said "we thought we had the fire knocked down pretty good. Then, everything let loose." Windows throughout the neighborhood were shattered by the concussion. Bystanders were knocked to the ground and injured by flying glass and other debris. Dishes and utensils in nearby homes and canned goods on store shelves were sent flying by the shock. Capt. Arthur Grover, who was ascending the stairs to check that no one was in the third-floor apartment said that the moment of the violent blast "the whole building lifted up and then went down." The structure collapsed around Grover but the stairs held long enough for him to climb out a hole in the collapsed roof and make his way down a slope of debris to the ground.

Firemen Who Made the Supreme Sacrifice

Ray Bordier, 28, joined the fire department one year ago.
William Sieger, 49, who had just 18 months to go to retirement.

Chronicles of Incidents and Response 121

> Thomas A. Walsh, 36, who was in the middle of his career as a
> fireman, carrying on a job his late father held before him.

Bordier was killed instantly. the bodies of Sieger and Walsh were recovered later by their comrades who dug with the aid of power shovels for hours in the smoldering ruins. Flames flared up occasionally setting the boots of firemen afire as gasoline seeped from ruptured tanks underground and ignited. The explosion hurled firemen off ladders some from the third-floor level. Others inside, including Murphy, were blown out of the building.

Daily Courier Connellsville Pennsylvania 1961-04-17

Houston, TX February 3, 1962
Gasoline Trucks Collide

Two tank trucks one of them carrying 8,100 gallons of high octane gasoline collided in a predawn fog today. Both drivers were killed as explosions and fire engulfed both vehicles.

Those killed were William T. Gregory, 40, and Thomas Bell, 58, both of Houston. Gregory had just started a trip to Bryan after loading his truck with gasoline at the Coastal Transport Co. terminal. Bell's truck was empty. It belonged to the Unico Co., Abilene. Accident investigator C. G. Wright said the trucks collided as the empty vehicle made a left turn off the Hempstead highway (U.S. 290) just inside the Houston city limits. The two vehicles, clinging together, then careened into a metal utility pole. "I saw a man crawling slowly away from one of the trucks," he said. "He was on fire. Then the fire surrounded him and I couldn't see him anymore." This was the driver of the empty truck. Gregory's charred body was found later in the twisted wreckage of his truck cab. Highway traffic in the area was blocked about four hours. The crash also knocked out a 12,000-volt power line, plunging the area into darkness. Firemen brought the fire under control two hours later with the use of chemicals.

Corpus Christi Times Texas 1962-02-03

Brandtsville, PA April 29, 1963 LPG
Cars Explode At Wreck Site

Three railroad fuel tank cars exploded nearly five hours after a train derailment Sunday, spewing flames and jagged metal over a wide area. At least 17 firemen and railroad workers were taken to hospitals for treatment of burns, injuries and inhalation of chlorine fumes. Two were detained for observation. An undetermined number of persons were treated at the scene.

The cars exploded separately within 10 minutes, the third being the most serious and the one which sent the mushroom of flame towering several hundred feet into the air (Figure 1.57). Authorities had ordered the sparsely populated area evacuated several hours earlier when chlorine fumes leaked from one of the cars.

It was feared for a time that the chlorine would contaminate the Yellow Breeches Creek, which supplies drinking water for West Shore suburbs of Harrisburg some 20 miles northeast of the wreck scene. The wreck occurred in open farm country near the village of Brandtsville in York County, about five miles south of Mechanicsburg. The derailed cars were part of a 45-car freight being operated by the Western Maryland Railroad over the Reading Railroad's east-west freight line between Hagerstown, Md., and Harrisburg.

Figure 1.57 The cars exploded separately within 10 minutes, the third being the most serious and the one which sent the mushroom of flame towering several hundred feet into the air.

A railroad official said the 10th to 32nd cars jumped the double track and two tank cars carrying liquefied petroleum gas caught fire at their safety valves. The fire appeared to be under control when the first car of liquefied petroleum gas exploded with a dull roar nearly five hours later. A spokesman for the Reading said in Philadelphia it appeared the derailment was caused by a broken wheel on one of the cars carrying liquid propane gas. He said the explosion followed a fire started by the sparks as a result of the derailment. Wreck crews from Reading and Rutherford were summoned, as was a team of chlorine experts from Nitro, W. Va. A tank car leaking chlorine gas had to be sealed before the immediate area was safe and residents could return, State Civil Defense Director Richard Gertstell said.

The Titusville Herald Pennsylvania 1963-04-29

Cleveland, OH August 13, 1963 Avis Rent A Truck Propane Truck Explosion

A Bobtail Propane truck carrying 50 to 100 gallons of propane gas turned into a bomb that killed two firemen, injured 20 other persons and reduced the Avis Rent-A-Truck South to rubble (Figure 1.58). Firemen were called to the building Tuesday on a report that the truck, inside the building for servicing, was leaking propane gas from a valve. A water heater installed

Figure 1.58 A Bobtail Propane truck carrying 50 to 100 gallons of propane gas turned into a bomb that killed two firemen, injured 20 other persons and reduced the Avis Rent-A-Truck South to rubble. (Western Reserve Fire Museum & Education Center.)

124   *Hazmatology: The Science of Hazardous Materials*

without a permit is suspected as the source of ignition. Propane is heavier than air and vapors moved across the floor and reached the hot water pilot light and ignited the propane. Firemen had just entered the evacuated building when one of them spotted a flame. A fire hose had just been turned on the truck, and firemen were approaching the vehicle to determine the source of the leak. A number of firemen yelled to get out of the building, but the gas exploded before all could get out. Five firemen were among the 20 injured taken to a hospital. Assistant Fire Chief John Eble estimated damage to the building at $100,000.

Firemen That Made the Supreme Sacrifice

Lt. Edward J. Hart, 65, Engine 3, 36 years on Department
Fireman John T. McKenna, 26, Engine 14
Fireman Robert Marquart, 42, Ladder 7
Fireman Robert H. Jones, 30, Engine 3
Mt. Vernon Register-News Illinois 1963-08-14
Western Reserve Fire Museum and Education Center.
Cleveland Fire Department incident reports/Fire Investigation Reports

Hammond, IN December 23, 1963 Indiana Storage Company Warehouse Explosion

An explosion described as "something like an atomic bomb" roared through a huge burning warehouse storing anti-freeze early today. One fireman was killed and 19 persons were injured. For a time it was feared several firemen might have been caught in the flaming debris but all were accounted for. The firemen were pouring water on the Illinois Transit Warehouse when the pre-dawn explosion blasted the building and collapsed a section of it. The explosion involving antifreeze was heard for miles One fireman was knocked about 20 feet. One firefighter was killed and 18 injured. A newsman, Gene Langle of Station WJOB, said he was knocked to the ground, and when he came to his senses he ran to a police car and shouted in the radio, "There's been a terrible explosion here. Send all the help you can."

The injured were taken to St. Margaret Hospital. All were firemen except one. Aides said seven of them were treated. Most suffered from smoke inhalation and cuts. The dead man was identified as Fire Pvt. Francis Volk. He was standing by an engine and was killed outright by a flying object. The rig had pulled between the burning warehouse and an adjacent building and it was showered with flying debris. It was because of this that officials feared for some time firemen might have been trapped

Chronicles of Incidents and Response

under the rubble. The fire was brought under control about 8:30 a.m. EST. An official of the firm said the warehouse, a one-story building of steel and corrugated metal construction, contained an undetermined quantity of anti-freeze and soap. The official said he had no idea what caused the explosion or what the damage would amount to.

Fireman Who Made the Supreme Sacrifice

Bud Volk, 8 year veteran
Logansport Pharos-Tribune Indiana 1963-12-23

Anchorage, AK March 27, 1964 Great Alaskan Good Friday Earthquake

March 27, 1964 the largest earthquake to hit the United States and the 2nd largest in the world struck southern Alaska. The quake registered 9.2 on the Richter scale triggering massive tsunamis that wiped out entire villages. Landslides sent neighborhoods from suburban Anchorage into the ocean. Devastation was catastrophic ultimately killing 131 people with damage costs running upwards of 2.3 billion dollars. Large storage tanks of petroleum fuels were set afire by the quake and had to burn themselves out. Gas mains were ruptured and set afire as well (Figure 1.59).

A stunned population began to realize the economic ruin carried by the quake. In some communities, industry was as much as 95 per cent wiped out. "It might take a year and a half to two years to rebuild," said Bruce Woodford of smashed Valdez, "but we'll make it." Egan said his estimate of property damage was conservative. Other unofficial estimates were higher.

Egan had increased the figure after visiting Valdez, his home town. Information from many of the heavy hit areas was sketchy at best. In Kodiak, where a tidal wave washed out the waterfront, news reports said martial law had been proclaimed. Typically, Seward, 60 miles south of Anchorage, had only two known dead, but its business was 95 percent destroyed and few of its men still had jobs. The Alaskan Railroad, vital route from Seward to the interior, was a jumble of wrecked cars and twisted rails. A mile-long waterfront area collapsed into the sea.

All along the ring of the Gulf of Alaska where the great quake struck in fury at 5:36 p.m. Friday it was a similar story of low casualties but mighty ruin.

Anchorage, the metropolis of the state with an area population of 100,000, counted 12 dead. Its business district and its best residential section is where tottering heaps of awesome wreckage. Kodiak Island enumerated 12 dead. Its fishing fleet and canning plants were wrecked.

Figure 1.59 Large storage tanks of petroleum fuels were set afire by the quake and had to burn themselves out. Gas mains were ruptured and set afire as well. (Courtesy: U.S. Geological Survey.)

One hundred and five miles southeast of Anchorage, reports from the small town of Valdez, said many of the 32 dead were on a dock that collapsed when hit by a huge sea wave. The sea waves also worked terrible and deadly devastation thousands of miles away, killing at least 16 persons in California and Oregon. Worst of these sufferers was Crescent City, CA, more than 2,000 miles from the quake's epicenter. There 11 persons died and 15 were still missing. Anchorage, center of the Alaskan recovery effort, went soberly about its business, flinching at successive after-shocks.

One shake, felt strongly in Anchorage Easter evening, was rated at 7.3 on the Richter scale of energy by the University of Washington at Seattle 1,500 miles away. University scientists said it was a separate quake, in the Aleutian trench 600 miles northwest of Friday's epicenter, but Anchorage felt it with jittery apprehension. An earlier mid-afternoon shock led to a civil defense warning of a new tidal wave headed for Seward. It was called off quickly, but people who had lived through Friday evening's terror fled to high ground. The Friday evening quake was rated by experts at 8.2 to 8.7 on the Richter scale. This scale, measuring the release of energy, has never before rated a quake higher than 8.6 and then only rarely and

Chronicles of Incidents and Response

in unpopulated places. Anchorage wholesale grocers estimated they had about a 30-day stock of essential foods on hand mostly in wrecked warehouses, but still usable.

Electric current was being restored slowly. Many homes and buildings were without light, heat or power. Drinking water had to be boiled or melted from snow. Mass typhoid inoculations were arranged. Police, soldiers and National guardsmen, patrolling downtown Anchorage, reported no cases of attempted looting. Guardsmen, patrolling downtown's teetering four-ton slabs of the modernistic five-story J. C. Penny store and other dangerous ruins. Dynamiting was considered and rejected. A civil defense official said, "It is still possible some victims are in the rubble. We might not get to some for quite a while."

"It was as though the bottom had dropped out of the ocean." A steamship company representative thus described the violent earthquake and seismic waves that flattened and charred this seaport town and killed a reported 32 persons. The community of 1,000 on the Gulf of Alaska 150 miles southeast of Anchorage took one of the worst beatings from Friday night's violent shock. Mayor Bruce Woodford said it could be several months before residents evacuated to higher ground can return to their homes. Many of the dead were working or standing on the cityr's dock which collapsed with a roar. The freighter Chena, being unloaded at the dock, was tossed around by the wild wave action. "The water went down and then up," said steamship agent John Kelsey. "The ship hit bottom twice." Valdez Bay is 35 feet deep at dockside but the incoming tidal action covered the wharf and lifted the Chena above the normal shore level. Residents said the flip-flopping ship could be seen on the rise above housetops which normally obscure the bay.

The quake and wave also set five oil tanks on fire and damage, by official estimates, 90 percent of the town's buildings (Figure 1.60). Black smoke still curled from burning oil tanks. Mayor Woodford said property loss would total at least $20 million. "We are down on one knee," the mayor said, "but we're going to get back up." Only two bodies had been recovered but Woodford said 30 others were "surely dead." Included were five children who were watching the ship being unloaded."

Valdez, a salmon fishing and seaport community, was virtually deserted with surviving residents encamped at Gulkana, 117 highway miles to the north. The town was evacuated Saturday night except for security patrols. Cleanup crews moved in during the day Sunday. Residents said the town was hit by three tidal surges, the worst around midnight. "I was standing in water to about my knees," said ship agent Kelsey, "and it suddenly rose to chest height." The waterfront was relatively crowded with longshoremen and residents watching the ship unloading (U.S. Geological Survey).

Figure 1.60 In Valdez The quake and wave also set five oil tanks on fire and damage, by official estimates, 90 percent of the town's buildings. (Courtesy: U.S. Geological Survey.)

Marshalls Creek, PA June 26, 1964
Explosives Fire and Explosion

Marshalls Creek Pennsylvania is a small unincorporated peaceful community nestled in the Pocono Mountains of eastern Pennsylvania. The events of June 26, 1964 would change this forever. Shortly after 4 a.m. Friday, June 26, 1964, the Marshalls Creek Volunteer Fire Company was called to a tractor-trailer fire on the northbound side of Route 209, just southeast of where Regina Farms is now, in Middle Smithfield Township. Just as firefighters arrived to find the burning trailer, which at the time had been abandoned by the driver, the flames spread to the trailer's cargo of 26,000 pounds of nitro carbo nitrate (the oxidizing ingredient in fertilizer), and 4,000 pounds of partially gelatin dynamite and blasting caps and the truck exploded (Figure 1.61).

Chief McDonough responded driving Engine 2 and Engines 4-5 preceded him. He estimated it took 4 minutes to arrive at the scene. All three engines pulled up about 200 feet behind the burning trailer. There were no warning signs or placards visible on the trailer. Hose lines were deployed. As Chief McDonough passed between Engines 2 and 1 he heard a loud "spat" and could feel himself being propelled through the air. It was very difficult to breath. He did not remember hitting the

Figure 1.61 Just as firefighters arrived to find the burning trailer the trailer's cargo exploded. (Courtesy: Marshalls Creek Fire Department.)

ground, but found himself lying on his back partially pinned by a ladder from one of the pumpers. He said debris from the explosion was falling like rain.

Chief Miller was responding from his home 3 miles away when he heard the explosion. Without waiting for further word he radioed for help from Shawnee and Bushkill fire departments. Upon arrival he directed his men in caring for the injured and the responding mutual aid departments finished extinguishing the fires.

Three firefighters and two civilians were killed and Engines 1, 2, and 5 were destroyed. Leaving a large crater 10 feet deep and 40 feet wide visible from the air. Two firefighters and 11 civilians were injured. The explosion damaged the Regina Hotel, Middle Smithfield Elementary School and the Pocono Reptile Farm. Poisonous snakes were sent flying for miles to the point where it took weeks to find those snakes.

The driver of the truck Albert Koda 51, of Port Carbon was employed by the American Cyanamid Company. Koda had removed the explosive signs from the trailer and placed them under the seat before unhooking the trailer and going to find an open gas station for help. A second truck driver Joseph Horvath Jr. of Scranton saw the fire and drove to the nearest phone to sound the alarm. Horvath then returned to the scene to see if he could help.

The local fire company at the time of the fire had 50 firefighters and 5 pieces of apparatus. They had no idea that the truck contained explosives. They went to work to extinguish the tire fire when the explosion occurred. The blast killed three firefighters instantly before they knew what hit them. Killed were Francis Earl Miller, 50, Leonard R. Mosier, 48 and Edward. "Frank" Hines, "Hinesy" 48. Two other firefighters were seriously injured but survived Richard McDonough, 40 and Robert Heid, 33. In addition to the three firefighters, three civilians were also killed for a total death toll of 6. John Regina, 23, of Marshalls Creek, and passersby Joseph Horvath Jr., 24 of Scranton, and Lillian Paesch, 33, of Baltimore. In addition to the two firefighters injured, there were eight civilians also injured. Ruth Livingston, Marion Sherman, Theodore Regina and Anthony Monfredl, were all from Marshalls Creek. Passers-by included William Paesch of Baltimore (husband of Lillian Paesch who was killed), John Florio of Pleasantville, NY, Robert Shive of Sellersville and Robert Ruppert Jr. of Bethlehem. Three of the fire departments apparatus were destroyed in the initial explosion.

New York Times, 1964/6/27

50th Anniversary Memorial

During 2014 the fire company marked the Marshalls Creek explosion's 50th anniversary with a memorial ceremony at the firehouse, where Route 402 meets Marshalls Creek Road near Route 209. The three fallen firefighters' families, other area volunteer fire companies, local, county and state elected officials and the Monroe County Office of Emergency Services and Control Center were among those invited. "We have our old memorial to the explosion down alongside Route 402, but we'll be gathered at our new memorial for Saturday's event," said fire company safety officer, chaplain and former chief Eugene Berry. Costing more than $60,000 and expected to be completed by next spring, the new memorial consists of a brick walkway, surrounded by a landscaped lawn, leading to a circular gathering area. At the center of that circular area, which sits at the base of a U.S. flag pole, is the fire company's "Station 29" emblem bearing the words, "In memory of all who served."

More than a dozen fire hydrants were placed around the edge of the memorial area for Saturday's event, Berry said. A trellis consisting of old fire ladders were set up to bear three wreaths, one for each of the fallen firefighters. In the future, benches will be added to the memorial area, along with a three-section granite wall measuring 18 feet long by seven feet tall by eight inches thick. The three firefighters' names will be in the wall's center section while "The Firefighters' Prayer" and a brief history of the explosion will go on the other two sections. "This event we're having marks the anniversary of an incident that was not only tragic, but also changed federal safety regulations to now require all vehicles transporting hazardous or explosive materials to always have warning placards on them," said Fire Chief Joseph Quaresimo. "This way, the public is made

Figure 1.62 Edward Hines, 42, a welder who had helped start up the fire company in 1945 was killed in the explosion. (Courtesy: Anny Hines Jacoby Daughter.)

aware, as well as firefighters responding when one of these vehicles is on fire."

Firefighters That Made The Supreme Sacrifice

Edward Hines, 42, a welder who had helped start up the fire company in 1945, (Figure 1.62).
Leonard Mosier, 38, a carpenter, architect and local rod and gun club member.
Francis Miller, 50, a state highway department employee who also had helped start up the fire company and whose brother was chief at the time.

Pocono Record, Stroudsburg, Pa. Jun 23rd, 2014

Houston, TX August 10, 1965
Chemical Plant Explosion

An explosion almost destroyed the main building of a chemical company plant in east Harris County Monday killing two men and injuring two others. Four other persons were taken to a hospital but released when examinations showed they were not injured. Witnesses said the main building of the Retzloff Chemical Co. complex "just blew apart." Officials said the cause of the explosion at the plant which manufactures agricultural chemicals was not known. Police said the explosion was heard about five miles away and large sections of sheet metal were scattered over a 100-yard area. Other buildings were damaged as were about a dozen parked cars.

San Antonio Express Texas 1965-08-10

Chicago, IL Feb 7, 1968 Gasoline Tanker Explosion & Fire

Nine people are killed when a fire and explosion completely wrecks the offices of Mickelberry's Food Products Company at 301 West Forty-Ninth Place (Figure 1.63). Four of the deaths were firemen and one the President of the Company. A gasoline tanker truck, on site to fill the company's gasoline tanks, struck a garbage bin while driving through an alley, knocking the valve off of the tanker's discharge pipe. Gasoline poured out of the tanker, ran through an alley doorway, and into the basement of the sausage plant, where a boiler ignited the gasoline. The burning gasoline soon produced two small explosions that spread the fire and led to a buildup of gasoline fumes, which eventually caused a more powerful third explosion that destroyed the two-story general offices section of the building and demolished a portion of the sausage factory.

Figure 1.63 Nine people are killed when a fire and explosion completely wrecks the offices of Mickelberry's Food Products Company.

Chronicles of Incidents and Response 133

Chicago firefighters had just arrived on scene and were rescuing office and factory workers when the third explosion occurred. The explosion threw firefighters from their ladders and factory workers trapped on the roof fell into the rubble. Onlookers were showered with bricks, concrete, plaster, and glass, and windows as far away as three blocks were blown out. The explosion was so powerful that one section of glass block window was launched across the street, where it left an imprint in a brick building.

Following the explosion, the Chicago Fire Department issued a 5-11 alarm that brought 300 firemen to the scene. The massive response was not only due to the severity of the fire, but also because of the number of casualties. Nine individuals, including four firefighters and the Mickelberry's Company President, were killed and more than seventy people were taken to the hospital. Dozens more with injuries were treated at the scene. A check of hospitals indicated that at least 75 persons were injured, 23 hospitalized and three in critical condition. Thirteen of the injured were firemen.

Most of the injuries occurred while firemen were on the roof of a one-story section of the more than 400 employee plant trying to rescue some of the 20 frightened office workers who had poured onto the roof from the second-story windows of the adjacent two-story office section.

Two smaller explosions had occurred before the mass scramble on the roof and as firemen helped the employees descend ladders the third and most massive explosion sent them hurtling to the street below. After the blast there was only a door hanging from its frame leading to an abyss two stories deep. Authorities said the holocaust apparently started when a gasoline truck hit something and caught fire in the alley behind the plant. James Neville, second deputy fire marshal, said the gas apparently leaked from the 6,000 gallon tank of the truck into the basement of the meat-packing plant.

Chief Fire Marshal Harry Volkamer speculated that the fumes were touched off by hot water heaters or furnaces. Firemen were caught in the last and most powerful of the blasts as they battled flames and sought to rescue plant employees trapped on the roof by the fire. Many of the injured were spectators who came to watch the fire. The series of blasts some witnesses counted three, some said four some said six leveled the Mickleberry's Food Products Co. brick and concrete plant in a mixed industrial and residential area just south of Chicago's Union Stockyards.

The first explosion occurred about 4:30 p.m., a half-hour after 85 day-shift production workers had stopped work at the plant. Office workers and a night cleanup crew were still inside.

The arson squad was called in to help determine the cause of the series of blasts. Fire Commissioner Robert Quinn said investigators were attempting to determine whether gasoline leaking from a damaged

134 *Hazmatology: The Science of Hazardous Materials*

6,000-gallon tank truck in the alley behind the building triggered the disaster. Firemen prevented the gasoline trailer unit from exploding by pouring foam on it but the tractor unit was charred.

Firemen Who Made the Supreme Sacrifice

John C. Fisher Sr., Captain

Edward Liefker, Firemen

Charles Bottger, Firemen

Thomas Collins, Firemen

Capital Times Madison Wisconsin 1968-02-09

Richmond, IN April 6, 1968 Natural Gas and Explosives Fire and Explosion

Richmond Indiana's "Most Tragic Day and Finest Hour"

Saturday April 6, 1968, the day before Palm Sunday, by all accounts started out as an unusually beautiful spring day in east central Indiana. Large numbers of people were out shopping and the downtown area was bustling with activity. At 1:47 p.m. two tremendous explosions rocked the downtown area of Richmond killing 41 citizens including 7 children and injuring more than 127 (Figure 1.64). None of the injured people died, all of the fatalities occurred at the blast site. Five families suffered multiple loss of life including one family that lost a mother and two children.

The Explosion

Police officer John Ross was the first emergency responder on the scene of the two explosions on April 6, 1968 along with his partner Howard Crist. They had no idea what had happened. A body was found in the street, which they covered with a curtain torn from a broken business window. There were no massive fires at that time but they could hear ammunition popping like firecrackers in the distance. A car had flipped on its side from the force of the blast in front of the Hoosier Store. Using a heavy wooden door as a shield the officers rescued the occupant of the vehicle. They used the door as a stretcher and volunteers took the injured person to the hospital.

Ground zero on that fateful day in April was the Marting Arms Sporting Goods Store, located at the corner of 6th and Main Street in the heart of downtown Richmond. Guns, ammunition, black powder and primers were among the goods sold from the business. A gun firing range was also located in the basement of the store. Witnesses described the first

Figure 1.64 At 1:47 p.m. two tremendous explosions rocked the downtown area of Richmond killing 41 citizens including 7 children and injuring more than 127. (Courtesy: Richmond Fire Department.)

explosion as "muffled." It caused a severe vibration of the walls of Marting Arms and lifted the building slightly off its foundation. Investigation into the cause of the explosions determined that a gas leak occurred in the Marting Arms building and the leaking gas was ignited by a gunshot in the firing range.

The second explosion, the more powerful of the two, was touched off by the first, igniting gunpowder and primers stored in the basement. One person who was in the Marting Arms building at the time of the explosions survived. Jack Bales was standing shoulder to shoulder to his childhood and best friend Greg Oler at the rear counter of the store. He saw his friend fall through the floor and never saw him again, he died at the scene. Bales says "there were about one or two seconds between blasts. When the first one hit I moved to cover my head and I saw Greg go down. Then the second one hit. When I was in the hospital, I think the second day, my parents came and told me that they found Greg's body in the rubble in the basement of Marting's"

As a result of the two explosions, three buildings were completely destroyed, many others were damaged by the explosion and resulting fires. Window glass was broken for more than three blocks and the blast was heard over a mile away. Nothing but a hole in the ground remained where the sporting goods store had been. Adjoining buildings were ripped apart and fires raged through the business district. Twenty buildings had to be condemned, 125 buildings were damaged, estimated dollar loss 15 million, (over 102 million in today's dollars).

Fortunately for firefighters and police officers the explosions occurred before their arrival on scene. However, ammunition stored in the sporting goods store continued to go off as fire raged inside the building. Emergency responders faced hazards from fires, leaking gas, unstable buildings and heavy smoke reducing visibility. Hundreds of volunteers, ordinary people, joined police, firefighters and National Guard troops in searching through brick, glass and twisted metal for survivors and bodies.

Ordinary citizens helped firefighters man hose lines and performed other functions. Palladium-Item photographer Don Fasnacht captured Regernaild "Reggie" Webster helping with a fire hose (Figure 1.65). This photo has become an icon for the way citizens of Richmond and Wayne County joined together when two accidental explosions set off an inferno in downtown Richmond that day. The cooperation was all the more remarkable because in many cities, the assassination of the Rev. Dr. Martin Luther King Jr. only two days before had pushed racial tensions past the

Figure 1.65 Regernaild "Reggie" Webster helping with a fire hose April 6, 1968. (Courtesy: Richmond Fire Department.)

boiling point, resulting in riots and looting. Webster, worked at the Dana Corp's Perfect Circle division in Richmond when the photo was taken.

Firefighting efforts were directed at stopping the fire from spreading beyond the block between 5th and 6th Avenues. A stand was made at the State Theater on the East side that was successful in stopping the eastward travel of the fire. Given the massive volume of smoke and fire and the obstacles created by the explosions, fire crews did an excellent job of containing the inferno (Figure 1.66).

Construction companies from the community and personnel brought heavy equipment and worked for days without any compensation. Delucio Brothers Construction operated by six brothers brought staff and equipment from other construction sites. Mutual aid responded from surrounding areas of Indiana as well as Ohio. This aid included Civil Defense personnel, fire departments from as far away as Eaton and Dayton, Ohio, along with state and local police agencies. Fire departments responded from throughout Wayne County, Indiana including Boston, Centerville, Cambridge City, Greens Fork, Fountain City, Hagerstown, Dublin, Milton, Webster and Williamsburg. Other departments responded from neighboring counties and from Ohio to man now empty fire stations in Richmond, to handle any additional calls and provide assistance at the scene. These departments included, West

Figure 1.66 Given the massive volume of smoke and fire and the obstacles created by the explosions, fire crews did an excellent job of containing the inferno. (Courtesy: Richmond Fire Department.)

Alexandria, New Madison, Hollansburg, Trotwood and Dayton, Ohio; and Liberty, Brookville, Everton, Lynn, New Castle, Lewisville and Anderson, Indiana.

In 1968 there were no urban search and rescue teams (USAR) and few other organized collapse and confined space teams in Indiana and most other places in the United States. Little equipment or technology was available compared to today's resources, to assist rescue personnel in searching collapsed buildings and debris. However, a uniquely innovative rescue squad, Box 21 had been organized in Dayton, Ohio in 1935. Founded by seven Dayton, Ohio men visiting a firehouse decided to band together and form a rescue unit to aid firefighters. They asked which of the city's fire stations received the most disaster calls and were told that is was Box 21.

Thus the name of the organization was born. This totally volunteer organization responded to Richmond to assist in the search and rescue for victims of the explosion and were commended many times for their outstanding performance. Members are seen in several photos taken of rescue operations. Box 21 boasted 40 personnel, 8 boats, 1 canteen truck, 2 heavy duty rescue trucks, 3 ambulances and a child's ambulance donated by Kiwanis at the time of the response to Richmond.

Approximately a year before the explosion in Richmond, a Community Disaster Plan was created and exercised by the Wayne County Office of Civil Defense (Now Wayne County Emergency Management). Also involved in the development of the plan were local hospital officials, the Red Cross, the County Medical Society, City Police and the County Health Office. EMS across the United States in 1968 amounted to little more than first aid and transportation to the hospital. Because of the large number of injuries in Richmond and only a few ambulances available, private citizens used station wagons and other vehicles to transport victims to hospitals.

Most often ambulances were operated by funeral homes and hearses sometimes doubled as ambulances because of the vehicle design. (A hearse is a funerary vehicle used to carry a coffin from a church or funeral home to a cemetery.) In the funeral trade, hearses are often called funeral coaches. Many were fitted with connections that allowed for the easy installation of emergency lights and sirens to convert from hearse to ambulance. Often the ambulances were manned by funeral home workers with little or no first aid training.

Author's Note: When I was 18, I worked for a furniture store in a small town in southern Nebraska. The furniture store owner also owned and operated the funeral home. Because I was a volunteer firefighter, even without any real medical training, I was the one who went on ambulance

Chronicles of Incidents and Response

calls. It wasn't until the 1970's that EMS started to evolve with the implementation of Emergency Medical Technician programs.

In 1970, then President Lyndon Johnson's Committee on Highway Traffic Safety recommended the creation of a national certification agency to establish uniform standards for training and examination of personnel active in the delivery of emergency ambulance service. Richmond's first class of EMT's were trained in 1984 and certified in 1985. Today, all firefighters in the City of Richmond are required to have a minimum EMS Certification of EMT. In 2012 Richmond Fire Department was certified as an Advanced Life Support (ALS) provider.

Firefighter Bill Smith had been on the Richmond Fire Department for about a year and was assigned to Station 3. As his unit arrived on scene in front of Holthouse Furniture Store they were met by a police officer who told them there were people trapped in the basement. Firefighter Smith went to the basement and encountered a woman who was trapped who held his hand and begged him to save her. He tried to free the woman from the debris, but was unable to remove her on his own.

Firefighter Smith proceeded back outside and located two additional firefighters to assist him in rescuing the woman. (In 1968 radio communication capability on individual firefighters was not like it is today. While many emergency vehicles were equipped with radios, few portable radios were in use. Much of necessary communication occurred face to face or by the use of runners.) As firefighters opened the door to the basement they were met by heavy flames and smoke. Everyone who had been trapped in the basement died in the fire. Firefighter Smith retired from the Richmond Fire Department in 1992 after 25 years service to the community.

Twelve year old Jeannine Herron was at Virginia's Beauty Shop and had just finished getting a hair cut when the explosion occurred. She ended up on the roof of Sargent's Wallpaper Store, a couple of shops away. At the time it was believed she was blown there by the force of the second explosion. Jeannine still doesn't remember for sure how she got on the roof. She now believes she climbed onto the roof through burning debris. She remembers getting her hair cut, then "there was a fire and figured I better try to get up (from my chair) because I had debris all over me," she said. "A lady, she did not know who she was to this day, set me on a window ledge to get out." "As a 12 year old, I was screaming and crying for her to come with me. But she was calm and said, "No, you get help, I'll be fine." The next thing she remembers is Bobby Johnson coming to save her.

Firefighter Bobby Johnson was on his day off from the fire department and had been outside talking with his brother-in-law. He happened to look up in the sky and saw a mushroom shaped cloud of smoke darkening the sky. "I thought, maybe we had a downtown plane crash," he said. "It just looked like the picture of the atomic bomb." He hurried to the scene. Johnson saw Hiatt atop Sargent's Wallpaper Store, screaming. He grabbed a ladder from a fire truck and helped get her to safety. Hiatt told him more people were trapped in the beauty shop. Firefighters found the beauty shop engulfed in flames. Hiatt believes the woman who helped her and the woman who was cutting her hair both died in that fire. "I still believe I had a guardian angel," Hiatt said.

Firefighter Joe Perkins believes one other person survived. Off duty that day firefighter Perkins saw smoke from the explosion and headed downtown to Fire Station 1, the Headquarters Station, then located near 5th and Main Street. From there Firefighter Perkins walked the short distance to the explosion scene. He had just arrived at the scene outside Virginia's Beauty Shop "when somebody yelled," There's more people in there," Perkins said. He ran inside and "I got a hold of a lady and carried her on my shoulders." Firefighter Perkins attempted to return to the beauty shop to rescue more people, but was turned back by heavy smoke and flames.

After rescue efforts at the beauty shop, Johnson and Perkins continued to search for people in trouble. "There were so many people trapped," Johnson said. "You'd begin to help somebody, then see someone who needed it worse," Perkins said. Firefighter Johnson went on to the State Theater and helped rescue children who had been attending a movie there. Firefighter Perkins stayed on scene for 3 days. Firefighter Perkins rose through the ranks of the department to become a as a Battalion Chief before he retired.

According to former Police Chief Louis Gibbs, "We cashed it in that day." "Richmond was never the same after that." I believe Palladium Item Newspaper Reporter Bill Engle summed it up the best, "April 6, 1968 turned out to be the last normal Saturday that Richmond would see for a long, long time. The explosions destroyed a major portion of the downtown shopping district and burned a section of the city's heart out. It also brought out the best in the citizens, whose heroic efforts saved many lives (*Firehouse Magazine*)."

Finding Jack Bales

During a visit to Richmond, IN doing research on the explosion that killed 41 people in 1968 I found many stories of heroes and miracles. One such miracle occurred when one of the survivors that had been at ground zero then the explosion occurred, standing next to his best friend. His name was Jack Bales (Figure 1.67). He lived and his friend died. I tried to locate him during my visit or find information about his whereabouts, but to no avail. Then one day several months after the article was published in Firehouse, I got the following email from Jack. "I received a call from Mike

Chronicles of Incidents and Response

Figure 1.67 Nothing short of a miracle occurred when one of the survivors that had been at ground zero then the explosion occurred, standing next to his best friend. His name was Jack Bales, he lived and his friend died.

Stewart, from the Howard Colorado VFD. He stated that he was reading your article in the March 2015 Fire House magazine and wanted to know if I was the same Jack Bales, mentioned in the article."

I told him that I was the same person mentioned in the article. I had lived in Howard Colorado for 14 years and was a Member of the Howard VFD, plus on the Fremont County search and rescue. Mike said all the years that we worked together you never mentioned that you were in an explosion, but remembered that I was originally from Richmond, Indiana. In 2014 I moved back to Richmond, IN to be closer to my kids and grandkids, plus my 85 year old mother that still lives in her own home."

I am always amazed at how small the world really is, for someone, 1200 miles away, to read an article about an explosion that happened 47 years ago, then pick the phone and call me." Jack has invited me to come to his home to view the wall he has created in his basement of the Richmond Explosion articles. My wife and I were finally able to work this out in the fall of 2017 while visiting my daughter in Lawrenceburg, IN. We met Jack and his wife and had dinner at an old Richmond Fire Station that has been turned into a restaurant (Figure 1.68). Had a great time and was able to see the memorial wall he had created on the walls of his basement about the explosion. Since the explosion, Jack has served as a firefighter, rescuer, and Deputy Sheriff.

Jack sent me an email explaining what he remembered from that fateful day when he lost his best friend. "In the summer, that I was eight and Greg was 10, we played on a little league baseball team together. Greg

Figure 1.68 My wife and I were finally able to meet Jack and his wife in the fall of 2017 and had dinner at an old Richmond Fire Station that has been turned into a restaurant.

was the catcher and I pitched. That summer, we formed a very close bond that lasted for the next ten years. I should say that the bond has lasted a life time. I miss Greg, almost daily (Figure 1.69). For the next eight years we were together non-stop. We rode horses, showed horses, in the same 4-H club. Jr. leader group, FFA club, school functions, double dated many times, and even dated sisters. When Greg graduated and went off to Purdue, (Veterinarian School).

For the next two years I would go up there and stay, different times, when He came home we were together all the time. Two weeks before he

Figure 1.69 Greg Oler (Photo Courtesy: Jack Bales.)

Chronicles of Incidents and Response   143

was killed, in the explosion, he had come home and was in need of new arrows, (Greg was on the Purdue archery team). We went to Marting Arms (6th & main Street). We had to order the arrows he needed. They would be arrive in 10 days. We returned two weeks later, on Saturday April 6th 1968 at 1:45 pm. Approximately two minutes after we walked into the store, the first blast went off, followed by the second stronger blast. That was the last time I saw Greg alive. Greg's Mom & Dad asked my parents to go to the make shift mortuary to identify Greg's body. With several broken bones and severe burns, I was able to check out of the hospital and attend Greg's funeral. 10 days later. 50 years later I still maintain a very close relationship with Greg's, sister Sandy. We can't get together without talking about the old days and Greg. It still brings us both to tears."

Davenport, NE June 10, 1968 Propane Tank Fire

Shortly after noon piping underneath a 30,000 fixed facility propane tank was damaged when a Bobtail Propane truck pulled away from the tank without disconnecting the fill hose mid morning. Temporary repairs were implemented and the situation seemed to be under control for an hour or two. Propane gas started to leak around the repair and is heavier than air so it flowed along the ground aided by a gentle south breeze into the shop area of the Deshler Propane Company. The building housed a repair shop, appliance storage and the company's office. Once inside the propane gas found a pilot light on a hot water heater and explosive ignition occurred. The building was completely destroyed and the fire flashed back to the propane tank and ignited the leaking gas at the source.

There was 35% propane left in the tank, and the flame impingement on the underside of the tank was cooled by the liquid inside of the tank. While the propane kept the tank shell cool it heated the liquid inside, which eventually caused the relief valve on the tank top to activate with a loud roaring sound, sometimes reaching 40 feet in the air (Figure 1.70). Firefighters directed hose streams onto the tank shell to keep it cool. Plans were to keep the tank cool and let the propane burn off. It was approximately 12:55 p.m. on Monday when the fire started and the last propane was consumed about 11:00 a.m. on Tuesday.

Marvin Vieselmeyer, 51, the owner was in the building when the ignition and fire occurred. He received 2nd degree burns over 30% of his body. He was listed in fair condition at Mary Lnnning Hospital in Hastings NE about 50 miles away. Mr. Vieselmeyer was taken by private car to the hospital, because there wasn't much of an EMS system at that time Five area fire departments responded to Davenport to assist with the incident. Davenport is a small community in south central Nebraska with a population at the time of the fire of about 450. As is true in most small communities crowds gathered wanting to see what was going on. Fire

Figure 1.70 While the propane kept the tank shell cool it heated the liquid inside, which eventually caused the relief valve on the tank top to activate with a loud roaring sound, sometimes reaching 40 feet in the air. (Courtesy: Thayer County Museum.)

Chief Mark Bergt expressed concern for their safety. Fortunately, only one injury occurred, however it could have been a real disaster.

> ***Author's Note:*** *The author was just out of high school and would turn 19 in July of 1968. Family members lived in Davenport and I spent quite a bit of time during my teen years visiting. This summer I had obtained a job working for the owner of the hardware store in Davenport. Lunch had just ended and another employee and myself were heading out the south door to go back to work when we noticed the roaring column of fire just behind a house 3-4 blocks away. At first we thought the house was on fire and headed in that direction to see if we could help. As we got closer we realized the column of fire was emanating from a large propane storage tank. I am describing this with the knowledge of propane that I have today. Back then I didn't have a clue what was happening. Most of the afternoon I hung out at the fire scene with dozens of other curious townspeople. Wish I had a camera, cause I could have had some great photos. After a while with nothing much*

Chronicles of Incidents and Response 145

changing and everyone just waiting for the fire to go out, I went back to work. This was my first exposure to a hazardous materials incident really up close and personal.

When I look back on that day with the level of knowledge I have today, it scares me to death. Everyone there watching and all of the firefighters from six departments would have died had a BLEVE occurred with that tank. That's how close we were, most would have been the fireball. The fact that the tank only had 35% or less in it made the situation even more dangerous. There must have been dozens of Leprechauns in town that day with their full pots of gold.

Crete, NE February 19, 1969 Derailment and Anhydrous Ammonia Release

Shelter in Place Effectiveness Substantiated

February 18, 1969 set up as a "perfect storm" for a train derailment in Crete, Nebraska, a small college town of approximately 4,500 people. Temperature was 4°F, wind was calm, relative humidity was 90%, approximately14 inches of snow on the ground a temperature inversion was in place and ground fog. At 6:30 a.m. C.S.T. Chicago, Burlington and Quincy (CB&Q) Train #64, consisting of three locomotive units and 95 cars, were entering town at 52 mph on the single main line track. Eleven box cars were standing on a siding South of the main line. Train #824 with one locomotive and 49 cars was standing on a siding North of the main line. It contained three tank cars of Anhydrous Ammonia. As Train #64 passed the turnout leading to the Old Wymore main siding, the spread closure allowed the wheels of the 28th car to derail.

The wheels struck and broke the guard rail, then derailed and the car and train continued. The 72nd car derailed at the "frog" (switch) towards the side where the broken guard rail was located and a total of 19 cars of this train derailed. A collision occurred between cars on trains #64 and #824 and caused a tank car containing anhydrous ammonia to release its contents. The derailment site blocked completely Unona Avenue between Nebraska Highway 33 and 13th Streets. This left 13th Street East bound as only direct response route to the injured people and the only evacuation route out of the area.

Following an investigation the National Transportation Safety Board (NTSB) attributed the cause of the derailment to "movement of rail at the turnout due to lateral forces of the locomotive caused by surface deficiencies of track. The track was not maintained for 50 m.p.h. operation according to standards and irregularities contributed to the increase of lateral forces." In other words, the Train #64 was speeding! Speed limit for the area conditions was 35 m.p.h. Six previous derailments had occurred in or near Crete since 1954. One such derailment on Memorial Day 1963, occurred in almost the exact same location as this one. Crete police would set up radar at the location and clock the trains coming into the city. If traveling over 35 m.p.h. they would call the train dispatcher and warn them.

Tank car SOU263210, an ammonia car split into two pieces, releasing 29,200 gallons of ammonia almost instantly, the top 16 feet landing 200 feet over Highway 33 and landed in the front yard of a residence at 1109 Highway 33 (Figure 1.71). The rest of the tank with the sill intact was propelled 140 feet onto Unona Ave.

GATX 18120 (DOT112A), shattered completely, releasing 29,200 gallons of liquid ammonia, which almost immediately turned into ammonia gas. One gallon of ammonia liquid produces 877 gallons of gas volume. Portions of the shattered tank traveled to the yards of residences located North of the derailment at (1)813, (2)905, (3)907, 13th Street and one south of Highway 33(4) (Figure 1.72). The main group of people that died were located North and East of the pieces of tank car 18120. NTSB reported the shattering was caused by a heavy blow delivered to the head of the tank car by the coupler of another car and the brittleness of the metal at a very cold temperature, 4°F.

Figure 1.71 Tank car SOU263210, an ammonia car split into two pieces, releasing 29,200 gallons of ammonia almost instantly. (Courtesy: The Crete News.)

Chronicles of Incidents and Response

Figure 1.72 Portions of the shattered tank traveled to the yards of residences located North of the derailment at (1) 813, (2) 905, (3) 907, 13th Street and one south of Highway 33 (4).

Casualties

It appears the highest and ultimately lethal concentrations of ammonia were located on the West end of 13th Street on the South side of the street. All of the victims were in that area when the impact occurred. Three Crete residents died during the accident, three died later in the hospital. Three unidentified transients riding the train were killed by trauma during the derailment. Injury reports varied, however, the Crete News, the local paper, lists approximately 25, although the NTSB reported 53 in its final report. Of the injured were two train crew members, one, the train conductor, fell approximately 18 feet as he stepped from the train, going over a bridge West of the derailment. He was later transferred to Lincoln for treatment.

Hatchetts 725 W. 13th Street

"Ron Hatchett a 21 year old student at Doane College and star football player, his wife, Ethelene and 4 year old daughter Gloria apparently ran out of their house following the derailment. Mrs. Hatchett was found face down, unconscious in the driveway of the home. Her daughter was near her. Ron Hatchett had made it across the street to the Gottlob Rauscher home, 800 W. 13th Street. All of the Hatchett family was alive when they arrived at the Crete Hospital, however Ron and Gloria died a short time later. Mrs. Hatchett survived, possibly because she was face down when found, which might of limited the amount of inhaled ammonia she received."

Erdmans Between 1005 and 1045 W. 13th Street

"Louis Erdman and his wife Maxine operated the Crete Cleaners on West 13th Street Northwest of the derailment. Mrs. Erdman was talking on the phone to her daughter, Mrs. Don Wolverton at the time of the impact. Their home had been punctured by debris and they fled their home. Mr. Erdman made it between 25 or 30 feet before falling to the place where he was found. Mrs. Erdman made it to a neighbors home, and was later found dead.

Hoesche 1005 W. 13th Street

"Mr. Hoesche heard the crash and encountered the gas smell when he looked out the door of his house. He rushed into bedrooms to get his four children and covered their faces with wet towels. He then heard a woman screaming in the street, very likely Mrs. Erdman., and he ran out to help her into the house. Later Mrs. Erdman was found dead in the Hoesche home."

Safranek 905 W. 13th Street

Crete News reported, "The Lyle Safranek home, just to the rear and at the side of the Kovar home, was the closest to the impact." Firemen found Mr. Safranek in the street at the intersection with 13th street. He was dead. Mrs. Safranek and one-year-old son were found a short distance from the front of the house in a snow bank. Rescue worker Clarence Busboom said "we had come past this place maybe 15 minutes earlier, but because of the thick gas hadn't been able to see anything." Mrs. Safranek and the son were hospitalized. The son was transferred to Lincoln in very serious condition. Both survived.

Svarc 813 W. 13th Street

Crete News reported that "Frank Svarc and his wife," grandparents of Judy Svarc, and father of Firefighter Svarc, "heard the crash, looked out and saw the thick gas spreading towards them. He opened the porch door and got a bucket of water, got cloths and went to the basement to await rescue. They kept dipping their hands in the water and wiping their faces

Chronicles of Incidents and Response

and fanning the air with the cloths as well as holding them over their faces. They were rescued in about an hour by Crete firefighters and said it had seemed like an awfully long time."

Kovar 907 W. 13th Street

Information on the Kovar and Svare families was reported directly to the author by Roberta Kovar Strain and Judy Svare. "Mrs. Kovar, Roberta's mother and brother were away delivering newspapers when the derailment occurred, which may have saved their lives. They tried to get back to check on their family members, but were unable to. Parts of train cars crashed into the Kovar home puncturing the shell of the house and a window. Ammonia came into the dwelling through the openings. Mr. Kovar went out onto the porch to investigate what had happened and collapsed outside. He was taken to the hospital by rescuers, where he later died. Roberta Kovar was 19 at the time of the derailment and heard the noises from the crash. She was in her 2nd floor bedroom and covered herself with bed covers and was rescued about an hour later." The combination of covers and the fact ammonia is heavier than air likely kept the ammonia away from her and saved her life.

Svarc 915 Redwood Street

"Judy's father Firefighter Leonard Svarc lived at 915 Redwood Street, Northeast of the derailment site 3-4 blocks. He heard the fire whistle and responded to the fire house. When he found out what had happened, he tried to get back to his family to make sure they were ok, but he could not get there. Mean while, according his daughter Judy, "his wife Delma, sister Carol, brother Dale and cousin Ron were still in the home. Delma got rags and wetted them and had the children put them over their faces. They went out to get in the family car to try and escape. There was a strong smell of ammonia and it burned their eyes. Once in the car, they tried to drive away, but got stuck. By that time the father had returned and took them to the Armory. Judy believes that the wet towels saved their lives." Judy and "her mother, were hospitalized for two days, her mother burned by the ammonia and Judy experiencing recurring bloody noses."

Crete Fire Department Responds

Information on the Crete Fire Department response was provided to the author directly from firefighters Everett Weilange, Chuck Henning, Arnold Henning, Loren Henning, and Chuck Vyhniek, who all responded to the incident in 1969. First reports coming into Crete emergency responders indicated a propane tank had exploded. Firefighters and Fire Chief Don

Henning, were alerted by the outdoor fire sirens and Plectron alert radios. Crete firefighters Loren Henning and Chuck Vyhniek were on the first fire apparatus responding to the call, which was just a few blocks West of the fire station. The 1946 American LaFrance open cab pumper they responded in is still in the department inventory and used as a public relations apparatus.

Fog and anhydrous ammonia vapors made it difficult to see what was going on. At the time of the incident Crete fire apparatus did not have radio communications in there apparatus, so information initially had to be passed on person to person. With the very limited visibility, it was difficult to get a grasp of the scope of the incident and coordinate the response. As the incident progressed, radios from the Sheriff's office were pressed into duty, which improved communications. Turnout gear was limited as were SCBA. Henning and Vyhniek indicated the mixture did not move and just hung low in the air.

They did not know what the fog was, but some firefighters indicated there was a smell of ammonia in the air. (Anhydrous ammonia is heavier than air and tends to pool in low places and on the ground.) Henning and Vyhniek approached the site from the East. When they arrived on scene they smelled something, but didn't know what it was so they backed out and approached from a dirt road along the tracks from the North. They smelled something there as well. They drove by Douglas Manufacturing and stopped just West. Henning jumped from the apparatus at 17th & Main Streets and started evacuating people from the area. Henning then went door to door with SCBA and several other firefighters on 13th Street. The south and west sides of town were evacuated by firefighters.

Firefighters encountered bodies and injured as they made their search. Exposed skin on victims exhibited deterioration, likely from contact with ammonia or vapors. Firefighter Arnold Henning reported driving his apparatus into the cloud and finding body parts that were later determined to be transients. Firefighter Chuck Henning worked for Wanek's Furniture Store in Crete and used a furniture delivery truck to take bodies to the mortuary. While searching, firefighters located a car that had stalled because of lack of oxygen, which had been displaced by the ammonia gas. Mr. Alvin Rozdalousky got out of the car and ran 5 blocks East to Main Street and safety. Firefighter Everet Weilage used his own vehicle to alert people in the South side of Crete and move many from harm's way. Also helped out at the station. All of the firefighters I talked with indicated their first priority was to evacuate people to safety, and isolate the scene. Between 200 and 300 people were evacuated from the area. Crete firefighters had previously trained and participated in table top exercises for disasters. This preparation likely helped them as they responded to the train derailment.

Chronicles of Incidents and Response

Following evacuation, some people went to be with relatives; others to Doane College in Crete; the Armory; and others to the fire station. Food was brought in by residents and businesses and supported by the Red Cross. Nebraska National Guard from Lincoln assisted in a secondary search to determine if all had been rescued and evacuated. Later in the morning a helicopter was brought in to and helped to disperse the gas fumes with its whirling blades. In spite of the conditions at the scene and lack of protective equipment, only four firefighters were injured in the incident from ammonia exposure and were treated and released.

Firefighters remained on scene for three days. According to Wally Barnett, assistant state fire marshal, "If there had been a West wind and it had been a clear day that cloud of gas could have made a clean sweep of the town." Assistance from other towns fire and police came from Lincoln, Malcolm, York, Seward, Milford, Southeast Rural District in Lincoln, Hallam, Friend, Beatrice, Fairbury and Wilbur. Fourteen Crete firefighters who responded to the incident are still living at the time of this interview.

Crete's encounter with the train derailment and anhydrous ammonia occurred prior to the development of organized response to hazardous materials in the U.S. Fire Service. United States Department of Transportation (DOT) did not have an Emergency Response Guide Book or a placard and labeling system in place. No markings of the dangerous chemicals in transportation were required on the tanks of anhydrous ammonia. Requirements did not come until the early 1970's. D.O.T. had also not yet coined today's term for chemicals in transportation, Dangerous Goods or Hazardous Materials as they are normally called.

The idea of sheltering people in place inside of buildings against chemical exposure did not exist at this time. However, many people did in fact shelter themselves inside their homes, placed wet rags over their faces and some covered with blankets. Not only did they shelter in place and protect themselves from the ammonia, but saved their lives because of their inadvertent actions. Not a single person who stayed inside the entire time of the emergency and took self protective actions, died or was seriously injured.

Those who died from ammonia exposure, died when they left their homes and were overcome by the ammonia vapors outside. Some died on their driveways and one on a street corner, beyond the apparent safety of their homes. Curiosity, called them out to see what had happened and they paid the ultimate price. One person who was inside their residence and quickly went onto their porch, turned around and went back inside. They still had enough exposure to the ammonia that they died of complications later in the hospital. Unknowingly, even before the concept of Shelter in Place had been developed, victims of the Crete derailment confirmed the effectiveness of Sheltering in Place when hazardous materials are released outside of buildings or a vapor

152 · *Hazmatology: The Science of Hazardous Materials*

cloud travels to populated areas preventing an expeditious and safe evacuation.

It is interesting to note that the house number 1005 13th Avenue is located adjacent to the cleaners where two victims died and just West of 13th and Unona where one person died. The person at 13th and Unona who died on the street corner, lived at 813 13th Street. The house number "1005," is the United Nations/DOT designation for anhydrous ammonia in bulk quantities, which was not used at the time of the derailment. Had this incident happened at any other time of the year, it would have likely been much worse. In warmer weather, people would have had less clothing on, windows might have been opened in the dwellings more people may have gone outside to see what happened.

Liquid ammonia has a boiling point of –40°F below zero. Anhydrous means without water. Ammonia seeks water when released into the environment. However, at 4°F, there wasn't much water in the area of the derailment that was not frozen. It is unlikely that people exposed to the ammonia were sweating if they had been ammonia could have reacted with the moisture on the skin causing serious burns. Even though the air temperature and items in the environment were above the boiling point of ammonia, warmer temperatures would have caused even more gas to be formed quicker by the spilled liquid ammonia. Not everyone who went outside died, however no one who sheltered in place died. The snow on the ground may have someway created a barrier to the exposed skin and airways of those who survived.

During the following week after the derailment, the Crete Fire Department was dispatched to a fire at Doane College Merrill Hall for a fire. Firefighters laid hose lines and began to fight the fire, when the hoses failed and started to leak from exposure to ammonia at the derailment. One firefighter commented the hose looked like a soaker hose with all the leaks. Crete does not have a hazardous materials response team. All firefighters are trained to the hazmat operations level. If a hazmat incident occurs in Crete or the surrounding area teams are available from Lincoln and Beatrice (*Firehouse Magazine*).

Carpentersville, IL December 10, 1969
Chemical Plant Explosion & Fire

An explosion rocked through the plant about 04:50 a.m. CST with a force so sufficient enough to be felt by residents living near Algonquin three miles to the north. Three employees were injured in an explosion at the Alberto Culver Chemical Company in Carpentersville, about 40 miles NW of Chicago (Figure 1.73). The 60 x 100 foot building is one of a complex of small buildings that are not connected but with the same address.

Figure 1.73 An explosion rocked through Alberto Culver Chemical Company in Carpentersville about 04:50 a.m. CST with a force so sufficient enough to be felt by residents living near Algonquin three miles to the north.

Debris from the explosion smashed windows at the adjacent Burlington Manufacturing Company and the Chicago Furniture Mart Warehouse. Bricks and debris also were piled on two tank cars parked on a railroad siding west of the building. Ten families were evacuated from their homes across the street from the plant after firemen noticed pungent odors in the air when they arrived at the scene. All evacuees were allowed to return to their homes 2 hours later. No cause or damage estimates were available. The major explosion the blew out the buildings brick walls was followed by two smaller ones. The initial explosion blew out a large overhead steel door on a building 100 feet north. Firemen from East Dundee, West Dundee and Algonquin assisted Carpentersville Firemen in battling the blaze.

> ***Author's Note:*** *This was the first hazardous materials incident that I responded to in my career. In 1970 I did not have any hazmat training. In fact, the first hazmat team in the United States, Jacksonville, FL would not be formed for exactly 7 more years. This was my career hazmat incident and I didn't even know it at the time. To me this was just my first third alarm fire. My apartment was just across the Fox River and a few*

blocks southeast of the Alberto Culver Chemical Company. When the explosion happened, it woke me up, but I did not realize it at the time. Just thought it was a bathroom break. Then while sitting on the throne, I saw the blue light of another firefighter flash in my window. Next I heard the fire whistle sounding in the distance.

Needless to say, I put it into high gear and headed for Station 1 in "Old" Carpentersville, just a few blocks from the site of the explosion. My route to the station would take me along the west side of the Fox River, across the bridge and right by the scene. It was SOP to respond to the station and ride the apparatus to the scene. This was the first time I had ever driven by a scene on the way to the station. All I can say is WOW, broken glass everywhere, fire everywhere it looked like a bomb had gone off. Once at the station I passed on what I saw to the Lieutenant and the other two firefighters who would ride Engine 93 and be first in on the once in a lifetime incident of my career.

There was an alley between the main road where Engine 93 stopped and the fire building. The explosion had lifted the roof off of the building, laid the 4 brick walls out on their side scattered on the ground, and the roof sat back down on the top of the process tanks in the building (or what was left of it). Four of us advanced a 2 ½ inch supply line down that alley and used the corner of an adjacent building for cover as we poured water onto the fire. There were several small explosions that occurred while we were fighting the fire. It seemed like all four of us would jump off the ground together as each explosion occurred.

No one had any idea what they did in that building, what was burning, or what would happen next. I do not remember much more about the rest of the night. It is just a blur. We never did find out what the chemicals were that were involved. I don't think they ever found out what caused the explosion. There were rumors that military nerve agents were stored in the building. That was never confirmed or denied. I had no idea what they were about anyway. Wouldn't hear any more about them again for another 25 years. Three employees working in the plant when it exploded were injured. There were no firefighter injuries.

Crescent City, IL June 21, 1970 Derailment and LPG Rail Car BLEVE

Crescent City, IL, was the site of an accident involving LPG in transportation. On June 21, 1970 (Fathers Day), sixteen cars of the 109 car eastbound Toledo, Peoria and Western Railroad Company's Train No. 20 derailed in the center of town at approximately 6:30 a.m. Ten of the cars each contained 34,000 gal of LPG. Two additional LPG tanks remained on the tracks.

Chronicles of Incidents and Response

Fire Chief Orvel Carlson was awake at the time of the derailment and felt the heat of the explosion from his home three blocks away. Carlson and the 20 man volunteer fire department responded quickly with their two pieces of fire apparatus. Fire fighters arrived and tried to contain the fire which was burning intensely. Firefighting efforts were hampered by a lack of electricity that was knocked out by the derailment that prevented the city's water pumps from functioning. Firefighters took water directly from the city water tower to fight the fire.

Fire companies from 32 surrounding towns appeared with 53 pieces of equipment and 234 firefighters on the scene. Chanute Air Force Base in Rantoul sent a foam truck. They informed the local fire fighters that water would be unable to contain such a fire that including burning LPG.

Editor's Note: That was the thinking at the time, however, foam is not effective on LPG as it is not a liquid, it is a gas when released into the atmosphere.

An Illinois State Police Sergeant located at Watseka, Illinois about 6.3 miles East of Crescent City was notified of the derailment shortly after it happened and proceeded immediately to the scene. He arrived at approximately 6:45 a.m. and sized up the situation. When he determined that a tank car was being heated by the fire and contained LPG, he notified police officers in the area to evacuate the town and warned firefighters to move back to a safer location to fight the fires. His actions may very well have prevented serious injury and loss of life of firefighters, police officers and residents of the community.

During the derailment, one of the LPG tank cars was punctured by a coupler on another car causing a leak that ignited almost immediately. Flames reached several hundred feet into the air, impinging on the other tank cars of LPG. A nearby house and business were set on fire from the radiant heat, injuring several residents. Relief valves on the other tank cars began to open as the pressure built up from the surrounding fires (Figure 1.74). The first explosion (Boiling Liquid Expanding Vapor Explosion (BLEVE) occurred around 7:33 a.m., almost one hour after the derailment (Figure 1.75). In that first blast, several firefighters and bystanders were injured and some fire equipment was damaged. Additional explosions occurred at 9:20, 9:30, 9:45, 9:55, and 10:10 a.m. Parts of tank cars were propelled all over town, setting fires and damaging structures.

Sixty-six persons were injured by the explosions and 11 required hospitalization, but there were no fatalities. Twenty-five homes and 16 businesses were destroyed by fire and three destroyed by "flying" tank cars; numerous other homes received damage. There was over 2 million dollars in property damage as a result of the derailment, fires and explosions. Six fire trucks were damaged by the explosions and fires along with 3,050 feet

156   *Hazmatology: The Science of Hazardous Materials*

Figure 1.74 Relief valves on the other tank cars began to open as the pressure built up from the surrounding fires. (Courtesy: Crescent City Fire Department.)

Figure 1.75 The first explosion (Boiling Liquid Expanding Vapor Explosion (BLEVE) occurred around 7:33 a.m., almost one hour after the derailment. (Courtesy: Crescent City Fire Department.)

of 2 ½ inch fire hose, 500 feet of 1 ½ inch fire hose, several ladders, nine firefighter coats and seven firefighter helmets.

The remaining LPG tanks were allowed to burn, which took some 56 hours after the derailment. The exact cause of the derailment was not determined but a motorist spotted smoke coming from one of the train cars as the train was approximately ten miles west of Crescent City. Most of the responding firefighters had little training in dealing with propane fires. Some of the injured firefighters were not wearing their personal protective clothing (*Firehouse Magazine*).

Woodbine, GA February 3, 1971 Thiokol Factory Solid Rocket Fuel Explosion

Woodbine, Georgia was selected as the place to build the Thiokol Chemical Corporation in the mid-sixties, thanks to its close proximity to Cape Canaveral where the space race was in full orbit. Situated on 7,400 acres, the company built and tested solid propellant rocker motors for NASA. The Thiokol Chemical Plant, a sprawling complex of 36 buildings on 7,000, was working on a U. S. Army contract for trip flares (flares that are ignited by an external trigger, normally an enemy soldier approaching a camp's perimeter). Suddenly an explosion leveled one building and damaged three others. As a result of the explosion a forest fire, which would eventually destroy 200 acres of timber near the facility, was also set. Since the flares contained magnesium, many of the injured were severe burn victims, with second and third degree burns over more than 25% of their bodies.

The plant was comprised on 36 buildings that housed all types of fuels and chemicals produced here. On Wednesday morning, February 3, 1971, one of those buildings known as the Woodbine Plant (Building M-132) exploded killing 29 and injuring close to 50 workers, most of them women. Five were injured so severely, they died within days from their injuries. According to reports, the fire began in an area where ignition chemicals were added to other explosives including magnesium. The fire then spread to a storage area that contained 56,000 flares. The resulting blast leveled the building (Figure 1.76). Officials were not sure of the immediate death toll due to victims being dismembered from the explosion, and bodies being blown from the building into a nearby forest.

Three more buildings were heavily damaged. Nearby buildings sustained scorched and buckled aluminum walls, and charred utility poles. Another seven buildings received minor damage from the blast. One survivor said it was "like an atomic bomb" had gone off. Heavy smoke and dangerous fumes lay over the plant as a dismal rain began to fall on the wreckage. At the time of the tragedy, the plant had an order to produce 758,000 trip flares for the Army's use in Vietnam. The materials were originally given a Class 7 designation the highest ranking for hazardous chemicals and materials. But in 1967, the Army had downgraded the classification to a Class 2 which designated a fire hazard. The Army reissued the Class 7 designation in the fall of 1970.

Today, children of survivors are working to develop the Thiokol Memorial and Museum to honor those killed and injured in the incident - most of whom were women. Thiokol was one of the few places in the late

Figure 1.76 The fire then spread to a storage area that contained 56,000 flares. The resulting blast leveled the building, killing 24.

60s and early 70s where a woman could get a full time job and be paid the same wages as a man. Due to the Army contract, the plant's workforce at the time of the blast was close to 500 employees working round-the-clock to get the order filled. February 3, 2021 will be the 50th anniversary of the explosion. A memorial service will be held to honor those who died and were injured almost 50 years ago (Thiokol Memorial.com).

Waco, GA June 1971 Car-Truck Collision Sparks Dynamite Explosion

June 1971 a small foreign car-truck collision sparked an explosion that killed 5 and injured 35 others. The truck was carrying 20,000 pounds of dynamite and low grade explosives that sparked a "blockbuster bomb" that swept away spectators, cars and nearby houses. The blast created a 100 ft. wide and 20 ft. crater. Army demolition teams cleared the area when daylight came. The force of the blast severed the two lane U.S. Highway 78 black top road. U.S. 78 is the main route from Atlanta to Birmingham, Alabama and touched off woods fires for a quarter of a mile. Blast wave from the explosion collapsed the roof of a school gymnasium where 200 persons were attending a gospel song fest.

The truck driver A.W. Fielding of Birmingham foresaw a possible disaster when flames erupted after the collision of the Volkswagen with his truck shortly after 8 p.m. He told people to "get back there are explosives on the truck they may go off." He screamed at persons who began to crowd around the wreck. His actions likely saved lives as many headed his warning and survived the explosion that came several minutes later.

Chronicles of Incidents and Response 159

The truck owned by Baggett Trucking Co. of Birmingham, was carrying 10,000 pounds of dynamite and 10,000 pounds of a low-grade explosive called "slurry" (a water gel explosive) from Birmingham to a granite quarry in North Carolina. A water-gel explosive is a fuel sensitized explosive mixture consisting of an aqueous ammonium nitrate solution that acts as the oxidizer. Water gels that are cap-insensitive are referred to under United States safety regulations as blasting agents. Water gel explosives have a jelly-like consistency and come in sausage-like packing stapled shut on both sides. Water-gel explosives have almost completely displaced dynamite, becoming the most-used civil blasting agents.

The ensuing fire set off blasting caps, which ignited the explosives. According to Deputy Sheriff Gene Kirk, "It just blew down the onlookers and the biggest part of the part of the truck I could find was a wheel." The Volkswagen disintegrated along with its driver Talmadge L. "Skinny" Adams of Waco. Also killed were the wrecker driver, Cleve Heath of Bremem and a Bremen Firefighter David L. Smith. Cars were blown off the road like crumpled toys. Several nearby houses were demolished. Trees were snapped like broken tooth picks. "It looked like it just swept everything away," said Ivey chandler who was sitting in his yard when the collision occurred. Chandler a World War II veteran said the explosion "sounded like a blockbuster bomb." Fortunately, the shock only loosened the roof of the gymnasium, and the 200 persons got out safely before it collapsed.

Firefighter Who Made The Supreme Sacrifice

David L. Smith, Bremen Firefighter

Bucks County Courier Times Doylestown Pennsylvania 1971-06-05

Houston, TX October 19, 1971 Mykawa Road Rail Car BLEVE & Fire

Houston experienced a tank car incident on Mykawa Road following derailment on at approximately 1:15 p.m. Sixteen cars of an 82 car Missouri Pacific train derailed. Hazardous materials on the train included six cars total of vinyl chloride, one each of acetone, caustic soda, formaldehyde, plasticine and butadiene. Two of the derailed cars were involved in fire, a vinyl chloride car and the butadiene car. The butadiene tank car experienced a BLEVE approximately 40 minutes following the derailment and 20 minutes after the arrival of the Houston Fire Department. This BLEVE resulted in one firefighter fatality and 37 firefighter injuries (Figure 1.77). The primary reason so many fire fighters were hurt was that fire officials did not have information on the chemicals in the rail cars. They did not

Figure 1.77 The butadiene tank car experienced a BLEVE approximately 40 minutes following the derailment and 20 minutes after the arrival of the Houston Fire Department. (Courtesy: Houston Fire Department.)

fully understand the potential danger of the burning chemicals and tank cars. Firefighter Truxton Hathaway was assigned to the Fire Department Training Academy. When the explosion occurred, Hathaway took a camera to the scene to record the incident for future training purposes. He arrived on the scene and began recording the incident when the second tank car exploded, sending a wall of fire over Truxton and killing him instantly.

Water supply was an initial problem with the nearest hydrant ¼ mile away from the derailment site. Two alarms of fire equipment responded with 7 additional engines requested to supply water to the scene pumping in tandem from the nearest hydrant. The butadiene tank was 100 feet long and overturned during the derailment. This placed the tank car pressure relief valve in the ground and covered it with liquid which prevented the relief valve from functioning properly. As a result of the explosion, parts of the tank car traveled 400 feet from its original location. Video of this incident has been circulating through the fire service for years showing a firefighter on an aerial ladder engulfed when the explosion takes place.

Chronicles of Incidents and Response 161

That firefighter, Andy Nelson, was burned, but survived. District Chief V.E. Rogers who was today's equivalent of the Incident Commander (IC) was burned over 50% of his body and spent three months in the hospital. It was thought that the Mykawa incident was what motivated Chief V.E. Rogers to form a hazardous materials team in Houston. However, the team was formed as a result of Chief Rogers attending a chief's conference where a presentation was made by Ron Gore about the new Jacksonville, Florida Hazardous Materials Response Team. Chief Rogers returned home and directed District Chief Max H. McRae to organize Houston's team (HFD Pictorial History 1980 – 2014).

Firefighter Who Made The Supreme Sacrifice

Truxton Hathaway, Training Academy

Kingman, AZ July 5, 1973 Propane Rail Car BLEVE

On July 5, 1973 a propane tank car being off-loaded in Kingman, Arizona, caught fire, which resulted in a Boiling Liquid Expanding Vapor Explosion (BLEVE) that killed 11 Kingman firefighters and 1 civilian. Another 95 persons were injured by the blast and over $1,000,000 dollars in property damage occurred to surrounding exposures. Except for one career firefighter/engineer who was severely burned but survived, those injured were mostly spectators that had gathered along Historic U.S. Highway 66 to watch the incident. Most of those injured were approximately 1,000 feet from the explosion and ignored police warnings to stay back. Photographs of the spectacular BLEVE incident have appeared in countless articles, books and training programs over the years. Instructors have often referenced the Kingman incident when warning emergency responders of the dangers of flame impingement on the vapor space of propane tanks. When firefighters arrived they found a 1,000 gallon propane tank involved which exploded into a 400-500 foot fire ball approximately 20 minutes after the firefighters arrived.

The firefighters killed were reported to have been hit by the fireball but not burned, it is though they died from the concussion of the explosion. Two other firefighters were injured and medivaced to the hospital and were later released. Kingman, Arizona with a population of 7,500 in 1973 is a desert community located approximately 80 miles southeast of Las Vegas, Nevada, 184 miles northwest of Phoenix and 147 miles west of Flagstaff. At the time of the incident the Kingman Fire Department was a part career and part volunteer force operating out of two fire stations. There were 6 career firefighters with one on duty in each station at all times and 36 volunteers. Kingman's equipment in service at the time

162   *Hazmatology: The Science of Hazardous Materials*

of the explosion included 4 engines and 1 rescue vehicle. Station #2 was located just a half mile west of the site of the explosion.

Doxol Gas Distribution Plant

July 5, 1973 started off as a typical summer day in Arizona with temperatures reaching well above 100° F and a light westerly wind at 12 mph. Workers at the Doxol Gas Distribution Plant, located on the east side of Kingman, approximately two miles from the downtown area, were preparing to off-load a 33,500 gallon water capacity rail car of liquefied propane gas. The Doxol Plant consisted of an office and two above ground LPG storage tanks, one 30,000 gallon capacity and the other 18,000 gallons.

Located on the southwest side of Route 66, the Doxol Plant Office was approximately 70 feet from the highway. Fire protection features of the Doxol facility consisted of portable dry chemical extinguishers near the storage tanks and in the office. The nearest public water supply fire hydrant was 1200 feet north of the site on Hoover Street.

Two storage tanks were located at the rear of the office approximately 200 feet from the highway. Stationary propane tanks are generally un-insulated and flame impingement on the vapor space can cause metal failure in much the same way as the rail car did when the Kingman incident occurred. Usually, there were four employees present at the small facility, a clerk, the manager, and two delivery personnel. At the time of the explosion, there were three employees on duty at the plant. Just east of the Doxol plant 600 feet was the Double G Tire Company. The Country Kitchen Restaurant was located 800 feet away and the Phillips Truck Stop 900 feet east. Loading racks to off-load rail cars were located on the south side of the rail siding with piping running underground to the storage tanks. Kingman was serviced by the Santa Fe Railroad, which had just delivered the tank car to the Doxol Plant one month earlier on June 5, 1973. Delay in off-loading the rail car is believed to have occurred because fuel demand during the summer is low and the bulk storage tanks were full. The loading rack where the tank car was setting is located 30 feet across the Santa Fe Railroad main line southeast of the main Doxol facility, approximately 450 feet from the highway. Liquefied Propane tank cars at that time were un-insulated and all of the valves were located within the dome cover at the top of the tank. United States Department of Transportation (DOT) specifications required that the tank be constructed of 11/16" carbon steel with a hydrostatic test pressure of 340 psi and a rupture pressure of 500 psi. The pressure relief valve was set to expel excess pressure from the tank at 280 psi.

Around 1:30 p.m. workers began to connect the hoses to the rail car to start the off-loading process. Liquid lines are attached to the two liquid

Chronicles of Incidents and Response 163

valves located in the dome cover housing at the top of the rail car. Vapors are collected and routed into the vapor space of the tank. After all connections are made, the valves are opened slowly at first so as not to trip the excess flow valves. Connections are routinely checked for leakage and the valves fully opened when no leaks are present. Two Doxol personnel were involved in the off-loading operation. As the off-loading proceeded, one of the workers detected a small leak that was present in one of the connections.

Connections were typically tightened by striking them with an aluminum alloy wrench. In spite of the efforts to tighten the connection, the leak continued. The liquid connection was once again struck with the wrench. This time a fire erupted. It is thought that a spark was created as the wrench struck the steel fitting, because of magnesium being present in the alloy of the wrench. Both men fell from the top of the tank car with their clothing on fire, resulting in severe burns from the fire and extreme heat. One of them was to die from his burns. The second man ran back to the office building where he was driven to an Arizona Department of Public Safety Office, approximately one quarter mile away, to report the fire.

Kingman firefighters received the first call for help at 1:57 p.m. and arrived on scene 3 minutes later. Initially a call went out to the Haualapa Fire Department for mutual aid assistance. The fire spread quickly and was impinging upon the top of the rail car where the vapor space is located (Figure 1.78). Every liquefied gas container has an approximate 20% vapor space above the liquid in the tank to allow for expansion of the liquid to vapor within the tank during shipping and storage. This is the most dangerous place for flame impingement to occur because there is nothing to absorb the heat but the metal itself. Steel does not absorb heat well so when temperatures reach above 400° F the integrity of the tank is quickly compromised.

> *Hazmatology Point: National Fire Protection Association (NFPA) statistics show that pressure tanks can fail from flame impingement within the first 8 to 30 minutes of the first flame exposure, with 58% occurring in 15 minutes or less. It is estimated that the Kingman fire burned about 8 minutes before firefighters arrived and an additional 10 minutes before the first water was applied to cool the tank. It is estimated that the BLEVE occurred just 19 minutes after the first flame impingement began on the top of the tank.*

Flame impingement on the liquid level is a somewhat less dangerous situation because the liquid will absorb the heat and protect the integrity of the tank. However, the increased heat will cause the already boiling liquid to boil faster, causing the pressure inside the tank to increase. Firefighter's tactical objectives at Kingman were to provide water to cool the tank and

Figure 1.78 The fire spread quickly and was impinging upon the top of the rail car where the vapor space is located. (Courtesy: Kingman Fire Department.)

prevent an explosion. An engine with a 1,000 gallon booster tank was located approximately 75 feet from the rail car and two -1 inch booster lines were placed into service to cool the tank shell. (The water flow from a 1 inch booster line is about 30 g.p.m.) If water is applied effectively to the point of flame impingement the temperature of the shell cannot reach above 212° F, which is well below the failure temperature of the steel. This operation requires large quantities of water. NFPA recommends a water supply of 500 gallons per minute, uninterrupted, applied the surface of the tank for cooling.

While the first firefighters attempted to cool the rail car from the booster tank of the engine, others began laying two 2 1/2 inch lines to the hydrant 1200 feet away to supply a deluge gun located 50 feet from the burning tank car. The first 2 1/2 inch hose lay was completed, but the firefighters ran out of hose for the second supply line. The first line was being charged when the explosion occurred at approximately 2:10 p.m. (Figure 1.79). Twelve firefighters were within 150 feet of the burning rail car when the blast occurred. Eleven of them died from severe thermal burns; 1 career and 10 volunteer firefighters. A twelfth firefighter was taken to the hospital in critical condition, but survived.

Chronicles of Incidents and Response 165

Figure 1.79 The first 2 1/2 inch hose lay was completed, but the firefighters ran out of hose for the second supply line. The first line was being charged when the explosion occurred at approximately 2:10 p.m. (Courtesy: Hank Graham.)

Protective equipment typical for firefighters at the time was cotton duck with wool linings and helmets made of polycarbonate plastic, this was the type of protection worn by Kingman Firefighters. It is reported that at the time of the explosion some firefighters had full protective gear, while others wore only a coat and helmet or just a coat. Those firefighters killed had the coats and their street clothes burned off of their bodies by the fire and radiant created from the explosion. The tank broke into pieces from the force of the blast and one half of the tank bounced end-over-end westward along the tracks, landing approximately 1200 feet from its original location on the siding.

The other portion of the tank tore along the welds and flattened out on the ground. A ground level fireball ensued and extended 150 to 200 feet in all directions from the center of the blast. This was followed by a large mushroom cloud of flame extending several hundred feet into the air, measuring 800 to 1,000 feet in diameter. Fireball flame temperatures can reach well over 3,500 degrees F. The fireball and radiant heat resulted in setting five buildings on fire including a tire company, restaurant, truck stop, and gas company office building. Several brush fires were also started by the heat from the explosion.

Radiant heat was so intense that it caused the relief valve on the 30,000 gallon storage tank to activate. However, the released vapors did not ignite and once the pressure was relieved the valve closed and remained closed.

166 Hazmatology: The Science of Hazardous Materials

Following the explosion calls for mutual aid went out to the Lake Havasu City, Mohave Valley, and Bullhead City Fire Departments. Chiefs of the Lake Havasu City and Bull Head City departments set up a command post at Kingman Fire Station #2. Responding mutual aid companies were assigned to extinguish the numerous fires, the last of which was brought under control around 5:30 p.m. that same day.

Many of the photographs taken of the Kingman incident are from retired Santa Fe Railroad Conductor Hank Graham. Mr. Graham was working on a short train servicing the industries along the railroad in Kingman when the incident started. He had been notified of the fire at the Doxol plant and advised not to proceed into the area. While maneuvering several rail cars near East Kingman, he saw the smoke from the fire. Hank Graham pulled out his camera and started taking photographs of the incident. When the train stopped he proceeded to a point on Route 66 where a police officer was blocking traffic access from the incident.

He wanted to know how long the rail line would be blocked. Mr. Graham began photographing the burning tank car's relief valve and the fireball created when it opened up. While preparing to take another photograph of the flaming relief valve, the tank BLEVEed burning the hair off of his arms as he captured the now famous fireball/mushroom cloud from the BLEVE. In the years since the Kingman explosion, railcars have been manufactured with an outer layer of insulation and tank skin surrounding the inner tank to provide a period of time before a fire can reach the inner tank. This insulation adds approximately one hour to the 15 to 20 minute time of flame impingement before a BLEVE is likely to occur.

Kingman, Arizona Remembers 35 Years Later

At the time of the 35th Anniversary of the Doxol Disaster the Kingman fire department is was headed by career Chief Charles Osterman (who was just 16 years old when the explosion occurred) in command of 4 stations, 4 front line and three reserve engines, 1 – 100' Tower ladder (dedicated to the firefighters who died with each of their names inscribed on the sides of the tower bucket), 1 – new 2008 Light Rescue, a heavy support vehicle, an extrication truck and 2 brush engines with a response area that has grown to over 30 square miles. Each shift now has 9 career firefighters on duty supported by Chief Osterman, two assistant chiefs, five battalion chiefs, an EMS coordinator and a training officer. The career force is complemented by 10 part-time personnel and 3 volunteers.

Chief Osterman reported that the incident 35 years ago resulted in many relatives of the firefighters who were killed, including sons, nephews, and uncles becoming career and volunteer firefighters in Kingman and other communities. Assistant Chief Joe Dorner is a nephew of Butch Henry,

Chronicles of Incidents and Response 167

Battalion Chief Porter Williams is the son of Lee Williams, and Captain Bob Casson is the son of Bill Casson and are all currently career firefighters with the Kingman Fire Department. Chuck Casson, former volunteer captain, currently volunteer firefighter is the son of Bill Casson.

Chief Osterman's father John Osterman was also a volunteer firefighter in Kingman when the explosion occurred, but was at work at the Ford Proving Ground 25 miles away and missed the original call, which perhaps may have saved his life. John Osterman reported that his wife called him and told him "something was going on in Kingman but she didn't know what. She said the house just shook." John told her "he was on his way". He told his boss he had an emergency and needed to go back to Kingman. Gus Reichardt and Lawson Bradley also Kingman Firefighters worked at Ford as well and he picked them up and headed to Kingman. They arrived after the explosion had occurred and immediately went to work trying to do what they could to help.

Remembering Those Who Lost Their Lives

One of the most important things we can do as firefighters is to never forget the sacrifices made by those who went before us or the lessons we have learned from them. In particular we should not forget those who gave the ultimate sacrifice, their lives. It had been 35 years since 11 firefighters gave the ultimate sacrifice in a propane explosion in Kingman, Arizona. In spite of the terrible loss of life, no other hazmat incident has occurred in the United States that has had more of a positive impact on the fire service than the Kingman incident. Many changes in procedures and regulations occurred across the fire service as a result of this explosion.

We owe uncounted saved lives to those brave men who gave theirs in Kingman that fateful day in July 1973. Kingman firefighters; family members of those firefighters lost; brother firefighters from across the region and community residents gathered at Kingman Middle School July 5, 2008 at 10:00 a.m. to remember the 11 firefighters and 1 civilian who perished. Over 700 people attended including representatives from 35 fire departments and 7 law enforcement agencies. I had the opportunity to attend the service as well. It was one of the most moving and heartwarming experiences of my life. Chief Charles Osterman and his staff with assistance from the community at large did a remarkable job of honoring the 11 firefighters that died that day and making sure they are not forgotten. I am sure that those eleven men were looking down upon the service and where very proud.

The service began with opening remarks by Chief Charles Osterman followed by the Kingman Fire Department Honor Guard under the command of Fire Prevention Specialist Keith Eaton presenting the colors. Next a procession of all the visiting honor guard members and other uniformed firefighters who were present paraded through the aisles. Department Chaplin Dave Patriquin gave the Invocation. Brief remarks were made by Kingman Mayor John Salem, Bob Barger Director of Arizona Department of Fire, Building and Life Safety and Tim Hill President of the Professional Firefighters of Arizona. The speakers were followed by a video presentation showing scenes from the incident and an American Heat documentary prepared about the incident during the 25th Anniversary. Chief Osterman read a biography of each of the 11 firefighters as their photograph was displayed on the screen, followed by the traditional ringing of a fire department bell six times for each firefighter.

Biographical information about the firefighters revealed not only that the eleven men were dedicated to the Kingman Fire Department but also their community and nation as many of them participated in other community organizations and were veterans of United States Military Service. Following the bell ringing ceremony a bagpipe rendition of Amazing Grace was played by the combined bagpipes from Kingman and Glendale, Arizona and Henderson, Nevada followed by a closing prayer from Chaplin Patriquin.

Upon completion of the inside ceremony honor guard members along with uniformed firefighters formed two lines in the parking lot outside as family members passed through. Everyone gathered by the flag pole located in nearby Firefighter's Memorial Park. The five acre park was rededicated with the raising of a flag, lowering it to half staff and the playing of taps. An Arizona Department of Public Safety helicopter performed a fly-over to conclude the memorial service. Kingman Truck 34 was on display in the park with the tower extended and a very large American Flag attached creating a giant flag pole. The American Flag was waving in the breeze just below the tower bucket with the names of the eleven firefighters inscribed on the sides. With the blue Arizona sky and a few fluffy white clouds in the background; it was a breath taking sight (Figure 1.80). Each of the surviving family members were given a Maltese Cross with the fire department emblem on top KFD. The crosses were made from a remaining piece of the tank car that was saved by the department.

> *Author's Note: Chief Osterman also gave one of the crosses to me. It has become one of my most treasured items because of the sacrifice those 11 firefighters made that fateful day in Kingman.*

Eleven Kingman Firefighters Who Made the Supreme Sacrifice

Donald G. Webb

Donald Gene Webb 30, died at 3:20 a.m. Wednesday, July 18, in St. Joseph's Hospital in Phoenix. Donald Webb was married and had two children, a

Figure 1.80 Brick on memorial given to Author Robert Burke to add to the memorial. This was my thought about those brave firefighters that died.

daughter, 8, and a son, 5. Don was the owner and manager of the Eastside Shell Service Station in Kingman. He was a volunteer member of the Kingman Fire Department.

Arthur A. Stringer

Arthur Stringer 25, died Friday July 6, 1973 in Southern Nevada Memorial Hospital in Las Vegas, where he had been air evacuated following the explosion. His father, also an employee of the Kingman Fire Department was burned in the same conflagration and was hospitalized at Southern Nevada Memorial Hospital, but survived. He was also a military war hero. On returning home from Viet Nam he joined the 997th Aviation Company of the Arizona National Guard. Art had been employed full time by the Kingman Fire Department since June 1st of 1973. He was married and had one daughter and his wife was expecting another child.

Frank (Butch) Henry

Frank Stewart (Butch) Henry, 28, died in Good Samaritan Hospital in Phoenix on Tuesday July 10, 1973 of burns suffered in the fire and explosion in Kingman. At the time of his death he was the manager of ICX Inc. where he had been employed for seven years. A volunteer fireman, he had been on the official roster for five years. However he had attended fires all his life with his father, George A. Henry, who retired as a 30-year veteran of the Kingman Fire Department. He was married and had one son and a daughter.

Christopher G. Sanders

Christopher Sanders 38, a city fire Engineer with the Kingman Fire Department, died of burns suffered in the explosion and fire. Mr. Sanders died at 4 p.m., Saturday, at St. Joseph's Hospital Burn Center in Phoenix where he had been air evacuated by helicopter for treatment following the explosion. Mr. Sanders began working for the Kingman Fire Department April 16, 1971, and before that had served as a volunteer firefighter. The fire engineer has also been lauded by his fellow workers for his bravery at the scene of the explosion. Despite his burns Sanders worked to aid others and wanted to refuse treatment until his fellow firemen had been treated. Witnesses at the fire said Sanders was the first person to respond in helping treat other burned firemen and helped load several into an ambulance before being taken away himself. Christopher Grey Sanders was married and had a son; two step-sons; and two step-daughters.

Alan H. Hansen

Hansen 34, died early Thursday, July 19, 1973 from burns he received while rendering assistance in the July 5 explosion and fire. A patrolman with the Arizona Department of Public Safety, he was air evacuated to the Maricopa County Burn Unit in Phoenix immediately following the explosion and was under treatment there at the time of his death. He was a member of the Kingman Volunteer fire department. Mr. Hansen joined the Highway Patrol in October of 1961 and had been stationed in Wickenburg and Gila Bend before returning to Kingman in April of 1964. He was a veteran of the United States Army and was a member of the Arizona National Guard. He was married and had a son and a daughter.

John O. Campbell

John Odis Campbell, 42, assistant public works director for the City of Kingman, died early Monday morning July 9, 1973 at St. Joseph's Hospital in Phoenix. A member of the Kingman Volunteer Fire Department, he died of burns suffered in the explosion and fire which occurred there on July 5. Air evacuated immediately for treatment after the explosion, he was the 10th person to die in the explosion and accompanying fire. He was married had a daughter and two sons.

Joseph M. Chambers

Joseph Minter Chambers, III 37, owner and operator of Chambers Exxon Station and member of the Kingman Volunteer Fire Department with the rank of lieutenant, died of burns suffered in the explosion and fire

Chronicles of Incidents and Response 171

in Kingman. Mr. Chambers died at 7:50 a.m., Friday July 6, 1973, at the Maricopa County Hospital Burn Center where he had been air evacuated for treatment following the explosion. Joe's grandfather, Joseph Minter Chambers, was an avid firefighter too and was one of the first firefighters to serve in that capacity in this area. Joe joined the Kingman Volunteer Fire when he was 21 years old, as soon as he was old enough to serve. In 1963 he began operating the Chambers Exxon Service Station, a business in which he had been active until his death. He was married and had two sons.

M.B. (Jimmy) Cox
Jimmy Cox, 55 perished in the explosion and fire at Doxol Gas Company in Kingman Thursday July 5, 1973 company, he was a veteran of the Army Air Corps where he served more than four years, three of them at Kingman Air Base. He worked for Motor Supply, was a deputy sheriff, a parts manager for Coffman Motors, and operated a Mobil Oil Service Station at Second Street and Andy Devine. He had worked for the Kingman Bake Shop since 1960 and was in their employ at the time of his death. He had been a member of the Kingman Fire Department since 1951 and was serving as Assistant Chief when he was killed. He was married and had three children.

William L. Casson
William L. Casson 52 was district manager for Citizens Utilities Company electric division in Kingman and died in the explosion and fire while serving as a captain in the Kingman Fire Department Thursday July 5, 1973. He served in the Army Signal Corps during World War II from 1943-45. He joined the Fire Department in 1946 and was assistant chief under Joe Miller until Miller's retirement in 1959, at which time Casson took over the reins. Bill left the chief's post in 1964 when he became the utility company's top man here, and under Chief Charlie Potter has been Captain Number One. He was married had six children, four sons, and two daughters.

Roger A. Hubka
Roger Allen Hubka 27, a member of the Kingman Volunteer Fire Department, was one of the three firefighters killed in the explosion and fire Thursday July 5, 1973. Roger was employed by McCarthy Motors from the time he arrived in Kingman until a few months ago when he started to work ad Double G Tire Co. (which was destroyed by fire as a result of the explosion) as service manager. He was married to in February 1973.

Richard Lee Williams
Richard Lee Williams 47 was principal at Kingman High School, and a member of the Kingman Volunteer Fire Department, died of burns

suffered in the explosion and fire in Kingman. Mr. Williams died at 5:55 p.m. Sunday July 8, 1973 at the Good Samaritan Burn Center in Phoenix where he had been air evacuated for treatment following the explosion. He served in the U.S. Navy V-12 program during World War II. He was married and had three children, a son and two daughters. The high school athletic field was renamed Lee Williams Field in honor of the schools principle who died in the explosion on July 5, 1973 (*Firehouse Magazine*).

> *Author's Note: In July 2023 Kingman will celebrate the 50th Anniversary of the propane explosion. For information about the memorial event, contact the Kingman, AZ Fire Department.*

Houston, TX September 21, 1974 Englewood Rail Yard Collision Fire and Explosion

About noon on September 21, 1974, 2 loaded "jumbo" tank cars, cars 17 and 18 of a 145-car complement, were uncoupled as a unit at the crest of the gravity hump in the Southern Pacific Transportation Company's (SP) Englewood Yard at Houston, Texas. Englewood railyard is almost 4 miles long and is basically used to assemble railcars into trains for their final destination. The railyard is designed so that cars will be taken to the top of a hump, then released to roll down the hill to a selected track leading to the train they will be attached to.

Two cars passed through the hump master retarder and group retarder without being slowed and accelerated as they moved down the grade into bowl track 1. At a speed of 18 to 20 mph, the two tank cars impacted an empty tank car. Upon impact, the coupler of the empty tank car rode over the coupler of car 17 and punctured the tank head. Butadiene spilled from the car and formed a vapor cloud, which dispersed over the area. The car contained 38.000 gallons of Butadiene. After 2 to 3 minutes, the vapor exploded violently; as a result, 1 person died and 235 were injured. Total damages amounted to about $13 million, which included the destruction of 231 railroad cars substantial damage to 282 others. Houston Fire Department responded with over a dozen ambulances (Figure 1.81). It quickly escalated from a two alarm fire to 4 alarms, requiring still more fire equipment. Smoke could be seen in the FM 1960 area of Northwest Harris County.

The National Transportation Safety Board determined that the probable cause of the over speed impact was the failure of the retarding system to slow the two coupled tank cars and the absence of a backup system to control cars which pass through the retarders at excessive speeds. The failure of the retarding system was caused by foreign substances on the wheels of the two cars that preceded the two tank cars through the retarders. Contributing to the accident was the failure of the Southern Pacific

Figure 1.81 Houston Fire Department responded with over a dozen ambulances. It quickly escalated from a two alarm fire to 4 alarms, requiring still more fire equipment. Smoke could be seen in the FM 1960 area of Northwest Harris County. (Courtesy: Houston Fire Department.)

Transportation Company to enforce procedures to exclude cars with a foreign substance on their wheels from the humping system, and the Shell Oil Company's failure, after notification of the hazard, to eliminate spilled epoxy resin from the flange ways of their track (HFD Pictorial History 1980 – 2014).

Philadelphia, PA August 17, 1975
Gulf Oil Refinery Fire

It was Sunday August 17, 1975 just before dawn. A sea going tanker in the Schuylkill River at Girard Point was off-loading crude oil at the Gulf Oil Refinery in South Philadelphia. The refinery was constructed in 1905 on 723 acres of land located on the east bank of the Schuylkill River at Penrose Avenue near the airport. The Penrose Avenue Bridge (now the George Platt Memorial Bridge) connects the east and west sides of the Schuylkill River and passes directly over the refinery. At the time of this fire, the refinery produced 180,000 barrels per day of refined petroleum products.

Suddenly and without warning accumulated vapors from off loading the tanker were ignited starting a fire that threatened 600 storage tanks at the refinery tank farm on shore, many with a capacity of 80,000 gallons of crude oil (Figure 1.82). Hydrocarbon vapors, emanating from

Figure 1.82 Suddenly and without warning accumulated vapors from off loading the tanker were ignited starting a fire that threatened 600 storage tanks. (Courtesy: Philadelphia Fire Department.)

Tank 231, accumulated in the area of nearby Boiler House #4 and were ignited. A flame front followed the vapors back to Tank 231 causing fire at the tank's vents and an explosion within the outer shell of the stack. These events began to unfold at 5:57 a.m. At approximately 6:02 a.m. in the wake of the first explosions and fire, the tanker terminated its pumping operations, left its Schuylkill River berth and relocated to the Gulf piers at Hog Island. Philadelphia's fire alarm office received the first report of the fire at approximately 6:04 a.m., upon receiving the report, they transmitted the refinery's fire alarm box; Box 5988, Penrose and Lanier Avenues. Firefighters in Philadelphia were no strangers to the Gulf Refinery. Prior to August 17, 1975 the refinery had been the scene of ten extra alarm fires since 1960, eight of which occurred since 1966. On September 9, 1960, several storage tanks were struck by lightning at the height of severe thunderstorms and resulted in an 8-alarm fire. On May 16, 1975, a six-alarm fire struck the Gulf Refinery. Following the disastrous fire of August 18, 1975, another six-alarm fire occurred just five months later on October 20, 1975.

Engine 60 was first due at the refinery and as they left their station firefighters could see fire and smoke conditions at a distance and before arriving requested a 2nd alarm at 6:09 a.m. Before the fire was over 500 firefighters would battle the blaze and 11 alarms would be transmitted, six Philadelphia firefighters would lose their lives, and an additional nine would be injured along with 4 Gulf firefighters. Two more Philadelphia firefighters would succumb to their injuries several days later. Not long

Chronicles of Incidents and Response

after the original ignition of the fire a second explosion occurred within Tank 231. Burning petroleum spilled from the tank's vents into a diked area surrounding the tank. Within the diked area a second tank (No.114) just north of Tank 231, containing No.6 grade fuel oil, also ignited as pipelines within the diked area began to fail. The initial explosion damaged the pipe manifold outside of the dike wall and petroleum pouring out under pressure ignited.

First arriving companies encountered large clouds of heavy black smoke billowing from Tank 231, fire on top of Tank 114, and fire showing from the 150-foot stack at Boiler House No.4. The third and fourth alarms were ordered in quick succession by Battalion Chief 1, Arthur Foley, at 6:11 a.m. and 6:14 a.m. Acting Assistant Fire Chief Dalmon Edmunds ordered the fifth alarm at 6:34 a.m. Engine 33 and Foam 133 responded on the fifth alarm from their station on the North side of Philadelphia. Firefighter Hugh McIntyre of Engine 56 had been detailed to Engine 33 on the day of the fire. He was the oldest firefighter to die in the inferno. The sixth alarm was ordered by Fire Commissioner Joseph Rizzo at 6:52 a.m.

Over the next several hours, firefighters utilized deluge guns and master streams to cool down surrounding exposures, and applied foam directly to the burning tanks and piping in an effort to extinguish the fire. By 8:44 a.m. it appeared that the fire was well contained and the situation sufficiently stabilized to declare the fire under control. Throughout the day Philadelphia's two foam pumpers, Foam Engines 160 and 133, along with the Gulf Refinery's foam pumper, continued to apply foam to the burning tank, piping and manifolds. Additional foam to support the operation was acquired from the fire department's warehouse and the nearby Atlantic-Richfield refinery. It was also obtained from the National Foam Company in West Chester, PA.

As the firefighting operation progressed, it became apparent that the refinery's sewerage system was unable to properly drain the foam, water and petroleum-naphtha product mixture that was accumulating on the ground along Avenue "Y," between 4th Street and 5th Street to the east, running in front of the refinery's administration building. These drainage problems were further exacerbated by a decision by refinery personnel to shutoff drainage pumps. These pumps were shutoff as part of a decision to de-energize overhead power lines that ran adjacent to Tank 231 along 4th Street. As the liquid mixture continued to build-up in Avenue "Y," Engines 16 and 40 were dispatched to Avenue "Y" and 5th Street, to draft from a sewer intake and pump the material to a diked area some distance away.

Foam Engine 133 was set up on the east side of Tank 231 at Avenue "Y" at 4th Street applying foam to the tank. Three members were attending to the apparatus and wading in the foam-water-petroleum mixture which was accumulating on the ground under the apparatus. The mixture

176   *Hazmatology: The Science of Hazardous Materials*

continued to get deeper as the battle with the fire continued and eventually reached the axils of the foam engines. Commissioner Rizzo and Gulf Refinery manager Jack Burk were on an overhead catwalk nearby observing the fire fighting operation. Without warning at approximately 3:30 p.m. that afternoon, and in full view of Commissioner Rizzo and Burk, the accumulating liquid surrounding Engine 133 ignited, immediately trapping firefighters Campana, Fisher and Andrews working at Engine 133.

Instinctively and without hesitation other nearby firefighters dove into the burning liquid to rescue their comrades, not aware of the danger to themselves. Five more firefighters would be consumed by the advancing fire. Firefighters McIntyre, Willey and Parker died during the attempted rescue of their fallen comrades. Lieutenant Pouliot and Firefighter Brenek were gravely injured during the rescue efforts and would die several days later in the St. Agnes Hospital burn center in South Philadelphia.

At approximately 4:41 p.m. a fire storm was developing as the fire quickly spread eastward along Avenue "Y" towards 5th Street. Viewing the unfolding horror before him, Commissioner Rizzo ordered two more alarms, five additional rescue squads, and the recall of all companies which had previously been released from the fire ground throughout the day. On these orders the fire alarm office transmitted the seventh and eighth alarms simultaneously. As the fire had been placed under control nearly eight hours earlier, firefighters in stations across the city knew that the unthinkable had occurred as these additional alarms were struck. At 4:46 p.m. Commissioner Rizzo ordered the ninth alarm and notification of Philadelphia Managing Director Hillel Levinson as a major disaster was now unfolding at the Gulf Refinery.

As the fire swept rapidly eastward along 5th Street, Philadelphia's foam pumpers 60 and 33, and the Gulf Refinery foam pumper, were rapidly destroyed in the fire's advance. At 5th Street, where Engines 16 and 40 had been assigned to improve drainage, their pieces were also destroyed in the fire's path, although their pump operators were able to escape. Upon reaching 5th Street, the fire traveled two city blocks north along 5th Street, now threatening four additional storage tanks and the 125-foot Penrose Avenue Bridge. At 5:37 p.m., Commissioner Rizzo ordered the tenth alarm as the fire was now traveling southward and engulfing the refinery's administration building, which was located on the south side of Avenue "Y" between 4th and 5th Streets. The tenth alarm companies were ordered to report to Gate 24 at Penrose and Lanier Avenues, to set up deluge guns and leave the area.

As the situation continued to deteriorate at the Gulf Refinery, Commissioner Rizzo ordered all "D" platoon members from the day shift held over, and at 6:01 p.m., he ordered the 11th alarm. By seven o'clock, the involved tanks and pipelines were gushing flames and nearby streets in the complex were burning streams of oil and other petroleum products.

Chronicles of Incidents and Response

For a period of time, it was far from certain where the fire would be stopped. Burk was quoted as saying that contingency plans had been made for a retreat through the refinery, street by street, tank by tank. But far from retreating, the courageous men of the Philadelphia Fire Department attacked, retaking 5th Street, 4th Street, and finally Avenue "Y." At 1:00 a.m. Commissioner Rizzo left the fire ground relinquishing command to Deputy Fire Commissioner Harry T. Kite who placed the fire under control at 5:38 a.m. on Monday, August 18, 1975.

The original cause of the fire was the overfilling of Tank 231. While no crude oil escaped from the tank as a result of being overfilled, large quantities of hydrocarbon vapors were trapped above the surface of the tank's crude oil. As the quantity of crude oil increased, these hydrocarbon vapors were forced out of the tank's vents and into the area of the No.4 Boiler House where the initial flash occurred. The overfilling of the tank, in turn, resulted from a failure of the tanker's personnel to properly monitor the quantity of crude oil being pumped to the tank. It is believed that the second fire and subsequent explosions were triggered when the hot muffler of Foam Engine 133 came contact with the hydrocarbon vapors above the flammable liquids floating on the water under the truck (Figure 1.83A)

Firefighters Who Gave the Supreme Sacrifice

John L. Andrews, Engine 49

Hugh McIntyre, Engine 56

Ralph J. Campana Ladder 19

Joseph R. Wiley, Ladder 27

Roger T. Parker, Ladder 27

Robert J. Fisher, Engine 33

Lt. James J. Pouliot, Engine 20

Carroll K. Brenek, Engine 57

Philadelphia Pays Tribute to Gulf Oil Refinery Fire Fallen

In Philadelphia, the Gulf Oil Refinery fire is far from forgotten. In August 2007, about 200 people gathered at the Fireman's Hall Museum in Philadelphia as plaques were unveiled to honor the firefighters lost in the refinery disaster. Paying tribute to the fallen, were IAFF General President Harold Schaitberger, Philadelphia Local 22 President Brian McBride, Local 22 executive board officers and many Local 22 fire fighters. "Since the refinery was all but destroyed in the fire, we felt it fitting to place the plaques outside the museum where many passersby can see them and remember what happened," says McBride. "We will never forget the sacrifice our brother fire fighters made that day," says IAFF General President Harold

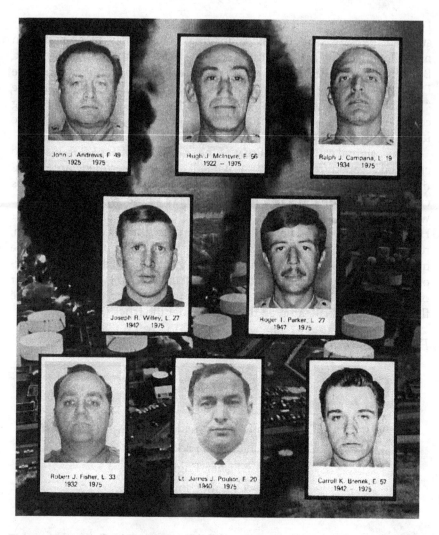

Figure 1.83A Firefighters who made the supreme sacrifice at the Gulf Oil Fire. (Courtesy: Philadelphia Fire Department.)

Schaitberger. "These eight plaques will serve as a permanent reminder of their dedication to the fire service and to the citizens of the great city of Philadelphia." The eight memorial plaques were sponsored by broadcast pioneer, publisher and philanthropist Kal Rudman who also awarded scholarships for the children of the eight fallen fire fighters (Figure 1.83B).

The plaques placed in front of the Firemen's Hall were the 19th through 28th in an ongoing series of memorials to be placed honoring the 285 members of the Philadelphia Fire Department that died in the line of duty

Figure 1.83B The eight memorial plaques were sponsored by broadcast pioneer, publisher and philanthropist Kal Rudman who also awarded scholarships for the children of the eight fallen fire fighters.

since 1871. The plaques are placed on the sidewalks in the city or as near to the location of where the firefighters died as possible. This will allow citizens, friends, family and fellow firefighters to remember these fallen heroes far into the future. Thanks to Philadelphia Fire Commissioner Lloyd Ayers, Michael F. Roeshman Battalion Chief, Hazardous Materials Administrative Unit and curator of Philadelphia Firemen's Hall Harry McGee for their assistance in the gathering of information and photographs for this section (*Firehouse Magazine*).

Niagara Falls, NY December 15, 1975 Hooker Chemical Chlorine Tank Explosion

Clouds of deadly chlorine gas killed four persons and injured 87 others after a railroad tank car at the Hooker Chemical plant exploded and unleashed the gas over this honeymoon resort city Sunday night. Authorities said persons three miles from the blast scene were injured by the toxic fumes. Company officials said the explosion of unknown origin occurred in one of three tank cars used for storage of recovered chlorine. Firemen extinguished a brief fire that followed the explosion, but continued to stand by early today to check a small leak in one of the other tank cars, which was ruptured in the initial explosion. Each tank car contained 30 tons of liquid chlorine, company officials said.

The explosion sent a cloud of chlorine gas over the area, affecting residents and motorists driving through the region. But a company official said the cloud dissipated after about two hours. Niagara County Coroner Oscar A. Bell listed the cause of death as inhalation of chlorine gas. Fifty-three other persons were also treated at the Medical Center. The spokesman said 40 of the injured were treated and released and 13 were admitted to the hospital. Thirty-four other persons were treated at Mount St. Mary's Hospital in nearby Lewiston. Ten of those persons were admitted to the hospital. A hospital spokesman said most of those injured were either employees of the plant or were in the immediate area of the chemical plant at the time of the explosion. He said most complained of feeling ill after inhaling the toxic chlorine fumes.

The Daily Messenger Canandaigua New York 1975-12-15

Houston, TX May 11, 1976 Anhydrous Ammonia Tanker Accident

Worst Accident in Houston History

The worst transportation incident to occur in Houston happened almost 43 years ago on May 11, 1976 at approximately 11:08 hours. This incident occurred a little more than 2 years before the hazardous materials team was formed. An MC331 Tanker truck hauling 7,509 gallons of liquid anhydrous ammonia struck and penetrated a bridge rail, then struck a support column and fell approximately 15 feet to the Southwest Freeway below. Great clouds of ammonia fumes billowed over the area, turning spring foliage to a golden brown, and chasing stunned motorists from their cars (Figure 1.84).

Hospital emergency rooms were jammed with persons suffering from burning eyes and lungs. All of the ammonia was released from the tank resulting in 6 fatalities, 78 hospitalizations, and approximately 100 other persons treated for exposure to the ammonia. Had this incident occurred at another time or location the results could have been catastrophic. If the location had been in downtown Houston next to office buildings, apartments or a congested residential area there could have been many more injuries and fatalities. As a result, the City of Houston created designated routes for hazardous materials transported within the city to avoid such vulnerable areas.

"I saw a big whole tank fly up as high as the (40-story) Humble Building," "said cab driver Robert Galindo, 32, about 150 yards from the explosion. "The smell was horrible. I couldn't breathe. But I saw people on the inside of that cloud of smoke." Another witness said several motorists were trapped in their cars. "I stopped and my car was rocking from the shock waves," the witness said. "I started crossing over to the grassy area.

Figure 1.84 Great clouds of ammonia fumes billowed over the area, turning spring foliage to a golden brown, and chasing stunned motorists from their cars. (Courtesy: Houston Fire Department.)

People stopped to stare at the wreckage like they were in a trance. I got off the road and my heart was still in my throat. It was terrible."

The National Transportation Safety Board dispatched a team of investigators from Washington to Houston to investigate the accident. The owner of the truck, Transport Co. of Texas, said the tank normally carried 4,000 to 5,000 gallons of ammonia under great pressure.

The truck crashed through a guardrail on Loop 610 and fell onto U.S. 59. A Texas Highway Department spokesman said 345,000 cars daily use the interchange, the "heaviest traveled in the state." "Cars and people were scattered everywhere," said Shelby Hodge, a reporter for the Houston Post, whose offices are adjacent to the interchange. All that was left of the truck was four tires and a heap of metal. Miss Hodge was sitting at her desk when the blast shook the four-story building and engulfed the area in smoke.

Playground Daily News Fort Walton Beach Florida 1976-05-12

Westwego, LA December 22, 1977
Grain Elevator Explosion

Weary rescue crews working without rest today recovered another dozen bodies from the steel and concrete rubble of a Mississippi River grain elevator, forcing officials to raise their estimates of the explosion death toll to 35 in the nation's worst grain industry accident (Figure 1.85). Shortly before dawn, the 22nd body was recovered from the wreckage of the $100 million Continental Grain Co. plant in suburban New Orleans and Jefferson Parish Sheriff Alwynn Cronvich said he feared another 13 victims were buried inside. Eleven other persons, including a sheriff's deputy, were hospitalized with injuries. "It's almost more than we can bear," said Jefferson Parish Sheriff Alwynn Cronvich."

Louisiana Agriculture Commissioner Gil Dozier said the strike, which slowed shipping at ports from Maine to Texas for two months this fall, may have forced Continental officials to hold on to their grain supplies too long. "If you put any grain in an enclosure it's like using the grain for mash and gasses are created," Dozier told UPI. "It's not the grain that explodes, it's the gasses that build up. It could have been from spontaneous combustion." Dozier said he was uncertain how long the current supplies of grain had been stored at the Continental plant 10 miles upriver from New Orleans. He said investigators would be asked to look into the question. "This is the worst grain disaster in the history of the nation," he said. "They had one on the Great Lakes a few years back, but we already have enough deaths to exceed that one. Apparently this plant is a total loss."

"There was terrific blast that shook all the houses and then there was this huge black smoke pouring out," said P. J. Angell, a retired railroad who lives four blocks from the plant. "By the time I got there, I still had to duck and dodge, concrete was falling everywhere."

The blast destroyed 48 of the 73 giant silos used to store soybeans, wheat and oats at the plant. The damaged silos leaned against one another

Figure 1.85 In the nation's worst grain industry accident, 35 people were killed in an explosion of a grain elevator in Westwego, LA.

Chronicles of Incidents and Response 183

like dominoes ready to fall. The view was the same from a distance or up close. Everyone thought an atomic bomb had been dropped. "I was driving and it was just an unbelievable sight looking to the west," said Walt Pierce, who was 10 miles away in New Orleans when the Continental Grain Co. plant exploded Thursday.

"I saw this giant mushroom cloud and I couldn't believe it. It looked like a hydrogen bomb." The explosion destroyed a grain elevator at the plant along the Mississippi River and set off what officials called the nation's worst grain industry accident in history. Twenty-five persons were believed dead. "Flames came out of the top of the elevator for a little while. It looked like a gas burner." The top half of the grain elevator was destroyed. Its concrete and steel spewed into the air and crushed a nearby two-story office building where most of the victims died.

The Ruston Daily Leader Louisiana 1977-12-23

Galveston, TX December 29, 1977
Grain Elevator Explosion

The bodies of six persons were found Wednesday beneath tons of debris from the explosion of the Farmers Export Grain Elevator, pushing the death count to 15 in the disaster (Figure 1.86). "You have drier, combustible materials" when they humidity is low, said L. E. Bartelt, administrator of the Federal Grain Inspection Service. Bartelt arrived in Galveston late Wednesday morning. At least four of the victims of the disaster were identified as members of the Federal Grain Inspection Service and three more remained unaccounted for. The humidity in Galveston about 8:30 p.m. Tuesday was 50 percent, lower than normal for the island.

The humidity was also lower than normal last Thursday at a grain elevator explosion in Louisiana. Bartelt likened the volatile nature of grain to gasoline. "Handling grain is not too different than handling gasoline. We've got to learn to do it better." Despite the speculation that the conditions for an explosion may have been made more favorable by the low humidity Tuesday, investigators said they still have no clear indications what set the holocaust off. "It's still undetermined at this time (late Wednesday afternoon) as to the cause," said Galveston Fire Chief Hugh O'Donohoe.

Statements taken by Galveston police from survivors of the holocaust show the explosion was apparently touched off in or near the rail car dump bin south of the elevator and then spread through an underground connecting tunnel into the elevator Head House Officials feared that the bodies of the persons still unaccounted for would be found beneath the ruins of the rail dump bin. Cutting torches and bulldozers were being used late Wednesday night to untangle the mass of twisted steel that once

Figure 1.86 The bodies of six persons were found beneath tons of debris from the explosion of the Farmers Export Grain Elevator, pushing the death count to 15 in the disaster.

was the bin and railroad cars. Most of the bodies located Wednesday were in the Head House area where heavy cranes had to be used to move the pile of rubble. A brother of one of the victims watched the body search calmly for most of the day but then openly cried when workers found his sister's body in the Head House area.

Robert Steen, the first police officer on the scene, told of a grim scene in his official report of the explosion aftermath Tuesday night. "At first," Steen said, "no one could be found, but when we (Steen, with Officer T.

Cantela and Lt. D. K. Lack) started going through the rubble, bodies could be seen on the outskirts of the explosion." Lack pulled back to take over command of the rescue operations and Steen and Cantela were joined by Officer Milton Strickland. Steen's report continues:

"We began to yell to locate survivors. After several minutes, I heard a faint voice. He had been sitting in a chair when a cement wall fell on him. We then found two other survivors and began digging them out. These three survivors were buried under several feet of twisted metal, cement and grain." Galveston officials received such an overwhelming response from out of town aid that many emergency units had to be turned back when the scene began to be clogged.

Galveston Daily News Texas 1977-12-29

Waverly, TN February 24, 1978
Derailment and LPG Explosion

The LPG explosion, which occurred in Waverly, Tennessee on February 24th, 1978, was the high water mark of hazardous materials incidents in the United States. The Waverly incident also resulted in many changes in both tactics for dealing with LPG fires in containers, and safety equipment on railroads. Some years after the Waverly incident, a similar derailment occurred in Weyuwega, Wisconsin and the incident commander there used his knowledge of the Waverly incident to formulate tactics in Weyuwega. This may have had a direct impact on the Weyuwega incident in terms of safety to emergency personnel and residents. There was not a single serious injury or death as a direct result of the derailment in. I think we can all learn from lessons from past incidents such as Waverly and be able to protect personnel and the public from harm. Unfortunately, sometimes our memories are very short.

Waverly, Tennessee is located 56 miles West of Nashville and about 142 miles East-Northeast of Memphis. Waverly is the county seat of Humphries County and had a population of 5,000 in 1978. It was a typical winter day on Wednesday in Northwestern Tennessee. Temperatures hovered in the mid 20's with about ½ inch of snow on the ground. At approximately 10:30 p.m. a Louisville and Northern (L&N) train heading from Nashville to Memphis derailed in this small community. Investigators determined that a wheel on a gondola car, overheated from a handbrake left in the applied position, broke apart East of Waverly. A wheel truck damaged by the breaking wheel managed to remain with the train for seven miles before it finally came loose from the car causing the derailment. Twenty-four of the ninety-two cars of the train left the tracks in the center of downtown Waverly.

Two of the derailed tank cars, which contained liquefied petroleum gas, played a major role in the incident that unfolded over several days.

186   *Hazmatology: The Science of Hazardous Materials*

Varieties of Liquefied petroleum gases that are bought and sold include mixes that are mostly propane, mostly butane, and most commonly mixes including both propane and butane. LPG sometimes contains a variation of butane called isobutane. LPG is a colorless gas with an odor of Natural Gas (methane). Flammability wise, it doesn't matter what the combination of gases are, they are all very flammable. Mixing the gases changes their physical characteristics like boiling point. Natural gas and LPG do not have an odor naturally, so an odorant is added for leak detection purposes. LPG vapor, as most hydrocarbon based compound vapor, is heavier than air.

The primary hazard to be concerned about with liquefied petroleum gases is flammability. When liquefied petroleum gases are in a container, a boiling liquid, expanding vapor, explosion (B.L.E.V.E.) can occur under certain conditions. Those include direct flame impingement on the vapor space of the container, over pressurization of the container, or physical damage to the container shell or a combination of factors.

Incident commanders should be well aware of these potential effects on the integrity of liquefied petroleum gas pressure containers. The explosion in Waverly was unusual in the fact that it did not involve flame impingement on the tank. When the derailment occurred in downtown Waverly, there were no leaks, no fires, no explosions; nothing overly exciting resulted beyond the derailment of the cars themselves. This may have been one of the primary factors, which allowed response personnel and others let down their guard leading to disaster two days later.

Volunteers from the Waverly Fire Department along with the Waverly Police Department were first on the scene of the derailment. Responders did not have monitoring devices to check for leaks and relied upon their own sense of smell to determine there was no immediate danger. LPG has an odor threshold of 1800mg/cuM. Using the human senses is not recommended as a means of detection for hazardous materials because of the potential danger it presents to response personnel. Without monitoring equipment, there is no way to know what amount of vapor may be present. Responders could become asphyxiated by LPG or other gases without proper respiratory protection. They could find themselves in the middle of the flammable range without a meter to determine the amount of vapor/air mixture that is present. This could lead to serious injury or death if ignition occurs.

Following the initial scene survey, a nearby single-family residence and custodial care facility were evacuated as a precaution. The state civil defense agency was notified the next morning of the derailment. Initial reports to the agency indicated no hazardous materials were involved. It was not until 05:10 on February 23rd that civil defense authorities were told there were in fact hazardous materials in some of the derailed train cars. With that information in hand, a state hazardous materials team was dispatched to the scene and arrived at 06:30 hours on February 23rd.

Chronicles of Incidents and Response

Once the hazardous materials team was on scene, an additional evacuation distance of ¼ mile was implemented and all electrical and natural gas sources were shut off in the hazard area. Local firefighters had already placed heavy hose streams in place for cooling the derailed tank cars.

> **Hazmatology Point:** *Remember that the boiling point of LPG is usually −42° F to 0° F but it is kept as a liquid by the pressure in the tank. The actual temperature of the liquid is ambient and any operations of applying water to tanks which are not on fire can actually cause the LPG to be heated, increasing pressure inside the tanks. Water from booster tanks that was in heated fire stations can be in the 70's or higher. At some point, the water pressure will equal ambient pressure and about that point the water will act as a coolant to the tank and the liquefied gas inside. It is a questionable tactic to cool tanks that are not on fire. We do not need to do it just because it is in our Hazmat Tool Box. We need to evaluate the situation and make sure there is a good reason to cool the tanks.*

Railroad personnel began the process of clearing the right-of-way to get rail traffic moving again as soon as possible. They complained the mud created by the "cooling" efforts of emergency responders was making it difficult for their workers, and the hose lines were shut down but left in place. By 2:15 p.m. on the 23rd of February, the rail line had been cleared of all derailed cars. One derailed tank car #83013 (which was to B.L.E.V.E. later) had been moved some 12 feet from its original resting point underneath several other rail cars (Figure 1.87). The L&N line was once again opened to limited rail traffic at 20:00 hours on the 23rd. Up to this point, no efforts had been made to deal with the LPG still in the tank cars.

At the direction of the L&N Railroad, crews were dispatched to off load the LPG tank cars and they arrived on scene about 13:00 hours on the 24th of February. By this time the sky had cleared and the sun had come out raising the temperature into the mid-50's.

> **Hazmatology Point:** *One of the possible contributing factors to the incident indicated by the National Transportation Safety Board (NTSB) was an increase in pressure in the tank. Ambient temperature increases can cause the pressure in a tank to increase. Also unknown at the time was that a portion of tank #83013 had been damaged and weakened by the derailment. Keep in mind that tank cars piled up on each other or banged into each other may have sustained damage. Moving them before offloading the product can lead to further damage or catastrophic tank failure, which occurred in Waverly.*

The combination of the two factors may have resulted in the B.L.E.V.E. involving tank #83013). Before the off loading process was to begin, air monitoring of the area with combustible gas indicators revealed no leaks

Figure 1.87 By 2:15 p.m. on the 23rd of February, the rail line had been cleared of all derailed cars. One derailed tank car #83013 (which was to B.L.E.V.E. later) had been moved some 12 feet from its original resting point underneath several other rail cars. (Courtesy: Waverly Fire Department.)

or LPG in the area. Because of the lack of catastrophic events surrounding the derailment, the evacuation was relaxed and Waverly was pretty much back to business as usual by the time the off loading was to begin. In fact, persons were observed smoking in the area of the derailed propane tanks. The fire chief, police chief, a fire crew, and two representatives of the Tennessee Civil Defense were on scene along with workers from the L&N Railroad along with personnel from the Liquid Transport Company.

Prior to initiation of the off loading process at approximately 14:58 hours someone noticed LPG vapors leaking from tank car #83013. Before anyone could react to the leaking LPG (a matter of seconds), a B.L.E.V.E. occurred (Figure 1.88). The resulting fires and effects of the explosion killed five people instantly, and severely burned the Waverly fire chief (he would die later at the hospital). Also killed was a Waverly Civil Defense Officer. Injured were taken by ambulance to local hospitals and critically injured were flown by helicopter to burn centers in Tennessee, Ohio, Kentucky and Alabama. Ultimately 16 people died as a result of the explosion. Most of them were in businesses next to downtown, located right next to the derailment. The explosion compromised most of the Waverly Fire Departments on scene firefighting capability. Hoses left in place in case they were needed, were shredded by the explosion, leaving no immediate means of fighting fires. Parts of the tank car, burning LPG, and other debris were scattered over a wide area.

Figure 1.88 Prior to initiation of the off loading process at approximately 14:58 hours someone noticed propane vapors leaking from tank car #83013. Before anyone could react to the leaking propane (a matter of seconds), a B.L.E.V.E. occurred. (Courtesy: Waverly Fire Department.)

One piece of the tank car was propelled 330 feet by the explosion. Noise and blast pressure from the explosion were felt several blocks from the scene. Numerous large buildings were set ablaze by the heat from the fireball, as well as vehicles, by standers, and other railcars. A second LPG tank car was also set on fire by the explosion, but did not B.L.E.V.E. Fortunately, flame impingement on the second LPG tank car was below the liquid level and not on the vapor space. Liquid in the tank even though flammable, will absorb heat from flame impingement and protect the tank shell from thermal damage.

As you can imagine at this point the small town of Waverly was in chaos. Many emergency responders including fire alarm police personnel and their equipment were directly affected by the explosion and were unable to assist those injured. They had become victims themselves. Calls went out statewide for assistance and over 250 emergency vehicles from 39 counties responded to Waverly's call for help. Once incident command was established, the evacuation area was extended to one mile around the derailment scene where the second LPG tank car was burning.

Thirty Medevac helicopters were dispatched from the Fort Campbell, KY and Tennessee State Police Army Post in Kentucky to airlift burn victims to burn centers in Nashville, Louisville, Kentucky, Birmingham, Alabama, and Cincinnati, Ohio. In all, 16 people died from the explosion

190   *Hazmatology: The Science of Hazardous Materials*

and resulting fireball and 43 people were hospitalized with injuries and numerous others treated as outpatients for their injuries. Sixteen buildings were totally destroyed and another twenty were damaged. Total cost of the damage in 1978 dollars was estimated at $1,800,000. A team from the National Transportation Safety Board (NTSB) investigated the incident. Excerpts from their report are provided in the next paragraph.

"The National Transportation Safety Board determined that the probable cause of the loss of life and substantial property damage was the release and ignition of liquefied petroleum gas from a tank car rupture. The rupture resulted from stress propagation of a crack, which may have developed during movement of the car for transfer of product or from increased pressure within the tank. The original crack was caused by mechanical damage during a derailment, which resulted from a broken high-carbon wheel on the 17th car which had overheated.

Those Emergency Responders Who Made the Supreme Sacrifice

Fire Chief Wilbur J. York, 65

Police Chief Guy Oakley Barnett, 45

Civil Defense Officer, Donald M. Belyen, 24 (*Firehouse Magazine*).

Youngstown, FL February 27, 1978 Train Derailment with Chlorine Gas Cloud

A cloud of chlorine gas that escaped from a derailed freight train and has already killed eight people floated north, triggering a string of evacuations in its path. "We don't know how far we're going to have to go with the evacuation," a spokesman for the Bay County Sheriff's Department said. "The chemists are telling us when we have an unsafe area." At least 2,500 people with a five-mile radius of Sunday's derailment were evacuated, Civil Defense Director Ron Johnson said propane leaking from one of the tank cars could mix with the billowing chlorine and explode, turning the 115-car freight train into a chain of fire.

Chemical specialists from the Environmental Protection Agency planned to spray the train wreck area with detergent today to cut off the spewing chlorine and propane gas and blot out fires or explosions. Forty-three cars of the Atlanta and St. Andrews Bay freight train jumped the tracks about 2 a.m. CST Sunday. One of the chlorine tank cars cracked open on impact and the toxic gas boiled out. Shifting winds spread the chlorine, stalling cars and running them off the road as the gas choked off oxygen from automobile engines. Some motorists were trapped in their cars and died. Others jumped free and fled to the nearby swamp, where they were caught by the searing fumes. At

Chronicles of Incidents and Response

least 80 people were treated at hospitals for the effects of the gas, which turns to acid in the lungs and leaves victims bleeding from the nose and mouth. "It was instant death," said Al Smith, an EPA emergency coordinator. "The kind of death we're talking about is horrible. It literally burns your lungs up."

Eight teen-agers packed into one car for a return trip from a party in a nearby community were caught in the cloud. They tried to hide from the fumes in the swamp. Four died. One of them was not found for 14 hours, lying dead in an open field near the spewing tanker car, a jacket wrapped around her head in a futile attempt to thwart the gas. At least 45 other people, including four sheriff's deputies, were hospitalized, nine in critical condition. All of the train's crew members, trained to run upwind from a chemical spill, escaped unharmed. But engineer Ray Shores of Panama City ran so far into the swamp that he wasn't found until eight hours after the crash. He guided a search helicopter to himself with a walkie-talkie he had taken from the train. "It happened so fast I really don't know what happened," Shores said. "There was no noise or anything like that. It just jumped the tracks and derailed. I finally got the train stopped, but it just kept derailing."

Most of the train carried lumber flats, but nine tanker cars held toxic chemicals. Two of the tanks contained chlorine used mainly in bleaching and water purification; one held liquefied petroleum gas and began leaking late Sunday; four were filled with caustic soda; one had ammonia nitrate, and one had sodium hydroxide.

Kingsport Times Tennessee 1978-02-27

Brownson, NE April 2, 1978 Train Derailment Phosphorus Fire & Explosion

Brownson is located is an unincorporated village about 5 miles west of Sidney NE on US Highway 30. Union Pacific 142 was eastbound with 101 cars when the derailment occurred. 31 cars left the tracks including a liquid tank car carrying white (yellow) phosphorus. Phosphorus is an air reactive material that ignites spontaneously in air at 86° F. The phosphorus car turned over and approximately 1/2 of the water covering the phosphorus leaked from the tank. Immediately the phosphorus ignited and started to burn vigorously (Figure 1.89). No water supply was available so the decision was made to allow the fire to burn. Consultation with NFPA action guides, and CHEMTREC they were advised that there was no hazard of the tank car exploding. White phosphorus is very toxic and should not be allowed to come in contact with the skin. Dense white clouds of toxic fumes evolved from the burning tank which can attack the lungs.

Figure 1.89 The phosphorus car turned over and approximately 1/2 of the water covering the phosphorus leaked from the tank. Immediately, the phosphorus ignited and started to burn vigorously. (Courtesy: Sidney Fire Department.)

Firefighters were given three options by CHEMREC to deal with the burning tank car. They could deluge the tank with large volumes of water. This was not possible because of the remote location of the derailment and no water source readily available. They could bury the car with sand and keep the sand wet, which would block the phosphorus from getting to the air. Lastly, and the option chosen was to let the tank burn. Wind speed and direction changed around 11:00 a.m. Winds were gusting to 30 miles per hour, blowing the toxic cloud towards Sidney. Evacuation of the north side of Sidney was undertaken as a precaution. The fire continued to burn until 11:38 a.m.

Without warning, the tank car exploded and parts of the tank car were propelled 2,400 feet away. Even though fire fighters were in close proximity, no one was killed in the blast. Six railroad employees were injured (*Firehouse Magazine*).

> **Hazmatology Point:** *Everyone was correct in terms of the phosphorus not causing an explosion in the tank car. Then what caused the explosion? The very thing they didn't have enough of to fight the fire ended up causing the explosion. Phosphorus being air reactive is shipped under water in the tank thus keeping the phosphorus from spontaneously combusting. When the water leaked off, the phosphorus ignited and continued to burn all the while heating the remaining water in the tank.*

Chronicles of Incidents and Response 193

>When water is heated to the point of boiling it becomes a gas, steam. Since this was a liquid tank it was not designed to hold pressure. When the steam pressure built up to the point the tank could no longer hold it, the tank came apart, explosively. This was a simple boiler explosion. When dealing with hazardous materials incidents, be sure to consider all of the actors. In this case in addition to the phosphorus, the tank and the water were also actors; Actors that precipitated the explosion.

Gettysburg, PA March 22, 1979
Phosphorous Truck Fire & Explosion

During the night of March 22, 1979 a truck transporting phosphorus though the town of Gettysburg, PA stopped on Buford Avenue just west of the downtown business district when the driver noticed smoke coming from the cargo space of the truck (Figure 1.90). The truck was hauling white phosphorus in 55 gallon drums, (the same white phosphorus that was in the tank car in the previous incident). Phosphorus being air reactive is shipped under water to keep the phosphorus from igniting. One of the barrels developed a leak and water covering the phosphorus dropped below the level of the material in the drum. At that point the phosphorus ignited and burned, impinging on other drums on board. This repeated like a chain reaction until the entire truck was ablaze on the streets of Gettysburg (Figure 1.91). Gettysburg Volunteer Fire Department responded and were told by dispatchers not to put water on the fire. So,

Figure 1.90 During the night of March 22, 1979 a truck transporting phosphorus though the town of Gettysburg, PA stopped on Buford Avenue just west of the downtown business district when the driver noticed smoke coming from the cargo space of the truck. (Courtesy: Gettysburg Fire Department.)

Figure 1.91 At that point the phosphorus ignited and burned, impinging on other drums on board. This repeated like a chain reaction until the entire truck was ablaze on the streets of Gettysburg. (Courtesy: Gettysburg Fire Department.)

exposures were protected and the fire allowed to burn itself out. A couple of buildings received surface damage from the radiant heat.

By the next morning the fire had gone out and clean-up crews were on site to clean up the remaining phosphorus. As the phosphorus would be uncovered it would again ignite and sand had to be brought in to keep the phosphorus from reaching the air. Several persons are injured when 55 gallon drums of white phosphorus exploded during the clean-up from a fire that destroyed the truck that carried them. Parts of the drums were rocketed several hundred feet. It is likely the water heated up in the drums from the heat of the fire turned the water to steam and caused the drums to open up explosively from the steam pressure build up inside. Drums of the type the phosphorus is shipped in are ordinary 55 gallon drums meant for liquids. They have no ability to withstand pressure increases inside the drum and will fail rather easily which is likely what happened that day. Several area fire departments provided ambulances and other apparatus through mutual aid (*Firehouse Magazine*).

Hazmatology Point: *Remember that 55 gallon drums are used for many chemicals including acids, poisons, flammable liquids, and solids. They do not carry gases because they are liquid containers. Anytime 55 gallon drums are exposed to fire, expect pressure build-up in the containers, and they will rocket when the pressure causes the container to fail.*

Chronicles of Incidents and Response

Collinsville, IL August 7, 1978
Propane Tank Car Explosion

The 200 residents who evacuated their homes when a tank car full of propane exploded have been told to stay away from their houses until at least Tuesday. Two persons were injured early Sunday when the tank car derailed, exploded and threatened to explode four other tankers.

Police evacuated everyone within a half mile of the explosion site on the outskirts of town while firefighters attempted to keep the fire from spreading to the other propane cars, three of which also derailed.

Evacuated residents were asked late Sunday to remain away from their homes at least until Tuesday, when the fire was expected to burn itself out. After the area cools down, the propane would be pumped from the remaining tanks. The blast and flash of red light came were seen 40 miles from the scene in west St. Louis County. One caller said the red fireball was "like a little atomic bomb," according to the Illinois State Police. The explosion burned trees in a wide area around the site, charred railroad ties and a crossing warning sign and melted asphalt at the crossing.

After the explosion, police blocked nearby Interstate 55-70 for a brief time as a precaution. Firemen continued wetting down the burning car and neighboring tanks throughout the day Sunday. Collinsville firemen were assisted by departments from Maryville, Long Lake and Troy. Civil Defense units and auxiliary police helped Collinsville police and Madison County deputies seal off the area and set up a command center in Collinsville. The American Red Cross set up a center for the evacuees at the Collinsville VFW Hall.

The Daily Leader Pontiac Illinois 1978-08-07

Mississauga, Ontario, Canada November 10, 1979 Derailment Fires Explosions

The "Mother of All Hazmat Evacuation"

Mississauga Train Derailment, Mississauga, ON, 10 November 1979 A 106-car train Canadian Pacific Railway was eastbound near Mississauga, Ontario when a wheel bearing on the 33rd car began to overheat due to a lack of lubrication. Shortly before midnight a wheel/axle assembly on the car fell off of that car causing 23 cars to derail. The derailed cars contained a variety of cargo including styrene, toluene, propane, sodium hydroxide and chlorine.

Several of these cars, including the car carrying chlorine, ruptured, and spilled their contents. Propane cars exploded and burned (Figure 1.92). The force of the explosions knocked emergency responders to the ground

Figure 1.92 Propane cars exploded and burned. The force of the explosions knocked emergency responders to the ground and hurled one propane car over half a mile. (Courtesy: Mississauga Fire Department.)

and hurled one propane car over half a mile. An hour and a half after the derailment occurred officials ordered an evacuation. Additional evacuations over the next two days caused more than 218,000 of the 248,000 Mississauga residents to leave the area. Firefighters initially concentrated on cooling cars to allow the fire to burn itself out.

Six hours after the derailment responders discovered a 3-foot hole in the chlorine tank car. This prompted further evacuations. After three days the fires were brought under control. Responders found that most of the product in the chlorine car had been released, however, the gas continued to evaporate at the rate of about 35 pounds per hour. Responders spent the next three days removing the remaining 14 tons of chlorine. There were a total of eight evacuations ordered in the first two days of the response. The evacuees were allowed to return five days later. This response has since become a standard case study in how to conduct large-scale evacuations. This is the largest evacuation for a hazmat incident. There were no fatalities.

Chronicles of Incidents and Response 197

Saturday, November 10, 1979

This was to be one of the many dull, uneventful runs. A 106-car train carrying a mixed cargo, including dangerous chemicals, rolling through Ontario's rich farmland and heavily-populated areas to rail-yards in the northeast end of Metropolitan Toronto. And so it was, until just before midnight, November 10, 1979. Canadian Pacific Railway train 54 began its fateful journey in early afternoon at Windsor in the southwestern part of Ontario. It stopped in Chatham 90 minutes later where it picked up cars from a train arriving from Sarnia. Some of these cars were carrying caustic soda, propane, chlorine, styrene and toluene, a cargo which causes environmentalists to shudder and warn of horrendous derailment nightmares.

After connecting the tank cars, the train left Chatham at 6 p. m. to travel east to London where crews were changed before it continued its journey towards Toronto. As the train passed the Milton area, about 40 kilometers from the outskirts of Metro Toronto, lack of lubrication in a wheel bearing apparently started to spell trouble. On one car, the journal box at each end of the axle, where friction builds up between the moving axle and the car above, was an old-fashioned type, needing lubrication by oil. Modern freight cars have roller bearings, which don't heat up as do the older friction or journal bearings. However, when the journal box lacks lubrication, tremendous heat builds up. In trainmen's vernacular, the overheated journal box becomes a "hot box".

Residents living beside the tracks later reported seeing smoke and sparks coming from the middle section of the train. As the train progressed, people living closer to Mississauga, a community just west of Toronto, thought part of the train was on fire. Friction burned the journal bearing causing the stub of the axle to break off. Immediately after the train passed the Burnhamthorpe Road level crossing, the 33rd car - the one with the hot box and with a cargo of toluene - lost one of its four axles, complete with glowing wheels. The set of wheels crashed through a fence and landed in the backyard of a house, about 15 meters from the tracks and three kilometers from the Mavis Road crossing.

The train went past a further residential section of apartment buildings and suburban homes with its undercarriage hanging until reaching the Mavis Road crossing in a light industrial area about 30 kilometers from downtown Toronto. The dangling undercarriage left the track three minutes after losing the axle. Twenty-three other cars followed the tanker, causing a deafening crash and squeal of iron as cars collided at the Mavis Road crossing. On impact, some propane cars burst into flames. That was 11:53 p.m. - the beginning of a tense week for thousands of Mississauga residents.

As the derailed train's tank cars became twisted and tangled, tankers containing styrene and toluene were punctured, spilling their chemicals on to track beds. Within a minute, flammable liquids and vapors ignited, causing a massive explosion of a tank car. The yellowish-orange fire rose to a height of 1,500 meters and could be seen 100 kilometers away. The fire was fed by six dangerous ingredients - 11 tank cars of propane, four with caustic soda, three with styrene, three with toluene, two box cars with fiberglass insulation and one with chlorine. While chlorine is non-combustible in air, most combustible materials will burn in chlorine as they do in oxygen. Liquid propane, styrene and toluene are flammable while caustic soda is not combustible, but in solid form and in contact with moisture or water, it may generate sufficient heat to ignite combustible materials.

As the flames erupted, trainman Larry Krupa jumped out of the engine and ran towards the derailed portion of the train. He closed a cock on the 32nd car which permitted engineer Keith Pruss to drive the front part of the train eastward along the tracks out of danger. Citizen reaction was immediate. Police and fire department switchboards lit up with a flood of calls alerting them of the derailment. Officers on patrol and at the station closest to the derailment saw the fire. Within minutes, firefighters began connecting hoses and police were setting up roadblocks at the derailment site. Both reported to their headquarters a similar message - more help was needed urgently.

Sunday, November 11, 1979

As firefighters made preliminary plans to battle the fire, a violent explosion at 12:10 a.m., caused by a propane tanker blowing up, showered the surrounding area with large chunks of metal. The force of the explosion knocked police officers, firefighters and curious onlookers to the ground. Near the explosion, a green haze was seen drifting in the air. Along a kilometer stretch, windows were shattered and three greenhouses and a municipal recreational building destroyed. Between five and 10 minutes later, a second explosion erupted. A BLEVE (boiling liquid expanding vapor explosion) in another propane tank car hurled the car in the air, spewing fire and landing in a clear area. It tumbled across a field before coming to rest 675 meters northeast of the Mavis Road crossing. Five minutes later, another BLEVE in a propane car occurred with one end of the car travelling about 65 meters.

By now, CP Rail dispatchers' offices in London and in Agincourt, northeast of Toronto, the ultimate destination of train 54, were notified of the derailment through CP's radio system.

Meanwhile, hasty telephone calls were also placed to Mississauga Fire Chief Gordon Bentley, Police Chief Douglas K. Burrows, and the Ontario Ministry of the Environment. Mayor Hazel McCallion telephoned the police after her son climbed on the roof of the family house in nearby Streetsville to describe the blazing fire.

Chronicles of Incidents and Response 199

Police and fire officials acquired the train's manifest, a description of the cargo and emergency procedures, from the conductor, but it was unintelligible. Another copy was subsequently requested from the CP Rail dispatcher in Toronto. The front part of the train, which had the other copy of the manifest, arrived in Cooksville, about six kilometers from the derailment. At 1:30 a.m., a readable copy of the manifest was delivered to a makeshift command post, which has been established shortly after the derailment in a building just south of the fire.

Peel Regional Police and other emergency services established on-site emergency command posts just south of the site. Peel Police Chief Burrows and his Deputy Chief William Teggart assumed control of the police command centre. Members of neighboring police forces, fire departments and ambulance services had been alerted or volunteered services.

After obtaining the manifest, senior officials gathered for a meeting to evaluate the situation. This meeting involved the police chief, Deputy Fire Chief Arthur Warner, Chief Fire Inspector Cyril Hare, various CP Rail officials, two officials from the Ontario Ministry of the Environment and some local chemical experts. On checking the serial numbers of the derailed cars with the manifest, some worst fears were confirmed. The derailed cars were carrying a mixed cargo of dangerous chemicals. In the command post, officials discussed the possible chlorine gas threat.

Chlorine, a deadly chemical, forms a greenish-yellow cloud when released and is so heavy that it often hovers close to the ground. Since its mass weight is roughly 2.5 times that of air, a cloud of chlorine will slump following the terrain as it drifts and disperses. This feature led to its use as a weapon in the First World War at Ypres, Belgium, where thousands of Canadian soldiers were killed as a result of the gas release. Once chlorine gas is breathed, it saps the fluids in the linings of lungs and blood and starts a chain reaction that ends with slow suffocation.

At the site, it was quickly deduced that the chlorine tanker was indeed close to a filled propane tanker in continuing danger of exploding. After consulting with Fire Chief Bentley, the police chief made the first tough decision of the long week. He ordered 3,500 residents living closest to the derailment to leave the area for their own safety. With the yellow and red fire background, police officers using loud hailers or knocking on doors alerted sleepy residents of the evacuation notice. This evacuation - the first of 13 in a 20-hour period - began about two hours after the car went off the tracks.

Later, as winds shifted and more information about the fire and the train's cargo became known, areas of evacuation were widened. Shortly after 2 a.m., Metropolitan Toronto Police sent sound trucks to assist in telling residents of the evacuation. Police arranged for the selection and establishment of evacuation centers for those who could not stay with friends and relatives outside the area. The Mississauga section of the Canadian Red Cross Society began organizing for registering and feeding evacuated residents at reception centers. Square One, a huge covered shopping centre 2.4 kilometers northeast of the derailment, was selected as the first centre.

Other preparations started. The provincial Ambulance coordinating Centre sent a general call for ambulances in the surrounding area. One hundred and thirty-nine ambulances and 300 ambulance workers arrived in the area within six hours of the accident from as far south as Niagara Falls (130 km) and as far east as Kingston (275 km). Twenty-seven other vehicles were also provided, including buses from the Toronto Transit Commission, Oakville Transit and Mississauga Transit.

Throughout the night and early morning as machinery arrived and plans developed, experts to handle the dangerous substances also entered the scene. Experts from chlorine emergency plan, (CHLOREP), arrived armed with their equipment headed by Stu Greenwood and his team from Dow Chemical Co. in Sarnia, owners of the chlorine in the tanker. Experts agreed that it would be impossible to seal the chlorine tanker leak until the propane fires had burnt themselves out. Firefighters continued to increase the spray of water, and more water lines were added. Eventually 10 master streams were applied through about 4,000 meters of hose. After an hour at the scene, it was decided that firefighters would cool the cars and not extinguish the flames. This would allow a controlled burn of escaping gases and avoid possible explosions.

Shortly after another area was ordered evacuated, Chief Burrows moved the police command post because of a shift in winds. The mobile communications trailer of the Ontario Provincial Police joined other trailers at the new command centre in a Bell Canada building less than a kilometer north of the derailment. Just before 5 a.m., officials at the scene decided that the seriousness of the situation was not diminishing and that provincial officials, including Solicitor General Roy McMurtry should be notified. Under provincial government guidelines on such emergencies, the solicitor general was the provincial official in charge as chairman of emergency planning committee of the Ontario Cabinet. Other local officials such as Frank Bean, Chairman of Peel Region Council, and Mississauga city councilors, Peel Region's social services department and Mississauga city employees were also alerted.

As dawn broke on a dull cool day, the first four members of "think tank" group, the decision-making committee, which would operate during the next six days, met at 7:30 a.m. It included Bean, Mayor McCallion, Chief Burrows and Fire Chief Bentley. About an hour later, Chief Burrows issued another evacuation notice. But the later decision to evacuate Mississauga General Hospital and two adjacent nursing homes

Chronicles of Incidents and Response

would be most dramatic and tense. With the arrival of Solicitor General Roy McMurtry and Deputy Minister John Hilton, further meetings were held and the evacuated areas were increased as the threat to public health and safety became apparent. By 1:30 p.m., the boundaries were further extended south to Lake Ontario and Square One, the first evacuation centre, was closed although it was just north of the evacuated border. Evacuees were transferred to other centers.

As winds shifted, new dangers were presented, forcing more and more residents to join the exodus - some with packed luggage and others with Sunday dinner abandoned on the stove.

At the day's end, about 218,000 persons had left their homes, six nursing homes, and three hospitals including Oakville-Trafalgar Hospital, just outside the western border, and Queenway Hospital, just beyond the eastern boundary. The southern part of Mississauga, Canada's ninth largest city with a population of 284,000 was a virtual ghost town.

Monday, November 12, 1979

It truly was a closed city. Commuter traffic to Toronto was rerouted around the evacuated area, causing massive traffic jams for the rest of the week. The Queen Elizabeth Way, the busiest stretch of highway in Canada, which runs through the central part of the Mississauga core, was closed at its eastern and western entrances to Mississauga. Officials feared that a propane tanker might explode during the rush hours or that chlorine might waft over the highway, trapping thousands of commuters in their cars. By 10 a.m., three or four propane cars continued to burn but these fires were under control. Firefighters were sticking to the strategy of permitting the fires to burn themselves out. Because of various explosive vapor-producing substances, firefighters were ordered only to confine and control the flames.

Meanwhile, Proctor Ltd., of nearby Oakville, a major manufacturer of railway tank cars, prepared a steel patch to cover a one-meter hole in the tanker. Photographs revealed the hole, and it was surmised that at least some chlorine had escaped. But at this point, it was not known how much remained in the tank. The task then involved attempts to find out how much chlorine was left in the tanker and to cover the hole to prevent more leakage.

During the day, railway crews removed box cars and tankers, which had not been derailed, attempting to clear as much debris as possible without disturbing the chlorine tanker and propane tankers piled around. Chemical experts worked to devise ways of eliminating the chlorine threat while staff of the Ontario Ministries of the Environment and Labor was constantly monitoring air in the area. Most samples showed no hazard for healthy adults but a few pockets of chlorine gas had collected

202   *Hazmatology: The Science of Hazardous Materials*

in low-lying areas near the site. However, there were enough chemicals in the air to cause discomfort over a significant area.

In reception centers, volunteer groups, Red Cross and St. John Ambulance, supervised the settling of displaced residents and overall co-ordination of food and health services. Meanwhile, police patrolled deserted streets and checked all vehicles entering the area for possible stolen property. Officials could not consider lifting the evacuation until the fire was out and the chlorine danger had ended.

Tuesday, November 13, 1979

A sigh of relief was breathed by anxious experts and officials when the propane flame finally went out at 2:30 a.m. Most firefighting equipment was removed from the site. The all-out effort was now concentrated on patching the tanker. In late morning, patients were being returned to Queenway Hospital and Oakville-Trafalgar Hospital, which were just outside the fringes of the evacuated areas, but were closed as a precaution. In early afternoon, the command post committee set new boundaries, after air sampling tests indicated the situation was stable in those areas. At 3:30 p.m., Solicitor General McMurtry announced new borders on the eastern and western boundaries. Five hours later, a further eastern section was opened. The two announcements meant 144,000 persons returned home. However, the rest of the evacuees, living closer to the derailment and in the path of the prevailing winds carrying the deadly gas, would have to wait.

Patients from the Extendicare nursing home in Oakville and the Sheridan Villa nursing home were also returned to their residences. At the site, workers had been hampered in completely sealing the tanker by another tanker blocking access to it. As a result of an incomplete seal, a small amount of chlorine continued to escape. However, crews registered success in being able to drain the contents of one propane tanker at dawn. It was later hauled away. Meanwhile at the command post, officials were worried that there was a chance that a propane tanker, which has caught fire, might flare up again.

Wednesday, November 14, 1979

Weary workmen struggled to seal the chlorine tanker and decided to take a calculated risk. In trying to lift and drain another half-empty propane tanker before tackling the chlorine tanker, they gambled that the propane would not explode and further tear open the chlorine tanker. To complicate matters, firemen and command post officials were concerned when a large white cloud of chlorine vapor and water vapor wafted from the derailment site. Pockets of chlorine gas monitored in

Chronicles of Incidents and Response

the deserted area still presented a health hazard for infants, the elderly and anyone with respiratory problems. During the day, resentment and frustration grew among some evacuated residents, who wanted to return to their homes. The 25-square-kilometre area remained closed, including the two entrances to the Queen Elizabeth Way through Mississauga.

In an effort to alleviate some bitterness, CP Rail offered to pay for hotel rooms for about 1,000 displaced residents, thus relieving the reception areas of some strain and tension.

Thursday, November 15, 1979

As crews worked throughout the night and early morning to patch the leak, an estimated 20 to 30 kilos of chlorine were escaping each hour. The steel patch could not be fitted tightly over the rupture. It was supplemented by a neoprene air bag pressed over the opening by a timber mat and secured by chains. This virtually sealed the tanker, and officials could announce that there was little leakage. Between 7 and 10 tons of liquid chlorine remained in the tank since most of 90 tons of chlorine had apparently been sucked up into Sunday's giant flames, and the resulting chlorine gas had been dispersed harmlessly over Lake Ontario.

Technical experts explained to the "think tank" meeting that a slushy ice mixture of chlorine and water had built up inside the tanker from water poured in by fire hoses. The mixture formed a layer over the liquid chlorine, complicating the removal of any remaining chlorine and delaying this phase of the operation. Scientists worried that this layer of ice might break up and fall into the liquid chlorine, exposing it to the air. However, it was decided that pumping would not start until favorable winds prevailed. The pumping started at 11 p.m. Earlier in the day, Solicitor General McMurtry announced on behalf of the command team that the remaining 72,000 could not return that night. The end would depend on the removal of the chlorine.

Friday, November 16, 1979

Generally, the transfer proceeded smoothly. As a precautionary measure against the spread of small amounts of chlorine emitting from the tank after the patch was fitted, firefighters set up monitors in a fog mode downwind from the chlorine tank to ensure that any remaining chlorine in the air would be captured by water and drawn to the ground. The problem involving the layer of ice was resolved by applying a liquid line below the ice and a vacuum line above it. X-rays were taken to measure the levels of the tank car and truck during the pumping operation. By noon, most of the chlorine had been pumped into trucks and shipped safely away.

Figure 1.93 Police removed road blocks. Only the derailment site remained out of bounds. (Courtesy: Mississauga Fire Department.)

Throughout the pumping, air monitoring continued. Their tests showed no dangerous pockets of chlorine. By 3 p.m., 37,000 persons of the remaining 72,000 were permitted to return home. But the 35,000 residents living closest to the derailment and the first to evacuate, waited another four hours. Finally the boundaries were lifted.

While CP Rail under the supervision of the Canadian Transport Commission, a federal government agency, removed wreckage, the chlorine tanker was not disturbed until the liquid chlorine has been removed and the empty car purged. At 7:45 p.m., the city was reopened. Police removed road blocks. Only the derailment site remained out of bounds (Figure 1.93). By late evening, the last reception centre was closed and by midnight, Metro Toronto police, the provincial police and RCMP had finished their duties (Mississauga Fire Department).

Somerville, MA April 3, 1980
Phosphorus Trichloride Release

April 3, 1980 at 9:10 a.m. a minor collision between a locomotive with 38 cars traveling 4 mph and a tank car in a railroad switching yard in Somerville, MA resulted in the release of highly toxic and corrosive phosphorus Tri-chloride (PTC) (Figure 1.94). The tank car had approximately 13,000 gallons in the tank. The puncture created a spillage of 100 gallons a minute and immediately upon contact with the moisture in the air a white fog was generated. The engineer, front brakeman, and the rear brakeman abandoned the site and headed north. Liquid PTC quickly flooded the ground adjacent to the track and followed the ground contours and flowed west under the locomotive to the left side. Next, the spilled PTC flowed south 30-50 feet along an adjacent track before flowing down a 10 foot embankment into a recessed fill area, adjacent to the tracks and bounded by joy Street. The PTC flowed through this recessed area and fanned out to about a 20 foot pool.

Somerville is a city with a population of 80,596 people living in 4.9 square miles. That is a population density of 16,448 persons per square mile. The product is classified as a corrosive liquid by the United States Department of Transportation (DOT). Initially, a 1 1/2 square mile area containing 23,000 people was evacuated. During the first 48 hours of the release 418 persons were treated at the Somerville Hospital.

PTC is a clear colorless or slightly yellowish, highly volatile liquid. The chemical is used in the manufacture of phosphites, gasoline additives, plastics, and dyestuffs and is used as a chlorinating agent for water purification. Vapors have a pungent odor and are extremely irritating to the eyes, skin and mucous membranes. Skin contact may produce burns, especially when moisture is present from perspiration. PTC is heavier than water and vapors are 5 times heavier than air. This was the first ever PTC spill ever reported to the DOT.

Figure 1.94 April 3, 1980 at 9:10 a.m. a minor collision between a locomotive with 38 cars traveling 4 mph and a tank car in a railroad switching yard in Somerville, MA resulted in the release of highly toxic and corrosive phosphorus trichloride (PTC). (From NTSB Investigative Report.)

206   *Hazmatology: The Science of Hazardous Materials*

EPA requested the fire department to use water spray downwind of the pit of PTC to leach out the escaping hydrochloric acid vapors from the air. Firefighters were reluctant to do this because they had seen earlier the accelerated vapor production caused by water spraying. The fire department strongly opposed the procedure but reluctantly complied with the EPA recommendations. During the operation, the wind shifted and as a result the spray hit the pit directly where it mixed with the PTC and created additional massive vapor clouds. Streams of water were terminated immediately. As the clouds spread, the Somerville Central Hills was evacuated.

Officials met with experts from Commonwealth, Federal, and industry advisors who suggested to the fire chief three different approaches for disposal of the two inch residue in the pit.

1. Back fill the pit with sand.
2. Drowning and diluting the remaining product with water.
3. Neutralizing the pit with limestone or soda ash.

The advisors could not agree on which was the best plan to follow. Faced with these disagreements and the continuing chemical cloud, the Mayor of Somerville designated a technical advisory committee with representatives from the shipper Monsanto, the carrier (BM), and the Commonwealth of Massachusetts and instructed them to reach unanimous agreement on how to dispose of the remaining chemical in the pit. Almost 10 hours after the spill, the committee of advisors selected backfilling with sand and gradual addition of water as the best, because the other two alternatives contained more uncertainties (NTSB Investigation Report).

> **Hazmatology Point:** *Neutralization and dilution both carry significant risks in terms of reactions and side effects. While both may on the surface seem to be potential fixes, they should be fully understood before taking these actions. Neutralization is a chemical reaction, which can cause splattering, creation of vapor and heat.*
>
> *Neutralization is basically changing the pH of the material until it is as close to neutral as possible. You may need hundreds if not thousands of pounds of material to accomplish the neutralization. You also need a means to monitor the pH. Neutralization on a large scale is not an easy accomplishment. Dilution sounds easier than it is. Many corrosive materials are water reactive from mild to violent. Once again, you are trying to change the pH to a point where it is near neutral by applying more water than there is spilled product. This may require thousands of gallons of water, so you need to be able to contain the mixture. Neutralization and Dilution work better on smaller amounts of spilled materials, several gallons. There are also alternatives even for small amounts such as absorption or solidification.*

Corpus Christi, TX April 7, 1981
Grain Elevator Blast

An explosion and fires ripped the top off several silos at the Corpus Christi Public Grain Elevator at the city's seaport, on Texas' middle Gulf coast. This was the worst explosion of its kind in Texas history. Nine people were killed and 32 injured, many critically (Figure 1.95). Kevin Saunders, a Federal Grain Elevator Inspector was working in his office along with his supervisor Albert Trip. He had inspected the elevator within the past week and found dangerous levels of dust in the elevator. Mr. Saunders had a meeting with elevator management and told them of the dangerous conditions. He was told the elevator could not be shut down for repairs because it was in operation 24 hours a day to keep up with the movement of grain. Management further advised that there was no money for the needed repairs.

Firefighters found thick smoke billowing from the silos as they first arrived on scene. Inside the gates of the Corpus Christi Public Elevator bodies were scatted everywhere (Figure 1.96). A single ambulance arrived with two attendants. They moved from victim to victim to check the conditions of the injured. Workmen trying to assist the ambulance attendants hurried to find makeshift splints and stretchers to take victims to the ambulance and private vehicles for transportation to hospitals. A parking lot next to the silos looked like a battle ground. Just as they had attended to those on the parking lot, more were found under debris. Bulldozers and a huge crane were brought in to remove massive chunks of concrete and twisted metal to locate missing victims. The force of the explosion tore gaping holes in 10-story-high grain silos, hurled glass and huge chunks

Figure 1.95 This was the worst explosion of its kind in Texas history. Nine people were killed and 32 injured, many critically. (Courtesy: Kevin Saunders.)

Figure 1.96 Firefighters found thick smoke billowing from the silos as they first arrived on scene. (Courtesy: Kevin Saunders.)

of concrete and left mangled sections of the elevator walls dangling from twisted girders and reinforcement rods. One official estimated the damage at $30 million. Some of the bleeding and burned casualties stumbled out of the rubble while others had to be dug out from beneath the debris by rescue workers. An emergency medical technician, who asked not to be identified, said there were bodies lying all over the ground. "It looked like a 'Dam clearing station," he said. "Nobody had time to scream."

Flames ignited by the blast and fanned by 35 mph winds sent thick, black clouds of smoke billowing 500 feet into the air and troubled firefighters who warned bystanders about the possibility of additional explosions. Irwin said "the crane would be used to fight the fires as well as move debris in the search for the missing persons. As soon as we get the crane going, we can put some of our people in the basket and raise them to where these fires are still burning," Chief Irwin of the Corpus Christi Fire Department said. "That's the only way we'll get them out." "That structure (elevator) just can't hold people. There are no floors left." He said "50 firefighters kept watch on the smoldering fire overnight to prevent any other explosions," which was caused by an accumulation of grain dust, authorities said. Shortly after dawn, fresh crews and about 20 employees arrived to begin sifting through the debris, clearing away huge chunks of concrete, mangled metal supports, glass and rubbish lying around the area. One of the missing was a guard, an elevator spokesman said. He arrived about 10 minutes before the blast. One of his first duties was to check the controls and safety equipment on the basement level. It is thought the blast originated here.

Eighteen of the injured remained in Memorial Hospital today, said spokeswoman Kay White. Six were in critical condition, one in poor

Chronicles of Incidents and Response 209

condition, and 11 in fair condition. Seven of the victims were in the burn unit, she added. She said 12 others had been treated and released.

Spokesman Vince Hefley of Spohn Hospital said two people remained hospitalized there today.

Manuel De Los Santos was in critical but stable condition with head, neck and abdomen injuries and burns; and Jesus Lopez was in good condition after surgery for a compound arm fracture and he also had burns.

The fire still was burning early the next day, although firefighters said they had isolated the flames to one section of the silos. Acting Fire Chief E. E. Irwin said the fire would be allowed to burn itself out. Officials had not determined what sparked the blast. "The cause will probably remain hidden in all that rubble for some time. Grain dust is very volatile, anything can ignite it," said Don Rodman, a public affairs officer for the port of Corpus Christi.

The explosion came during a shift change and Rodman said officials estimated 50 to 51 people were near the elevator when it exploded. He said the three missing men included two elevator employees and a security guard. An explosion at the same elevator in 1968 killed one person and caused $2 million in damage, officials said. Tuesday's blast damaged 75 per cent of the structure built in 1952-53 and would cause an estimated $30 million to repair, said engineering director Nolan Rhodes. "It will probably have to be torn down and rebuilt it is that extensively fractured," said Rodman. He said $3 million to $4 million had been spent in the last three years to modify and improve the dust collection and emission control mechanisms. The facility had added new safety equipment as late as March of 1980. Everything was working the way it was supposed to be. Everything seemed to be okay," said Rodman, who added that officials were at a loss to explain the explosion.

Pete Garcia was in his late fifties and had been doing this job for over 40 years. With that level of job familiarity, he had incorporated shortcuts as he had never experienced an accident after so many years on the job. That day, he wore his rubber knee pads to kneel down by the 100 lb. iron manhole cover at the top of the silo to apply a pesticide. However, with the grain constantly in motion, and the dust collectors not working, grain dust was suspended in the air all the way up to the top opening. When Pete opened the can, it flashed a flame into the highly combustible dust in the air that caused the first explosion and took Pete's life that day as he was the first person killed. When equipment malfunction and human error both occur at the same time, it causes a catastrophe!

Corpus Christi Caller Times 1981-04-06
The Del Rio News-Herald Texas 1981-04-08

210   *Hazmatology: The Science of Hazardous Materials*

Meet Kevin Saunders

Author's Note: While searching for historical incidents, I came across the Corpus Christi Grain Elevator Explosion. It sounded interesting so I contacted my friend Jim DeVisser who is a battalion chief with the Corpus Christi Fire Department to see if he knew anything about the incident. He sent me an internet website for Kevin Saunders. I was about to read one of the most incredible stories I have ever seen. There were several photos and news clippings of the incident on Kevin's website. I emailed Kevin and asked if I could use the photos in my book and could I speak to him about the incident. Several days later I received a phone call from Kevin and we talked for almost 3 hours about the explosion and how it had affected his life. He made me feel like we had been friends for life.

I wanted to meet him and he said he was coming to Grand Island Nebraska February 12, 2019 to provide a motivational presentation for the Aurora COOP and its customers and stockholders. This was a private meeting and he said I would need to get permission to be there. So, I called the COOP and told them my story, and they said they would get back to me. A few days later they called back and told me I was welcome to attend and they provided me with a guest pass. On February 12, 2019, I met Kevin in person, the most amazing man I have ever known. Kevin's presentation was very moving and you could hear a pin drop in the room. You could tell by the laughs and see by the looks on faces that they too were amazed by Kevin's story. This has to go down as one of the most special things that I have experienced in my life. When the presentation was over, we talked for a while and I met his wife and came away that day with the pledge to myself that I will never again complain about things that happen to me the rest of my life. If Kevin can endure and rebound from the things that have happened in his life, and accomplish what he has, I can do anything I set my mind to.

Kevin Saunders was just like any other young man from the Kansas countryside. Fresh out of college and starting a family, he worked long hours as a Federal grain elevator inspector. Touring facilities day after day in the heat, he knew the job wouldn't be easy but he didn't know it would nearly take his life. But then, on a busy afternoon like any other, Kevin heard the sound he would never forget, it sounded like an earthquake the government building Kevin was in was rattling and shaking with the floor and building things were falling off the wall in the government building that Kevin and his supervisor Albert Trip were in. Kevin glanced out the window and saw chunks of concrete the size of a vehicle some weighting more than a ton being blown hundreds of feet through the air and it was coming right at him towards the government building he was in.

Chronicles of Incidents and Response 211

The experts said that there were 12 explosions that ripped through the grain elevator at 1,500 feet per second. So as Kevin saw those 2 foot thick concrete walls of the grain elevator being blown apart like paper coming right at him and his supervisor. In a split second the earthquake like rattling grew with so much intensity that the cracking and popping were so loud it was like it was going to split his head wide open. At the same instant Kevin caught a glimpse of his supervisor Albert Trip out of the corner of his eye and all the blood had drained out of his face and he had turned pale white and he had this look of absolute terror on his face.

Kevin's supervisor didn't utter a word he eyes were glazed over but his eyes said it all: "We're not going to make it!" and before Kevin could even take another breathe the biggest and final explosion blew out where Kevin and his supervisor were. That last and most powerful explosion completely destroyed the government building he and his supervisor Albert Trip were in there was nothing left of that government building but the concrete foundation.

Kevin saw his supervisor Albert Trip hit the floor just before the wall of that government building blew out in Kevin's face and Albert was one of those lives that were needlessly killed that day. Kevin was knocked out when the wall blew out in his face and he was blown through the roof of the building he was in and over a two story building over 300 feet through the air onto a concrete parking lot. Kevin hit the concrete parking lot hard. First the back of his head hit that concrete parking lot then his shoulder blades hit and shattered in pieces.

If it wasn't the last blast that blew the wall out in Kevin's face, killed his supervisor Albert Trip and completely destroyed the government building Kevin was in it looks like it was pretty close. The force of the blast blew his legs over his torso and broke his ribs, collapsed his lungs and severed his spinal cord at chest level. His body was broken over at the chest like people bend at the waist. Paramedics found Kevin lying in a pool of blood with blood and cerebral spinal fluid oozing out of his nose, ears and mouth. His Body was broken over at the chest like people bend at the waist. After the paramedics took his vital signs they black tagged him because they didn't think he would survive the ambulance ride back to ICU of Memorial Medical Center.

Kevin was unconscious and would have died in that parking lot, as no ambulances or stretchers were available while rescue personnel scrambled to get the injured to hospitals. However, a paramedic who didn't want to leave him recruited a fire and rescue guy, and together they found a door lying in the debris, which they used as a makeshift stretcher and carried him to safety. They didn't know it at the time but they saved Kevin's life. After a month in a coma, Kevin finally woke up in the hospital. He was in excruciating pain, face down in a hospital bed with massive internal and external injuries.

Doctors told Kevin there had been an explosion. The explosion propelled him through the air and landing broke his ribs, collapsed his lungs and severed his spine at chest level. Doctors thought these injuries would kill him, but he survived. However, Kevin would be paralyzed for life. Kevin told me in a phone interview that "his faith in God helped him to hang on when others may have given up." He said "I know that God was there because the one paramedic stayed even after they had run out of ambulances and stretchers". We talked about football, his sports career, life, his adventures and some of mine and that day that forever changed his life. Kevin cheated death doctors gave him a 1% chance to survive his injuries. Not only did he prove them wrong but launched an incredible career that I believe would not have happened if he had not gone through that terrible accident.

When I came across the news clipping of the Corpus Christi Grain Elevator Explosion, I never dreamed there would be such an extraordinary story that would rise from the death and destruction of this incident. There is not room here to fully cover the incredible life of Kevin Saunders and all of the people's he has touched. I would encourage readers to go to his website and find out more about Kevin. https://www.kevinsaunders. com.

For over 30 years, Kevin Saunders has been recognized as one of the top Motivational Speakers in the United States and the World, helping groups like NASA, State Farm, AT&T and many others achieve breakthrough success. Based on his experience as a Para-Olympic Track and Field champion and world record holder in multiple events, Kevin knows exactly what it takes to overcome obstacles, achieve peak performance, and develop a champion's mindset in sports, business and in life. In 1991, Saunders became the first person with a disability to be appointed to the President's Council on Fitness, Sports, and Nutrition by President George H. W. Bush.

The President selects only 20 individuals nationwide to serve on this Council, making this an exclusive honor and personal achievement. In 1994, Saunders became the only person to be reappointed to the Council by President Bill Clinton. He served on the Council until 2000. Under President George W. Bush, Kevin was named to a Commission by the President's Council on Physical Fitness and the Department of Health and Human Services to come up with a plan to improve fitness and health for people with disabilities. He was also commended for his help in the creation of the National Initiative on Physical Fitness for Children and Youth with Disabilities, or the I Can Do It, You Can Do It Program. Saunders also served as an International Ambassador for health, fitness and proper nutrition, a role in which he met with government leaders at the city, state and national level in numerous European countries.

Chronicles of Incidents and Response   213

In 1991, driven by the desire to improve his race times, Saunders spent time at his alma mater, KSU. There, he collaborated with the College of Engineering to prototype faster and lighter wheelchair designs. Mutual friends introduced him to the newly hired Head Football Coach Bill Snyder, who was impressed by Saunders' story and asked him to serve as the honorary motivational coach to the team. In 1993, Coach Bill Snyder created the "Kevin Saunders Never Give Up Award" for the Kansas State University football team. The award was given to the player who displayed the most courage, determination, dedication, and perseverance in the pursuit of team goals. Many of the award winners have gone on to professional NFL careers, including placekicker Martin Gramática, a member of the 2002 Super Bowl Champion Tampa Bay Buccaneers and 2001-02 Pro Bowl starter (Kevin Saunders.com).

Bellwood, NE April 7, 1981
Grain Elevator Explosion

On the same day as the explosion that occurred in Corpus Christi, TX, another grain elevator exploded in Bellwood, Nebraska killing two, injuring one and causing severe damage. 1981 saw 21 grain elevator explosions in 12 different states, 5 of those occurring in the State of Nebraska. All of the deaths occurred in Nebraska and Texas, a total of 13 deaths and 62 injuries. The injuries occurred in 9 states, with Nebraska and Texas leading the way with 7 and 33 respectively.

Explosion at the grain elevator in Bellwood, Nebraska caused all of the bins at the top to pop open in the 100 ft. tall elevator. One bin was ripped open from top to bottom, with a large piece of concrete left hanging. Other pieces of concrete were hurled 3 blocks away. One person was killed by the initial blast, two others were hospitalized with critical injuries. One person had minor injuries. The explosion severely damaged one of the largest businesses in this small community of 450. Some residents were evacuated and taken to the high school auditorium in nearby David City. Fire departments from several neighboring communities responded to assist the Bellwood Fire Department. One of the employees was missing underneath the rubble. Rescue workers spent 6 hours in what became a body recovery. They had to stop on one occasion because they though the structure was unsafe.

Two 20,000 gallon propane tanks had to be off loaded because of the fear part of the elevator might fall on them creating another serious hazard. Following the draining of the tanks, firefighters lit off a spectacular 30 foot flame to burn off the residue in the tanks. Nebraska Governor Charles Thone toured the site to get a first hand view of the devastation. He was also concerned about so many explosions in grain elevators. He visited to

214   *Hazmatology: The Science of Hazardous Materials*

see if there is anything that could be done from his office. After the visit Governor Thone set up a task force to investigate the problem.

Columbus Times, Columbus NE, April 8, 1981

> **Hazmatology Point:** *One of the positive things that directly resulted from the Bellwood Explosion was the hiring of three grain elevator inspectors for the State Fire Marshal's Office to inspect all grain elevators in the state at least once annually. Over the next decade this program greatly reduced the frequency of grain elevator explosions in Nebraska.*

Columbus Telegram, Columbus, NE April 8, 1981.

Thermal, CA January 7, 1982
Derailment Radiation Container

About 9:50 p.m., P.S.T., on Thursday, January 7, 1982, Southern Pacific Transportation Company freight train No. 01-BSMFF-05, derailed 14 cars at Thermal, California, while traveling about 57 miles per hour on the tangent single main track. Four transients riding on the train were seriously injured, a fifth transient died as a result of injuries. No crewmembers were injured as a result of the accident. A County Sheriff's Officer was in the vicinity responded to the scene shortly after 9:55 p.m. and requested all available ambulances be sent to the derailment site. The officer then began searching for injured people.

Emergency Medical and Fire Department personnel began arriving about 10:15 p.m. The Riverside County Fire Chief stated he went to the caboose about 10:20 p.m., and asked the conductor "what cargo the train contained, and if the cargo was hazardous to the firefighters." The conductor told the Fire Chief that the profile he had showed the train didn't have no hazardous materials. After the chief left the caboose, the conductor continued reviewing waybills. Shortly before 11:00 p.m., the conductor found the waybill that, although not known at this time, erroneously identified the RAM on the 48th car in the train, the TPAX car, as "Fissile Class III" He immediately took the way bill to the Sheriff's Officer in charge, who relayed the information on the waybill to the police dispatcher. Police dispatch notified the Radiation Health Unit of the OSHA of the State of California and was told to have "everyone stay back 100 yards, since there was no fire, and keep people there until a determination of radiation danger could be made.

About 01:00 hours CHEMTREC assisted in establishing contact between the Hazardous Materials Control Supervisor and the originating shipper of the RAM. The originating shipper subsequently advised the Hazmat Supervisor of the actual quantity and form of the Americium. Radiological monitoring devices were brought to the scene by emergency

Chronicles of Incidents and Response

response forces about 02:00 hours. Following a survey of the area, no indication of radioactive contamination was found.

The presence of radioactive material in the derailed Trailer-On-Flat-Car train was discovered about 1 hour after the accident occurred, resulting in the handling of the emergency response effort as a serious radiological emergency. Contributing to misdirected emergency response efforts was erroneous and conflicting information concerning hazardous material on the train. Accurate information regarding the precise nature of the radioactive material shipment was not available at the accident site until about 5 hours after the derailment occurred; at that time radiological emergency procedures were terminated. Damage was estimated to be about $1,015,350. The National Transportation Safety Board determines that the probable cause of this accident was the inadequate company evaluation of defect data which should have indicated that the rail in the vicinity of the derailment was approaching service life limit for main track use and the consequent failure of the company to initiate an accelerated inspection program to detect incipient fatigue fractures of the rail (NTSB Investigation Report) National Transportation Safety Board Report.

Hazmatology Point: Transportation, storage and handling of radioactive materials are some of the most regulated hazardous materials in the U.S. Containers that carry radioactive materials in transportation are so well engineered and tested that to my knowledge there has never been a release of a radioactive material from those containers in an accident.

Livingston, LA September 28, 1982 Train Derailment, Fire and Vinyl Chloride Release

About 5:12 a.m., CDT, on September 28, 1982, Illinois Central Gulf Railroad (ICG) freight train Extra 9629 East (GS-2-28) derailed 43 cars on the single main track of the Hammond District in Livingston, Louisiana. Of the derailed cars, 36 were tank cars; 27 of these cars contained various regulated hazardous or toxic chemical commodities, 2 contained non-regulated hazardous materials, and 5 contained flammable petroleum products (Figure 1.97). A total of 20 tank cars were punctured or breached in the derailment. Fires broke out in the wreckage, and smoke and toxic gases were released into the atmosphere (Figure 1.98).

Thermally-induced explosions of two tank cars that had not been punctured caused them to rocket violently. About 3,000 persons living within a 5-mile radius of the derailment site were evacuated for as long as 2 weeks. Nineteen residences and other buildings in Livingston were destroyed or severely damaged. More than 200,000 gallons of toxic chemical product were spilled and absorbed into the ground requiring extensive excavation of contaminated soil and its transportation to a distant dump site. This has

Figure 1.97 Of the derailed cars, 36 were tank cars; 27 of these cars contained various regulated hazardous or toxic chemical commodities, 2 contained non-regulated hazardous materials, and 5 contained flammable petroleum products. (Courtesy: Livingston Fire Department.)

Figure 1.98 A total of 20 tank cars were punctured or breached in the derailment. Fires broke out in the wreckage, and smoke and toxic gases were released into the atmosphere. (Courtesy: Livingston Fire Department.)

Chronicles of Incidents and Response 217

resulted in long-term closure of the railroad line and an adjacent highway. Property damage has been estimated to be in excess of $14 million.

The National Transportation Safety Board determines that the probable cause of this accident was:

1. the disengagement of a worn air hose coupling when the train passed over a low track joint which initiated an emergency application of the train brakes,
2. an excessive buff force within the train resulting from the failure of the person at the locomotive controls to respond properly to the brake application, and
3. the placement of empty cars near the head of the train between heavily loaded cars Contributing to the cause of the accident were the impairment of the engineer's faculties by alcohol and his abandonment of the locomotive controls to an unauthorized and unqualified person, and the failure of Illinois Central Gulf to supervise train operations and operating personnel adequately, as well as to inspect and to maintain adequately its Hammond District main track Contributing to the contamination of the environment was tank damage resulting from the lack of shelf couplers on some tank cars and the inadequately protected bottom outlet valves on a number of other tank cars.

Tuesday September 28, 1982 began like no other in this small community fifty miles east of the State Capitol of Baton Rouge. Thirty-six cars of a 101 car Illinois Central Gulf (ICG) train derailed in the center of Livingston, LA shortly after 5:00 a.m. Derailments were nothing new to this stretch of Illinois Central Gulf (ICG) Railroad tracks built in 1908 from Baton Rouge to Hammond, LA. Previous derailments had occurred involving grain that was shipped from Illinois to Baton Rouge for river shipment to New Orleans and beyond. Near Holden, the first shipment of grain in the late 1960's derailed a total of nine grain cars. During the same trip, another derailment occurred in Baton Rouge Parish where 14 cars left the track when it collapsed under the weight of the grain cars. That same year, thirteen cars and chemical tankers derailed near Albany. In another incident 7 cars derailed outside the corporate limits of Denham Springs. This time a bridge over Gray's Creek collapsed.

Over the next several years' additional derailments continued to occur on the troubled stretch of track. Fifty families in Corbin, a small community near Walker were evacuated in 1970 after a tank car of petroleum and another containing vinyl chloride derailed. A year later, 300 families were evacuated in the same community because of a derailment when vandals broke a lock on a switch box and opened it causing the train to go onto a spur line. A car of vinyl chloride left the tracks, ruptured and caught fire.

In the same year, another grain train derailed causing 11 cars to leave the tracks near Amite. In 1978 ten cars derailed when the tracks collapsed near Walker. During 1980, more than 2,500 residents of Hammond, including Southeastern Louisiana University students were evacuated. The train derailed spilling styrene monomer chemicals. During the same year the same officials faced another derailment near an elementary school where it came to rest 15 feet from the school. Fortunately it was Saturday and school was not in session.

The derailment occurred in Livingston, LA Population around 1,000, across from the fire department and high school. A total of 20 tank cars were punctured or breached in the derailment. Chemicals released included Glycol; hydrofluorosilicic acid; Phosphorus; methyl chloride; sodium metal; tetraethyl lead; toluene; diisocyanate; perchloroethylene; perchloroethylene; styrene; vinyl chloride; fuel oil; lube oil; polyethylene. In addition to the spills, smoke and toxic materials were released into the air. Explosions and fires occurred almost immediately.

Emergency responders did not need to be called. However, Livingston Volunteer firefighters could only watch from a safe distance, as there was nothing they could do. Everyone in town heard the noised and saw the sky light up from the fires. Hannon Stewart, who owned Livingston's ambulance company, lived about 150 yards from the railroad, thought it was the end of the world. His fleet of ambulances was damaged beyond use by the derailment and explosions. Had he been able to use one, the only road from his house was blocked by the derailment. Fortunately, there was no immediate need for the ambulances. Residents in Denham Springs and Springfield heard the initial blasts when the train derailed. Livingston Mayor Allen Hunt and Fire Chief Darrel Jones decided everyone should withdraw to a safe distance.

Police cars with loud speakers went through the city notifying residents to evacuate immediately. The people left so fast that they failed to take extra clothing, pets, food, medicine, and other personnel property. Initial evacuation distance was one mile in all directions and later was extended. The evacuations went smoothly; in excess of 3,000 people were evacuated in a 5 mile radius of the derailment, some for two weeks. Most evacuations were completed in two hours. Those who did not have any place to go were housed in the Walker High School Gym.

Experts arrived quickly and the Fire Chief told me CHEMTREC was very helpful at the beginning of the incident. Railroad EPA Region 6 on scene coordinator was Frank Gordy. The Coast Guard Gulf Strike team responded as well. Louisiana State University sent a foam truck with a monitor to fight flammable liquid fires. When Louisiana State Police heard of the derailment they sent the State Hazardous Waste Team, who brought equipment and supplies. Explosives experts from the state police were also dispatched. In Louisiana, the State Police provide response

Chronicles of Incidents and Response

stabilization for hazardous materials incidents. A state police helicopter was dispatched for aerial surveillance of the scene. Video and still photographs were returned to the command post and studied for information on the tanks and status of the fires.

By nightfall on the first day environmental officials knew three-fourths of the overturned tank cars contained toxic chemicals. Managers from companies responsible for the chemicals, hazardous waste response personnel and railroad officials met to discuss the action plan. The original command post was determined to be too close and was moved two miles away near Interstate 12. A forward command post was retained where decontamination took place. All water from decontamination was collected and properly disposed of as hazardous waste. Twenty-four hour air monitoring was conducted by the Louisiana Department of Environmental Quality (DEQ). They also tested soil and ground water for contamination.

By day three initial fires had subsided to small lazy fires and identification of locations of tank cars began and damage assessments by engineers. The sodium metal car was found to be intact. The railroad began clean-up with vacuum trucks for chemicals that spilled and remained. They were hauled to a hazardous materials waste site. All personnel conducting the clean-up operations wore appropriate chemical protective clothing. On day four, there were not many tasks to be accomplished at the derailment site, so other needs were addressed such as feeding pets and livestock that were not evacuated and tests on water and vegetation for contamination. Contaminated water was being hauled to a disposal company and police officers continued their security patrols.

Plans were made to start extinguishing small fires in and around the derailed tank cars. At approximately 3:15 p.m. a spectacular explosion occurred causing a massive fireball of 400 feet in height and a mushroom cloud visible 15 miles away. The tank car of tetraethyl lead had been weakened by the flame impingement causing a boiler type explosion and the main part of the tank rocketed 600 feet. It caused damage ½ mile away. The tank car struck a mobile home causing it to catch fire. It burned to the ground before firefighters could extinguish the fire. Illinois Central Gulf sent a response team following the explosion.

By day seven fires once again began to subside and the process of clearing debris and contamination began again. During the clearing process a tank car of styrene ignited and burned. CHEMTREC provided a conference call between the officials in Livingston and the three top manufacturers of the product in the United States. Now all that remained were the 6 tank cars of vinyl chloride. Experts determined that it was too dangerous to try to move or off load the derailed tanks. They decided to blow them up and burn off the product: known as vent and burn. A pit was dug around the cars to contain the vinyl chloride spilling from the breached tank and allowing the vinyl chloride to burn. It took 12 hours to burn the contents of the six cars. During the sixth day of the incident the controlled explosions were set off at 5:30 p.m. Fires burned through the night making it easier for railroad workers to remove cars on day 6. State Health Department personnel used data from monitoring and testing to make a decision to allow residents to return. Finally, residents were allowed to return to their homes, offices and businesses and begin to decontaminate.

Mr. Stewart lost his home and other buildings on the property. His home was torn down after the railroad purchased the property. He did not replace his ambulances and retired. Nineteen homes and other buildings in Livingston were destroyed or severely damaged. Over 200,000 gallons of toxic chemical products were spilled and absorbed into the soil requiring extensive remediation and removal. This material had to be transported to a remote dump site miles away. Rail lines and the adjacent highway were closed for an extensive period of time. Environmental clean-up took several months. Remediation of the site required removing soil 50 feet deep and replacing with uncontaminated soil. Because of the exposures of residents to toxic chemicals prior to and during the evacuation, a clinic was established by the railroad in Livingston to test victims for the next 30 years following the derailment. Monitoring wells were installed to check for subsurface contamination and remained in place until 2016.

The National Transportation Safety Board (NTSB) conducted an investigation and issued a report in 1983. They determined there were several things that contributed to the derailment. Disengagement of a worn air hose coupling as the train passed over a low track joint, which initiated an automatic emergency application of the train's brakes.

- An excessive buff force within the train resulted the failure of the person at the locomotive controls to respond to the brake application.
- Placement of empty cars near the front of the train between heavily loaded cars.
- Contributing factors include impairment of the engineer's faculties by alcohol and abandonment of locomotive control to an unauthorized and unqualified person. Additionally, failure of the Illinois Central Gulf to supervise train operations and personnel adequately and a lack of inspection and maintenance of the Hammond District main track. Environmental damage was impacted by a lack of shelf couplers on some tank cars and inadequate protection of bottom outlets on the bottom of some cars.

Author's Note: During my visit to Livingston I learned from local officials that the train engineer. Permitted a female friend to ride on the

Chronicles of Incidents and Response

train and placed the unqualified person at the train's controls. It was also reported that both may have been under the influence of drugs and or alcohol at the time.

Following the Livingston derailment, railroad officials elected to install heavier rails and reconstructed the road bed, hoping it would stop future derailments. At the time the derailment in Livingston it was considered the grand-daddy of all train wrecks. Since the Livingston derailment in 1982 there have been no reported derailments along the 50 mile stretch from Hammond to Baton Rouge.

Constant inspection and prompt maintenance seem to have put an end to derailments. Trains still carry hazardous materials, but also seem to travel at safer speeds. The Livingston Incident was the worst in Louisiana history, but was also the best handled. In 1982 it was also the worst train derailment hazardous materials incident to date in the United States. There were no serious injuries or deaths when the incident occurred. During clean-up a railroad worker lost both legs when a cable broke during clearing operations. The first three state troopers arriving in Livingston have died from cancer. It is unknown how many others may have also died from cancer caused by exposure to toxic materials during the derailment.

Today Livingston still has a volunteer fire department with one paid driver during the day time and all volunteer at night. Present day population is 1850. Engine Number 1 is the only remaining apparatus from the 1982 derailment. The city has rebuilt many areas damaged by the derailment. However, the area in front of the fire station on Highway 190 has become a park. Only a concrete slab remains from a dentist's office. Hazardous materials response is still handled by the Louisiana State Police. The rail line is now owned by the Canadian National (CN) Railroad. Hazardous materials are still shipped regularly through Livingston (*Firehouse Magazine*).

Denver, CO April 4, 1983 Railcar Leak Nitric Acid

Denver and Rio Grande Western Railroad Company Train Yard Accident Involving Punctured Tank Car, Nitric Acid and Vapor Cloud.

Executive Summary

About 4:00 a.m. mountain standard time on April 3, 1983, a Denver and Rio Grande Western Railroad Company (D&RGW) switch crew was switching 17 cars in the D&RGW's North Yard at Denver, Colorado, when a coupler broke on the 4th car, leading to an undetected separation

222 *Hazmatology: The Science of Hazardous Materials*

of 150 feet between the 3rd and 4th cars. The engineer, responding to a hand lamp signal from the foreman, accelerated the locomotive, with a caboose, an empty freight car, and a loaded tank car coupled ahead. The loaded tank car impacted a fourth car at a speed of about 10-12 mph. Upon impact, the end sill of the fourth car (empty boxcar) rode over the coupler of the (loaded tank car) and punctured the tank head. Nitric acid spilled from the car and formed a vapor cloud which dispersed over the area. As a result, 9,000 persons were evacuated from the area; 34 were injured. Damage to railroad property was estimated to be about $341,000.

Probable Cause

The National Transportation Safety Board determines that the probable cause of the accident was the complete failure of a coupler of a box car leading to an undetected separation of cars being switched and the puncturing of a tank car by the end sill of the box car when the coupled cars overtook those which had separated. Contributing to the accident was the lack of a federal regulatory inspection or an industry practice for a periodic inspection to detect defects in hidden car components. Contributing to the severity of the accident was the nature of the released product, insufficient guidelines and absence of preplanning which led to the evacuation of about 9,000 residents and injury to 34 persons.

More than 2,000 people were evacuated from an area near central Denver before dawn this morning after nitric acid spilled from a railroad tank car and sent a plume of deadly gas over the city. Eight people suffered minor injuries after contact with the gas, and three firefighters sustained minor burns from the acid itself, according to the police. It took the Denver Fire Department nearly six hours to contain the 20,000-gallon spill and suppress the deadly clouds of vaporized acid. Until then, the Fire Department advised people to stay out of an area of nearly three square miles, including part of downtown, while firefighters worked to contain the acid spill.

Officials warned that the gas could cause skin irritation and respiratory problems. If inhaled in sufficient quantity, it can be lethal, they said. Still, many people appeared to ignore the hazard. Several churches in central Denver went ahead with Easter morning services as scheduled. The police said 25,000 people lived in the area. Officials had no precise count of the number who left their homes, but it was estimated that at least 2,000 fled or were evacuated. Many waited out the holiday in six Denver schools set aside by the Red Cross as emergency shelters. Stretches of Interstate 25 and Interstate 70, which intersect in the area, were closed until early afternoon. Switchyard Accident Officials said the accident took place about 4

Chronicles of Incidents and Response 223

A.M. in the Denver & Rio Grande Western Railroad switching yard when the coupler from one rail car punched a hole of 14 inches by 6 inches in the tank car while switching it.

About 20,000 gallons of nitric acid spilled onto the ground, where it vaporized into thick billowing clouds and was carried over the city by light winds out of the northwest. The police set off air-raid sirens in the area to warn sleeping residents of the danger, and went house to house advising people to leave. "The sirens woke me up and when I looked out the window, I saw this big black cloud," said Bertha Sperry, who lives in a public housing project within a block of the railyard. Mrs. Sperry said she woke her husband and two children, and the family began walking south, away from the smoke. Church Opens Its Breakfast

Like nearly 200 others, she ended up in a north Denver junior high school, where Red Cross volunteers dispensed juice and coffee. At the Highlands Lutheran Church, across the street, the annual Easter morning breakfast was opened to the evacuees. By midmorning, 300 people had been served. Fire Department workers managed to suppress the acid clouds by blowing soda ash onto the spilled acid to neutralize it. To spread the ash onto the acid, which had spilled over a wide area, firefighters borrowed blowers used to clear snow from the runways at nearby Stapleton International Airport (NTSB Investigation Report).

Houston, TX December 11, 1983 Borden's Anhydrous Ammonia Explosion

By Bob Parry

"The year 1983 had not been one of my better years. On January 4th, I witnessed District Chief Lonnie Franklin, a "second father" to many of us at 7/B, getting killed while in route to an arson house fire. Two months later, my own father died from complications from heart disease. We only had a couple of tours of duty to go before the end of the year and I was looking forward to the New Year.

It was a quiet Sunday morning at 7's and a hearty breakfast was being finished when Stations 1, 7, and 8 were being toned out for an ammonia leak at Borden's, located on Calhoun between Milam and Louisiana (Figure 1.99). Borden's was a two-story brick building (not counting the basement) known for their ice cream and other dairy products, occupying nearly the whole city block. Riding Acting Chief that day was Sr. Captain Ed Hauck and he was being driven by Fire Fighter Tim Jordan.

Don Sims was the Acting Captain on Engine 7 and Ricky Dail was one of the pipemen. I can't remember who was driving the engine. John Burleson had just left on a holiday just minutes before the box came in. The fourth man had to be a fill in. I was assigned to the right side of

Figure 1.99 Stations 1, 7, and 8 were toned out for an ammonia leak at Borden's, located on Calhoun between Milam and Louisiana. (Courtesy: Houston Fire Department.)

Ladder 7 that day along and our crew was Captain Royce Beck, E/O Delbert Burleson, and Fire Fighter Thomas Morant. In those days, we could stand up on the battery box because seat belts were an option and not a requirement. The death of fire Fighter Thomas Cooper in September 1981 removed us from the tailboard but it was not until Fire Fighter Robert Reyes, another good friend, fell off Engine 15 in December 1984 that the department finally required us to stop standing and be buckled up when the apparatus was moving.

While traveling north on Travis, we were a few blocks away from the Pierce Elevated when we caught our first strong smell of ammonia. The official name of the ammonia was anhydrous ammonia or as some called it, refrigerated ammonia. In fact, when we were crossing Pierce, I crawled in my seat to put on my mask. When we turned left on Calhoun and drove past the building, white clouds of ammonia were filling the sky and drifting to the southwest due to the northerly breeze. We staged at the corner of Louisiana and Calhoun with Engines 7 and 8 (The War Wagon), Snorkel 8, and District 7. Other companies, including Haz-Mat 17, were enroute as well. The Holiday Inn, located on the opposite corner on Calhoun, had several guests looking down from their balconies at the incident.

While the plan was being developed, a maintenance engineer named Melvin Machart came up to the command post and stated he knew what valve was leaking and he could help us shut it down. Though there may

Chronicles of Incidents and Response

have been a policy about putting the public in harm's way, it was important he direct us to the area of the leak so we could tightened the valve and shut off the leak. There was virtually no traffic, vehicles or pedestrian, in the area that may have complicated or delayed any tactic. The crews of Ladder 7, Engine 8 (Acting Captain Mike Miller along with Fire Fighters Mark "Dirt" Evans and Greg "Silky" Collins), and Acting Chief Hauck suited up. A quick class on the air pack was given to Mr. Machart and we proceeded toward the building. We were in the cloud and just about to enter the building when Mr. Machart became a little restless in the SCBA about his face piece. We went back to the staging area and assured him how the air pack would protect him from the fumes.

While walking back to the building, we felt the ground rumble and a fireball shoot across Calhoun. The force of the explosion sent bricks flying and crashing down on us in the street. In fact, popsicle sticks were found a mile away. I was covering my head with my arms and looked around and heard Acting Chief Hauck request a second alarm due to a major explosion at our location. Greg Collins vividly remembers the how a manhole cover flew over him and into the windshield of a car parked at an auto repair shop across the street. He, like the rest of us, can testify, "You can't run on moving ground"!! Larry Claxton, driving Engine 1 that day, recalls cleaning the Popsicle sticks off the pumper.

After we regrouped, Engine 7 laid a line from "the War Wagon" and pulled up in front of the building and operated their deck gun. The extra-alarm companies were arriving under the command of Deputy Chief D.E. Crowder. We were there most of the day sifting through the rubble. It wasn't until later that afternoon that relief crews began to arrive. Sid LaCombe, the officer on "the bucket", still laughs on how many of the "old head" fire fighters had "large ice cream spoons" in their shirt pockets when getting off their rigs and reporting to the command post. Yeah, nothing like listening to the Oilers with a gallon of your favorite ice cream! I did find an employee punch-in clock that had stopped at 9:04. I played that number for several months at a little corner bar "operation" in Youngstown but it never came up a winner- only that day though!! After we returned to service later that night and returned to quarters, the "what ifs" were discussed at the station.

The main one of course was "what if" Mr. Machart did not stop and continued to lead us in the basement where the leak was. It surely would have killed the crews of Engine 8, Ladder 7, the Chief, and Mr. Machart – 7 Fire Fighters and a civilian!! "What if" the wind was blowing from the south and we would have staged closer to the building on the Milam side of Calhoun? "What if" the leak would have happened on a regular workday with the area filled with spectators and employees – would the leak still would have occurred or prevented by the workers?? "Silky" Collins told me this little tid-bit that you can either laugh at or shake your head.

226 *Hazmatology: The Science of Hazardous Materials*

Around 2 a.m. the next morning, he was detailed to relieve FF Dennis Ganns on fire watch at the site.

In those days, a fire watch usually comprised of a couple of Fire Fighters with hose lines off the hydrant to hit the hot spots. When Greg pulled up, he saw Ganns from 17's covered in blood. Silky ran up to him and asked him "Are you OK, what the *^$@(!^& happened to you"? Ganns replied he was hungry and he crawled into the basement looking for ice cream. Though he found some, the ammonia was so strong it made his nose bleed. Ganns, lucky to be alive, is a state patrolman in Oklahoma. Mark Evans (78/D) probably said it best when "working one of his many debit day at 78/B". Someone said "that was close" referring to a close call at home plate during some downtime while watching an Astros game. "Dirt", without blinking, calmly stated, "close only counts in horseshoes, hand grenades, and ammonia leaks"!!

Anhydrous ammonia, as per the NFPA, was not supposed to ignite and explode. Though the cause of the leak was never determined, the agencies and investigators determined a single light bulb broke and that provided the ignition source. Everyone who made it on the "box" feels blessed they can talk about this event. Until I moved up to "Steeler Country", every time I passed the location on the Pierce Elevated or on the street I made the Sign of the Cross! Ironically, this location is the new home of Station 8. I just thank the Good Lord is not a park with a memorial plaque to the Houston 7!"

Bob Parry (Retired in Pittsburgh, PA)

Buffalo, NY December 27, 1983
Propane Gas Explosion

At 20:30 p.m. E.S.T. Buffalo firefighters responded to a reported propane leak in a four story radiator warehouse located the corner of North Division and Grosvenor streets. The building was approximately 50 feet by 100 feet. First arriving Engine 32 reported nothing showing, shortly after Truck 5 and Engine 1 and the Third Battalion arrived. Chief Supple assumed command. Thirty-seven seconds after the chief announced his arrival the propane tank experienced a BLEVE (Figure 1.100). The explosion heard 15 miles away, leveled the 4 story building and damaged others in a four block area. Seriously damaged buildings were noted over a half mile away. The fireball started buildings burning on a number of streets. A ten story housing project across the street had every window blown out.

Engine 32 and Truck 5's Firehouse over a half mile away had the windows blown out. The force of the blast threw Ladder 5, an aerial tiller, nearly 35 feet into the front yard of a dwelling, instantly killing all five crew members. Two civilians were also killed as they sat in their living room. Engine 1 was thrown across the street, injuring the captain and

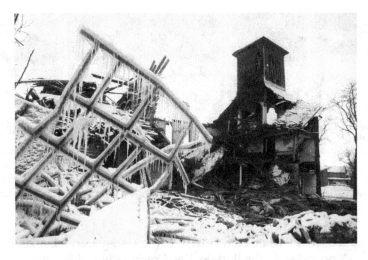

Figure 1.100 Chief Supple assumed command. Thirty-seven seconds after the chief announced his arrival the propane tank experienced a BLEVE. (Buffalo and Erie County Public Library.)

driver pinning them inside the cab. Engine 32 was slammed against the warehouse and buried in the rubble. Eleven firefighters were injured in the initial blast, several of them critically. During the rescue efforts 19 more firefighters were injured as a result of adverse weather conditions. A foot of snow that fell during that weekend hampered the rescue effort. Over 150 civilians were transported to hospitals suffered in the explosion, and many others were treated at the scene.

Investigation into the incident found the warehouse was housing an illegal 500 lb. propane tank. An employee was attempting to move the tank to another part of the building when it slipped off the forklift, breaking the valve. Leaking propane gas filled the facility and found an unknown ignition source.

This incident remains the largest number of firefighters killed in the line of duty in the Buffalo Fire Department history (Figure 1.101).

Firefighters That Made the Supreme Sacrifice

Firefighters Truck 5:

Mike Austin, 39

Mickey Catanzaro, 37

Red Lickfield, 43

Tony WaszkIelewicz, 37

Marty Colpoys, 47

Figure 1.101 This incident remains the largest number of firefighters killed in the line of duty in the Buffalo Fire Department history. (Buffalo and Erie County Public Library.)

Compiled from Buffalo News articles Wednesday December 28, 1983, provided by the Buffalo, Erie County Public Library.

Shreveport, LA September 17, 1984 Anhydrous Ammonia Fire Incident

In 1984, one firefighter was killed and a second was burned over 72% of his body in an anhydrous ammonia explosion and fire that occurred at the Dixie Cold Storage Company plant. Six other firefighters were injured. Five of the six were admitted to the hospital for observation from smoke inhalation. About 2:45 p.m. Employees discovered an ammonia leak inside a cold storage building and called the fire department for assistance. While firefighters were working inside in Level A chemical protection, the ammonia reached an ignition source and at 4:05 p.m. an explosion occurred that shook the building. The two hazmat team members were the only ones in the building. Both men were rushed to LSU Medical Center and were listed in critical condition. Chief Johnson and Captain Johnson went in to the building to shut off a valve. Assistant Chief Johnson received 3rd degree burns over 30% of his body and 2nd degree burns over 50%. Captain Johnson received 3rd degree burns over 95% of his body. Both were 10 year veterans of the Shreveport Fire Department. Chief Johnson is the Training Officer at the departments training academy. Captain Johnson is an instructor.

> *Hazmatology Point: Though it is listed as a non-flammable gas by DOT, ammonia burns inside structures and confined spaces; it is less likely to ignite out in the open. Precautions should be taken for ammonia leaks inside buildings and confined spaces just as for any other flammable gas.*
>
> *This is the second incident that I found in my research that involved anhydrous ammonia explosions inside buildings. The other was the*

Chronicles of Incidents and Response 229

> *Borden's building in Houston. Because the explosion in Houston occurred before the fire fighters went inside the building, there were no fatalities or injuries.*

The man who wanted to be a firefighter since the third grade, the man who went to court to become a firefighter, the man who became the first Shreveport Fireman to die in the line of duty in forty-two years was buried with full honors. He was not just a firefighter, but a fighter.

Captain Percy Rudell Johnson, a ten-year veteran of the City of Shreveport Fire Department, passed away early in the morning Wednesday September 19, 1984 as a result of injuries received while in the performance of hazardous duty as a firefighter on Monday September 17, 1984. Captain Johnson was a fallen hero, devoted public servant, a Christian gentlemen and a friend who was a credit to his community, church, and those with whom he worked. Captain Johnson was highly respected by all who knew him for his promotion of peace and harmony, as well as his extended efforts to assist others and for his obvious love of his job. Percy Johnson's many achievements are marked by his drive for social justice and willingness to be the first to step forward and accept obstacles inherent in promoting justice and equality for all.

Captain Johnson was the second black applicant to qualify for duty with the Shreveport Fire Department. He was one of the original plaintiffs in a lawsuit which resulted in the integration of the Shreveport Fire Department, leading to the hiring of blacks and women. Percy was the first black fireman for the city of Shreveport to be promoted to rank of Fire Instructor with the accompanying rank of Captain. The muscular, mustachioed Johnson spent much of his free time coaching the department's basketball team, running in competitions against his colleagues, and encouraging physical as well as mental exercises. He was also known to help struggling recruits after class. Captain Johnson was an innovator, helping to form the department's nine member color guard, which was present at his funeral. Percy was responsible for many unsung accomplishments. If a city is to be great, great spirits must walk among. Captain Percy Johnson was one such spirit. In the end, he gave his most precious gift while doing what he truly wanted to do.

Shreveport Fire Department was assisted by the Greenwood Volunteer Fire Department, Shreveport Police, Caddo Parish Sheriff's Department and the Louisiana State Police.

Shreveport Public Library.

The Times Shreveport-Bossier, September 18, 1984.

Firefighter That Made The Supreme Sacrifice

Captain Percy Rudell Johnson, 32, (Figure 1.102)

Figure 1.102 Captain Percy Rudell Johnson, 32, made the supreme sacrifice in the line of duty.

Norfolk, VA September 4, 1984
Gas Tank Truck Fire

A gasoline truck explosion that sent a river of fire through storm drains and forced residents to fell two apartment buildings left three men burned, one seriously, officials say (Figure 1.103). "We were really lucky we didn't come up with more injuries. The potential was there," Fire Chief Tom Gardner said of the explosion caused Monday by the collision between a garbage truck and an Exxon tank truck carrying 8,500 gallons of gasoline.

The Exxon truck driver, William Scott, 58, of Virginia Beach, suffered first and second degree burns on his face, legs and arms. He was listed in serious condition at Norfolk General Hospital. "He couldn't believe he was still alive," said Esther Danielson, a nurse who helped transport SCOTT to the hospital. Gardner said two firemen suffered minor burns while dousing the blaze, which destroyed the Exxon truck and 21 cars and created a flaming river that flowed through storm drains into the Hague, an inlet off the Elizabeth River.

Officials evacuated residents of the 168-unit Pembroke Towers and the 64-unit Hague Park apartment building by the accident site on Colley Avenue in the Ghent neighborhood.

Several elderly residents were rolled outside in wheelchairs. Five of those evacuated were treated for high blood pressure and breathing problems, said Don Haupt, supervisor of the city's paramedics. Another eight or nine people were evacuated from the Ronald McDonald House, which provides lodging for families whose children are being treated at nearby hospitals, said Greg LaRue, who manages the house with his wife.

Figure 1.103 A gasoline truck explosion that sent a river of fire through storm drains and forced residents to fell two apartment buildings left three men burned, one seriously. (Courtesy: Norfolk Fire Department.)

Firefighters using foam extinguished the fires within an hour of the 9 a.m. explosion and residents were allowed to return to their homes. Tralton Cooper, 32, said he was backing the garbage truck out of a driveway at the Ronald McDonald House when he collided with the Exxon truck. Cooper later was charged with reckless driving. "The tail of my truck caught the tail of the tanker," he said. Cooper said when he saw fuel flowing down the street, he pulled his garbage truck away from the tanker. "It exploded some five, 10 seconds after that," he said. LaRue said the Exxon truck driver jumped out just before the explosion. "Fuel started coming out of the side of the truck. He pulled over to the side and it went off," LaRue said. "It was a good-sized boom." Jane Hillman, a resident of Hague Park apartments, said she heard a big explosion and then "a series of explosions as the flames hit the cars along the back fence" of a parking lot.

Winchester Star Virginia 1984-09-04

Bhopal, India December 2-3, 1984
Release of Methyl Isocyanate

Background

December 2018 marked the thirty fifth anniversary of the massive toxic gas leak from Union Carbide Corporation's chemical plant in Bhopal in

232 *Hazmatology: The Science of Hazardous Materials*

the state of Madhya Pradesh, India. This incident was a "wake-up call" for the world in terms of illustrating the devastation that chemicals can cause when released into the environment. More than 40 tons of methyl isocyanate gas (MIC) leaked from a pesticide plant owned by the Union Carbide Chemical Company.

The company involved in what became the worst industrial accident in history immediately tried to dissociate itself from legal responsibility. Eventually it reached a settlement with the Indian Government through mediation of that country's Supreme Court and accepted moral responsibility. It paid $470 million in compensation, a relatively small amount of based on significant underestimations of the long-term health consequences of exposure and the number of people exposed. The disaster indicated a need for enforceable international standards for environmental safety, preventative strategies to avoid similar accidents and industrial disaster preparedness.

The Incident

At 11.00 PM on December 2 1984, while most of the one million residents of Bhopal slept, an operator at the plant noticed a small leak of methyl isocyanate (MIC) gas and increasing pressure inside a storage tank. The vent-gas scrubber, a safety device designer to neutralize toxic discharge from the MIC system, had been turned off three weeks prior. Apparently a faulty valve had allowed one ton of water for cleaning internal pipes to mix with forty tons of MIC. A 30 ton refrigeration unit that normally served as a safety component to cool the MIC storage tank had been drained of its coolant for use in another part of the plant. Pressure and heat from the vigorous exothermic reaction in the tank continued to build. The gas flare safety system was out of action and had been for three months.

At around 1.00 AM, December 3, loud rumbling reverberated around the plant as a safety valve gave way sending a plume of MIC gas into the early morning air. Within hours, the streets of Bhopal were littered with human corpses and the carcasses of buffaloes, cows, dogs and birds. Local hospitals were soon overwhelmed with the injured, a crisis further compounded by a lack of knowledge of exactly what gas was involved and what its effects were. It became one of the worst chemical disasters in history and the name Bhopal became synonymous with industrial catastrophe.

Around 12:50 a.m. local time a Union Carbide employee triggered the plant's alarm system as the concentrations of gas around the plant became difficult to tolerate. Activation of the system triggered two siren alarms: one that sounded inside the plant and the other that directed outward to the public and city of Bhopal. The two siren system had been decoupled from one another in 1982, so that it was possible to leave the factory warning system on while turning off the public system, and this was exactly

Chronicles of Incidents and Response

what was done: the public system sounded briefly at 12:50 a.m. and was quickly turned off, as per company procedure meant to avoid alarming the public around the factory over tiny leaks. Workers, meanwhile, evacuated the Union Carbide plant, traveling upwind.

Bhopal's superintendent of police was informed by telephone, by a town inspector, that residents of the neighborhood of Chola (about 2 km (1.25 miles from the plant) were fleeing a gas leak at approximately 1 a.m. Calls to Union Carbide plant by police between 1:25 and 2:10 a.m. gave assurances twice that "everything is OK", and on the last attempt made, "we don't know what has happened, sir." With the lack of timely information exchange between Union Carbide and Bhopal authorities, the city's Hamidia Hospital was first told that the gas leak was suspected to be ammonia, then phosgene. Finally, they received an updated report that it was "MIC" (rather than "methyl isocyanate"), which hospital staff had never heard of, had no antidote for, and received no immediate information about.

The MIC gas emanating from tank E610 petered out at approximately 2:00 a.m. Fifteen minutes later, the plants public siren was sounded for an extended period of time, after first having been quickly silenced an hour and a half earlier. Some minutes after the public siren sounded, a Union Carbide employee walked to a police control room to both inform them of the leak (their first acknowledgement that one had occurred at all), and that "the leak had been plugged." Most city residents who were exposed to the MIC gas were first made aware of the leak by exposure to the gas itself, or by opening their doors to investigate the commotion, rather than having instructed to shelter in place, or to evacuate before the arrival of the gas in the first place.

Over 500,000 people were exposed to the MIC. Initial fatalities are estimated to be 2,259, the government confirmed 3,787 deaths related to the gas release and 558, 125 injuries. Injuries included 38,478 temporary partial injuries and 3,900 severely and permanently disabling injuries. Estimates of the number of people killed in the first few days by the plume from the UCC plant run as high as 10,000, with 15,000 to 20,000 premature deaths reportedly occurring in the subsequent two decades. Several epidemiological studies conducted soon after the accident showed significant morbidity and increased mortality in the exposed population. These data are likely to under-represent the true extent of adverse health effects because many exposed individuals left Bhopal immediately following the disaster never to return and were therefore lost to follow-up.

Lessons Learned

The events in Bhopal revealed that expanding industrialization in developing countries without concurrent evolution in safety regulations could have catastrophic consequences. The disaster demonstrated that seemingly local problems of industrial hazards and toxic contamination are often tied to global market dynamics. UCC's Sevin production plant was built in Madhya Pradesh not to avoid environmental regulations in the U.S. but to exploit the large and growing Indian pesticide market. However the manner in which the project was executed suggests the existence of a double standard for multinational corporations operating in developing countries.

Enforceable uniform international operating regulations for hazardous industries would have provided a mechanism for significantly improved in safety in Bhopal. Even without enforcement, international standards could provide norms for measuring performance of individual companies engaged in hazardous activities such as the manufacture of pesticides and other toxic chemicals in India. National governments and international agencies should focus on widely applicable techniques for corporate responsibility and accident prevention as much in the developing world context as in advanced industrial nations. Specifically, prevention should include risk reduction in plant location and design and safety legislation.

Local governments clearly cannot allow industrial facilities to be situated within urban areas, regardless of the evolution of land use over time. Industry and government need to bring proper financial support to local communities so they can provide medical and other necessary services to reduce morbidity, mortality and material loss in the case of industrial accidents.

Public health infrastructure was very weak in Bhopal in 1984. Tap water was available for only a few hours a day and was of very poor quality. With no functioning sewage system, untreated human waste was dumped into two nearby lakes, one a source of drinking water.

The city had four major hospitals but there was a shortage of physicians and hospital beds. There was also no mass casualty emergency response system in place in the city. Existing public health infrastructure needs to be taken into account when hazardous industries choose sites for manufacturing plants. Future management of industrial development requires that appropriate resources be devoted to advance planning before any disaster occurs. Communities that do not possess infrastructure and technical expertise to respond adequately to such industrial accidents should not be chosen as sites for hazardous industry.

Since 1984

The Bhopal disaster could have changed the nature of the chemical industry and caused a reexamination of the necessity to produce such potentially harmful products in the first place. However, the lessons of acute and chronic effects of exposure to pesticides and their precursors

Chronicles of Incidents and Response

in Bhopal has not changed agricultural practice patterns. An estimated 3 million people per year suffer the consequences of pesticide poisoning with most exposure occurring in the agricultural developing world. It is reported to be the cause of at least 22,000 deaths in India each year. In the state of Kerala, significant mortality and morbidity have been reported following exposure to Endosulfan, a toxic pesticide whose use continued for 15 years after the events of Bhopal.

Aggressive marketing of asbestos continues in developing countries as a result of restrictions being placed on its use in developed nations due to the well-established link between asbestos products and respiratory diseases. India has become a major consumer, using around 100,000 tons of asbestos per year, 80% of which is imported with Canada being the largest overseas supplier. Mining, production and use of asbestos in India is very loosely regulated despite the health hazards. Reports have shown morbidity and mortality from asbestos related disease will continue in India without enforcement of a ban or significantly tighter controls. Some positive changes were seen following the Bhopal disaster. The British chemical company, ICI, whose Indian subsidiary manufactured pesticides, increased attention to health, safety and environmental issues following the events of December 1984. The subsidiary now spends 30–40% of their capital expenditures on environmental-related projects. However, they still do not adhere to standards as strict as their parent company in the UK.

The US chemical giant DuPont learned its lesson of Bhopal in a different way. The company attempted for a decade to export a nylon plant from Richmond, VA to Goa, India. In its early negotiations with the Indian government, DuPont had sought and won a remarkable clause in its investment agreement that absolved it from all liabilities in case of an accident. But the people of Goa were not willing to acquiesce while an important ecological site was cleared for a heavy polluting industry. After nearly a decade of protesting by Goa's residents, DuPont was forced to scuttle plans there. Chennai was the next proposed site for the plastics plant. The state government there made significantly greater demand on DuPont for concessions on public health and environmental protection. Eventually, these plans were also aborted due to what the company called "financial concerns" (Bhopal.org).

Author's Note: The incident in Bhopal did catch the attention of the U.S. Congress. They were concerned about such an incident occurring in the U.S. As a result the Emergency Planning and Community Right to Know Act also known as EPCRA and SARA Title III was passed. This act led to the first requirements for training of emergency responders to hazardous materials incidents. It also resulted in the creation of State Emergency Response Commissions and Local Emergency Planning

Committees to develop plans for response to hazardous materials releases and methods to notify the public about extremely hazardous materials used, stored and transported in an through their communities.

Miamisburg, OH July 8, 1986
Derailment Phosphorus Fire

July 8, 1986 at 4:25 p.m. a southbound Baltimore and Ohio Railroad Company (B&O) freight train Southland Flyer (FLFR) derailed near Miamisburg, Ohio. The train was made up of a locomotive and 44 cars traveling over the Bear Creek Bridge. The locomotive and 27 cars made it over the trestle, remained coupled together and as the next 11 cars moved over the trestle they derailed. Only one car on the train was carrying a considered hazardous material, and it derailed, turning on its side, it was the phosphorus tank car. Near the phosphorus car were a tank car of animal fat, which was leaking into the creek and a box car of rolled paper, neither of which were involved in the fire. Phosphorus is air reactive and shipped under water to keep it from spontaneously combusting. Several holes developed in the tank during the impact of the derailment and the water started leaking off causing the Phosphorus to be exposed to air and it ignited (Figure 1.104).

Figure 1.104 Several holes developed in the tank during the impact of the derailment and the water started leaking off causing the Phosphorus to be exposed to air and it ignited. (Courtesy: Special Collections & Archives, Wright State University.)

Chronicles of Incidents and Response 237

Fire Inspector Chuck Stockhauser was going to the west side of the river and a citizen stopped him and told him about the derailment. At 16:29, A General Alarm was sent out over pagers to call in all firefighters. Firefighter/Paramedic Steve Meadows, now a Battalion Chief, was off duty and came into town. He noticed the train had stopped and thought that was unusual. He saw Captain Sutton driving Engine 3, got in and heard there was a derailment. Captain Mike Sutton responded from the station with three firefighters. When they arrived on scene they began evacuations immediately of a few homes around the derailment site with the help of police officers and Fire Inspector Stockhauser.

Captain Sutton checked with the railroad conductor and was told there was only one car containing hazardous materials and that was the one on fire. There were minor explosions occurring, which may have been the gas tanks of automobiles on the two car carriers that derailed. Captain Suttons crew pulled lines and prepared to fight the fire. Mutual aid was requested from 30 area fire departments along with 107 medic units. Police agencies sent personal from Cincinnati, Dayton sent 25 police units along with other communities that sent help as well.

Meadows stayed with Engine 3 and ran the pump full throttle for 2 hours before he was relieved. Within two hours 3200 g.p.m. of water was being pumped through 4 master streams onto the fire. Water supply was an issue because hydrants were not close. Four inch hose was used with inline pumpers to supply water. Firefighters in or near the phosphorus cloud wore SCBA, which was refilled by the air cascade system. Air truck had to be taken to West Carrolton to be filled as they were using a lot of air.

Responders went to cartridge respirators, supplied by the Emergency Management Agency (EMA) on the second day or so. Many restaurants brought food for the firefighters including steaks. Steve Meadows helped deliver the food. None of the food came from Miamisburg, because the town had been evacuated. There were no security issues, and there was no looting during the evacuation. Miamisburg looked like a ghost town.

Returning from the hospital from and EMS call, Andy Harp and his partner Leroy Kline heard the call on the radio and went to the scene in their medic unit. There was a 100′ column of white smoke rising from the derailment site. They parked behind Rausch Contracting building for protection. At the time of the derailment Miamisburg had a population of 16,000. There were 23 career personnel and 14 reserves. Seven personnel along with the fire chief and fire inspector were on duty that day. They operated from 2 fire stations with 2 engines, 2 Ladders, Air Unit, Brush Unit, 3 medics and two reserve engines. Robert Menker was the Fire Chief and Incident Commander (IC) at the time of the incident. He passed away in 2015.

When Chief Menker arrived on scene, he ordered an immediate evacuation of the west side of the Great Miami River. After evaluation of the incident status he additionally ordered the northeast section of town

238   *Hazmatology: The Science of Hazardous Materials*

evacuated as well. His amounted to about 10,000 people. Following consultation with other chiefs it was decided to mount an attack on the fire utilizing water cannons and direct hose streams. The Train Master arrived at the command post and challenged the Chief's authority to handle the emergency and was subsequently escorted from the command post under threat of arrest. The winds changed in the evening of July 9th and coupled with a wind inversion that kept the phosphorus cloud close to the ground, prompted a second evacuation effecting approximately 30,000 people.

During the first and second days of the emergency numerous proposals for handling the phosphorous tank were suggested and evaluated. Some of these included; direct hose stream attack; plugging; water flooding of the interior of the tank car; foam application; burial; opening the manhole to allow air injection to accelerate the burn rate; and use of explosive demolition (it was suggested that the Air Force could use a Fighter from nearby Wright Patterson to shoot he tank car with a missile). Chief Menker in consultation with the city manager, decided to proceed with the suggestion of opening the manhole to accelerate the burn rate.

Initially the Command Post was established on a bridge downwind, south of the derailment. The bridge provided an excellent vantage point for the incident because the tracks ran right under the bridge with a view north of the tracks and the derailment site. However, the wind changed to the North and the command post had to be moved. Representatives from federal, state and local agencies were at the IC. However, Chief Menker maintained absolute control over the incident. Radio communications was the biggest problem as many agencies were on separate frequencies, local mutual aid channels were overloaded with communications. So much so, that remote activation of evacuation sirens could not be accomplished when needed.

AT&T provided emergency phone bank at the command post. An emergency operations center was established at Station 1, where logistics and support was being provided. Firefighter Andy Sharp slept with his head on a rail the 1st night at the scene and at station 2, the second night on a roll of insulation. All but three firefighters lived in the city, so they could not go home to sleep. Those three went home to sleep on the 3rd night. Firefighter ID's were not used on a daily basis that became an issue when firefighters were sent home to rest upon returning they did not have any way to prove they were firefighters. Police would not let them back in because of the evacuation in town and security set up to protect the evacuation area. As a result, identification badges were created for personnel operating at the incident.

Over the five day duration of the incident, crews were rotated in and out, millions of gallons of water were flowed onto the fire. By the time the emergency was declared under control, local hospitals received 569 persons with nonfatal injuries, including 13 emergency response personnel.

Of these, 27 were hospitalized due to their injuries. The cost, excluding costs to evacuees, community disruption or business interruptions was estimated to be over 3.5 million U.S. dollars.

Steve Meadows and Mark Stockman took most of the up close photographs (Miami Valley Fire District).

Ord, NE July 15, 1987 Anhydrous Ammonia Leak

Author's Note: While with the Nebraska State Fire Marshal's Office, I lived in the town of Ord. While there I was a member of the Ord Volunteer Fire Department.

One day I was driving on the west end of town when the fire pager went off announcing an anhydrous ammonia leak on a farm south of town. It was just a short drive from where I was so I headed to the scene. When I arrived, I pulled into the first part of a long driveway leading to the buildings on the farm. There was a faint odor of ammonia in the air, but nothing significant. The first thing I noticed was there was browning of a section of corn in the corn field next to the driveway (Figure 1.105). It was June so the corn was about 3-4 feet high everywhere but the browned area. There the corn leaves were brown and the plants looked like they were stunted.

Soon other Ord firefighters started arriving and staged upwind near where I had stopped my car. My memory is fuzzy and do not remember much of the detail of the actions that were taken beyond that point. If it wasn't for the photographs I took most of the incident may have been lost to memory. There were two large horizontal tanks on concrete cradles

Figure 1.105 There was a faint odor of ammonia in the air, but nothing significant. The first thing I noticed was there was browning of a section of corn in the corn field next to the driveway.

Figure 1.106 Heat from the sun causes the ammonia in the hose to vaporize and the increase in pressure ruptured the hose and released the ammonia from the hose.

between the house and cornfields. About 50 feet from where the corn started turning brown. The tank had been turned off, but the owner had not drained the hose. Heat from the sun causes the ammonia in the hose to vaporize and the increase in pressure ruptured the hose and released the ammonia from the hose (Figure 1.106).

Ammonia has a very high expansion ratio from liquid to gas of 1 gallon of liquid ammonia creates 850 gallons of ammonia gas. The liquid expanded in the hose and when it ruptured it released the remainder of the liquid, which immediately turned to vapor. The vapor cloud traveled to the corn field and extended approximately 100 feet into the corn field before it dissipated. That was the end of the incident, the wind blew away any remaining gas and the emergency was over. No injuries were reported.

Henderson, NV May 4, 1988 PEPCON Explosions

The Pacific Engineering Production Co. facility in Henderson, Clark County, Nevada manufacturers ammonium per-chlorate (AP) for the National Aeronautics and Space Administration (NASA). AP has been mixed with finely ground aluminum and other combustible materials to create solid propellants for launch vehicles and military weapons. Following the Challenger disaster on January 25, 1988 all launch activities were halted by NASA. However, The stand down halted PEPCON's shipping of AP, yet did not affect their contract orders from Space Shuttle Rocket Booster (SRB) manufacturer Morton Thiokol. Over a period of

15 months PEPCON had accumulated a stockpile of over 4,000 tons of the oxidizer.

May 4, 1988 workers at PEPCON were repairing wind damage to the steel and fiber glass drying structure when at 11:30 a.m. P.D.T. sparks from a welding torch ignited some of the fiber glass building material. Workers tried to put the fire out with a hose line, but when they added a second hose line, water pressure was not adequate to fight the fire. Residue on the drying structure surfaces accelerated the combustion of the materials and the fire quickly grew out of control. The fire spread to 55 gallon drums of AP next to the drying structure. It is estimated that within 10-20 minutes from the start of the fire, the first explosion occurred. Of the 77 PEPCON employees, 75 escaped by running or driving away from the facility through the desert and the other two were killed by the explosion. Clark County Fire Department was notified at 11:51 a.m. PDT. Closest units were a little over 5 miles away. The fire chief of the city of Henderson, 1.5 miles away also saw the smoke and ordered his units dispatched to the scent.

At 11:53 a.m. PDT, multiple 55 gallon drums exploded into a giant fireball, the first of the two major explosions, approximately 100 feet in diameter (Figure 1.107). The shockwave shattered vehicle windows of the fire response units, halting their approach to PEPCON. Employees escaping in vehicles warned the responders of potential, larger explosions, sending the responders in the same direction as the escaping employees. Those warnings undoubtedly saved many responder lives then the second and biggest of the two explosions occurred. Four minutes later, at 11:57 a.m. PDT the fire reached the storage area of the facility that held large aluminum 5,000 pound shipping containers loaded with AP, resulting in

Figure 1.107 At 11:53 a.m. PDT, multiple 55 gallon drums exploded into a giant fireball, the first of the two major explosions, approximately 100 feet in diameter. (From National Aeronautics and Space Administration (NASA).)

242 *Hazmatology: The Science of Hazardous Materials*

the largest of the two explosions. Witnesses reported seeing the shock-wave traveling across the ground towards them. Many were temporarily blinded by immediate flash and others lacerated by the flying glass. Henderson responders were effectively incapacitated, although none were seriously injured. Of the 4,000 tons of AP stored at the plant, it is estimated that approximately 1,500 tons, just over 1/3 were consumed by the subsequent explosion.

Henderson residents reported windows blown out, doors ripped off hinges and cars overturned in the streets. Most of the city of 50,000 was evacuated because of concern that toxic gases from the explosions might drift through the community. By evening the evacuation was relaxed and allowed to return home and a curfew was put into effect and the National Guard was called in to help prevent looting. Explosions were felt 10 miles away at McCarran International Airport in Las Vegas. In downtown Las Vegas, 16 miles away, windows were broken and tall buildings swayed. Two of the blasts measured 3.0 and 3.5 on the Richter scale at the California Institute of Technology in Pasadena (NASA).

Kansas City, MO November 29, 1988
Ammonium Nitrate Explosion

November 29, 1988 at approximately 03:40 hours the Kansas city Fire Department received a call reporting a fire at a highway construction site. The fire was reported by a security guard at the site to be in a small pickup truck, however, a woman in the background, another security guard, could be heard saying "the explosives ore on fire". Pumper 41 was dispatched to the site with a captain and two firefighters (in the Kansas City area fire apparatus with a pump is called a pumper). Dispatch cautioned Pumper 41 that there may be explosives at the site. Pumper 41 arrived on scene at 03:46 and found there were two separate fires burning and a second pumper was requested. Pumper 41 also requested that dispatch warn Pumper 30 of the potential for explosives at the site. Pumper 30 was dispatched and arrived on scene at 03:52. Because there were two separate fires at the site the crew from Pumper 30 suspected arson and requested that the police be dispatched to the scene as well.

Approximately 5 minutes after the arrival of Pumper 30, Pumper 41 requested a battalion chief be sent "emergency" to the scene. There was a great deal of confusion at the site as to whether there were explosives on site or if the explosives were involved in the fires. Following the extinguishment of the pick-up fire, Pumper 41 proceeded to the other fires to assist Pumper 30. There was a truck, a trailer, and a compressor on fire at 04:02. None of the vehicles or trailer appeared to be marked. There were no indications that the firefighters suspected any explosives were involved in

Chronicles of Incidents and Response 243

the fires they were attempting to extinguish. As it turns out the "trailer" that was on fire was actually an explosives magazine.

At 04:04 a.m. Pumper 41 contacted Battalion Chief 107 who had been dispatched to the incident. They indicated that "Apparently this thing's already blowed up, Chief. He's got magnesium or something burning up here." At 04:08 a.m., 22 minutes after Pumper 41 arrived and approximately 16 minutes after Pumper 30 arrived, the magazine exploded killing all 6 firefighters assigned to Pumper 41 and Pumper 30 (Figure 1.108). Battalion Chief 107 and his driver were just arriving on scene and stopped about 1/4 mile from the explosion. They received minor injuries when the windshield of their vehicle was blown in.

Chronology of the Explosion

03:43 a.m.
   Caller on 911 reports a pick-up truck on fire near U.S. 71 and Blue River Road. Says explosives may be involved. Dispatchers send a company of firefighters, from Station 41 at 5700 Bannister Road.
03:48 a.m.
   Dispatchers send a second company firefighters from Station 30 at 7543 Prospect Ave.
03:57 a.m.

Figure 1.108 At 04:08 a.m., 22 minutes after Pumper 41 arrived and approximately 16 minutes after Pumper 30 arrived, the magazine exploded killing all 6 firefighters assigned to Pumper 41 and Pumper 30. (Courtesy: Kansas City, Missouri Fire Department.)

244   *Hazmatology: The Science of Hazardous Materials*

The fire department calls the police department for what was thought to be an arson fire.

04:07 a.m.

More than 30,000 pounds of ammonium nitrate explodes at a rock quarry near 87th Street and Blue River Road. Six firefighters are killed and two fire department Pumpers are destroyed.

04:09 a.m.

Dispatchers send two pumpers, three trucks and ambulances to the scene.

About 04:10 a.m.

Power outages are reported in an area near Bannister Road, U.S. 71 and Blue River Road.

04:29 a.m.

Evacuation begins. Residents are ordered out of surrounding area as officials fear another explosion.

04:49 a.m.

The second explosion; an additional 15,000 pounds of ammonium nitrate explodes.

06:06 a.m.

Several major thorough fares are closed.

06:20 a.m.

Police emergency lines are flooded with calls about property damage. Reports include, street lights blown out and windows broken in area businesses.

07:00 a.m.

Fire department personnel at scene confirm that six firefighters are dead.

Firefighters That Made The Supreme Sacrifice:

Captain Gerald C. Halloran 57

Thomas M. Fry 41

Luther E. Hurd 31

Captain James H. Kilventon Jr. 54

Robert D. McKarnin 42

Michael R. Oldham 32

Lessons Learned

- State of Missouri highway construction site with explosives in magazines and 50,000 lbs. of ammonium nitrate/fuel oil mixture (most with aluminum pellets) stored in two trailers/magazines.

Chronicles of Incidents and Response

- The Kansas City Fire Department had not been involved in blasting permit process and was unaware of explosives on the site prior to the incident Nor did the Department have jurisdictional authority over the site. State highway site is not under city control regarding permits or inspections, according to City Attorney.
- ATF has universal jurisdiction over explosives except during transportation, but does not ordinarily inspect or issue permits for sites.
- Local fire departments are almost always the first responder, have their personnel at risk, yet do not always have regulatory control or guaranteed coordination from other agencies.
- Dispatcher was told of presence of explosives, but not what was stored, nor where. Both pumper companies were told of explosives on the site by the dispatcher, but nothing specific.
- Trailers/magazines probably were not placarded nor marked to indicate contents. They were not required by ATF to be marked when parked on site.
- Material Safety Data Sheet (MSDS) clearly says flee this type of fire.
- While there is no official ATF policy against placarding there appears to be a generally accepted practice in the field of removing placards when not in transit.
- This was an arson fire and in 1997 five persons were convicted and sentenced to life in prison for the firefighters deaths. However, there is a great deal of controversy surrounding the convictions and a recent investigation by Kansas City Star Reporter Mike McGraw has resulted in a call for a re-opening of the investigation into the case (*Firehouse Magazine*).

> **Author's Note**: *On September 24, 1989 the Kansas City Hazardous Materials Team was placed in service. The team was formed following the explosion in this incident. Pumper companies 30 and 41 were lost during the explosion. Pumper company numbers 30 and 41 were added together to form the number for Hazmat Unit 71.*

Houston, TX October 23, 1989 Phillips 66 Houston Chemical Complex Explosion & Fire

At approximately 1:00 p.m. CST an explosion and fire occurred at Phillips 66 Company 1400 Jefferson Road at Hwy 225, Pasadena TX. The first of many secondary explosions registered 3.5 on the Richter Scale at Rice University and threw debris 6 miles away (Figure 1.109). The polyethylene plant was demolished in the explosion. The incident took 10 hours to control and killed 24 Phillips employees and injured 314. No firefighters were injured. The accident occurred when a massive gas release occurred in a

Figure 1.109 The first of many secondary explosions registered 3.5 on the Richter Scale at Rice University and threw debris 6 miles away. (Courtesy: Houston Fire Department.)

reactor, more than 85,000 pounds of the reactors contents were released instantaneously. Within 90 to 120 seconds the gas mixture found a yet to be identified ignition source and exploded with the force of 2.4 tons of TNT. All of those killed were within 250 feet from the point where the gas was initially released.

The cause of the release as determined by the FBI was two hoses that were used to open an close a valve had been reversed when last reconnected prior to the product blockage-clearing procedure in progress. As a result, the valve would have been in the open position when the actuator switch in the control room was in the valve closed position.

Initial Response

Initial response was provided by the Phillips 66 fire brigade, which was soon joined by the Channel Industries Mutual Aid Association (CIMA). This organization had 106 members in the Houston area at the time of the explosion. The mission of CIMA is to provide emergency assistance to members with regard to firefighting, search and rescue, first aid and equipment. Site command and coordination was vested in the incident commander who was the Phillips Company Fire Chief. Technical assistance was provided by a team from the U.S. EPA. Cooperating governmental

Chronicles of Incidents and Response 247

agencies were the Texas Air Control Board, the Harris County Pollution Control Board, the FAA, the U.S. Coast Guard, and OSHA.

Firefighting

Firefighting water system was a part of the process system. When the first explosion occurred, some of the hydrants were sheared off at ground level by the blast. The result was inadequate water pressure for firefighting. Shut-off valves which could have been used to prevent the loss of water from ruptured lines in the plant were out of reach in the burning wreckage. No remotely operated fail-safe isolation valves existed in the combined plant/fire-fighting water system. In addition, the regular-service fire-water pumps were disabled by the fire which destroyed their electrical power cables. Of the three backup diesel-operated fire pumps, one had been taken out of service, and one ran out of fuel in about an hour. Firefighting water was brought in by hoses laid to remote sources: settling ponds, a cooling tower, a water main at a neighboring plane and even the Ship Channel. The fire was brought under control in about 10 hours as a result of combined efforts of the fire brigades from other nearby companies, local fire departments and the Phillips 66 foam trucks and fire brigade.

Search and Rescue

All search and rescue operations were coordinated by the Harris County Medical Examiner and County Coroner. Search and rescue efforts were delayed until the fire and heat had subsided and all danger of further explosions had passed. These operations were difficult because of the extensive devastation and the danger of structural collapse. The Phillips 66 Company requested, and the FAA approved an implemented a 1-mile no-fly zone around the plant to prevent engine vibration and/or helicopter rotor downwash from dislodging any of the wreckage. The U.S. Coast Guard and City of Houston fire boats evacuated over 100 trapped people across the Ship Channel to safety. OSHA preserved evidence for evaluation regarding the cause of the catastrophe (HFD Pictorial History 1980 – 2014).

> **Hazmatology Point:** *This incident was well beyond the capability of even the Houston Fire Department to handle on their own. Few, if any, other jurisdictions around the country would have had access to the resources available in the Houston Shipping Channel area. This was a rare and complex incident caused by a simple reversing of two critical hose lines by human error. Lost water supply caused by a dual system usage between plant requirements and firefighting. Even back repetitive*

back-up systems failed. This is what I would classify as a career or once in a life-time incident that was handled by the resources available doing the best they possibly could under the circumstances. Amazingly, no emergency responders were injured.

Philadelphia, PA February 23, 1991 One Meridian Plaza High-Rise Fire

On what began as an uneventful Saturday night twenty-eight years ago, a fire on the 22nd floor of the 38-story Meridian Bank Building, also known as One Meridian Plaza, was reported to the Philadelphia Fire Department on February 23, 1991 at approximately 20:40 hours and went on to burned for more than 19 hours. The reason this fire is presented in a volume on the history of hazardous materials is because the fire was started by spontaneous combustion of linseed soaked rags that were being used in the remodeling of this building (Figure 1.110).

This type of spontaneous combustion is a chemical reaction that has caused a number of fires across the World and in this case resulted in the needless deaths of three Philadelphia Firefighters and injuries of 24 others. Jack Bloomer was the only survivor from his platoon. David

Figure 1.110 The reason this fire is presented in a book on the history of hazardous materials is because the fire was started by spontaneous combustion of linseed soaked rags that were being used in the remodeling of this building. (Courtesy: Philadelphia Fire Department.)

Holcombe, Phyllis McAllister and James Chappell perished in the Feb. 23 high-rise inferno. The resulting fire also caused irreversible contamination from products of combustion and the high-rise building had to be torn down. Grant it, there were numerous other factors that contributed to these firefighter deaths and injuries, if the fire had not started, they would have been a mute point.

The 12-alarms brought 51 engine companies, 15 ladder companies, 11 specialized units, and over 300 firefighters to the scene. It was one of the largest high-rise office building fires in modern American history completely consuming eight floors of the building and was controlled only when it reached a floor that was protected by automatic sprinklers. The Fire Department arrived to find a well-developed fire on the 22nd floor, with fire dropping down to the 21st floor through a set of convenience stairs. Heavy smoke had already entered the stairways and the floors immediately above the 22nd. Fire attack was hampered by a complete failure of the building's electrical system and by inadequate water pressure, caused in part by improperly set pressure reducing valves on standpipe hose outlets (Philadelphia Fire Department).

Firefighters Who Made The Ultimate Supreme Sacrifice:

David Holcombe

Phyllis McAllister

James Chappel

New York City, NY February 26, 1993
World Trade Center Bombing

A bomb explodes in the parking garage beneath the World Trade Center in New York City February 26, 1993. Six people died and 1,000 were injured by the powerful blast, which also caused the evacuation of thousands of people from the Twin Towers (Figure 1.111). An informant later identified a group of Serbians in New York as the culprits. However, when the FBI conducted surveillance of the gang they found not terrorists but jewel thieves, putting an end to a major diamond-laundering operation.

Fortunately, investigators at the bomb scene found a section of a van frame that had been at the center of the blast. The van's vehicle identification number was still visible, leading detectives to the Ryder Rental Agency in Jersey City, New Jersey. Their records indicated that Mohammed Salameh had rented the van and reported it stolen on February 25. Salameh was already in the FBI's database as a potential terrorist, so agents knew that they had probably found their man. Salameh compounded his mistake by insisting that Ryder return his $400 deposit. When he returned to collect

Figure 1.111 Six people died and 1,000 were injured by the powerful blast, which also caused the evacuation of thousands of people from the Twin Towers. (Courtesy: Fire Department New York.)

it, the FBI arrested him. A search of his home and records led to two other suspects.

Meanwhile, the owner of a storage facility in Jersey City came forward to say that he had seen four men loading a Ryder van on February 25. When this storage space was checked, they found enough chemicals, including very unstable nitroglycerin, to make another massive bomb. Investigators also found videotapes with instructions on bomb making that led to the arrest of a fourth suspect.

Other evidence showed that one of the terrorists had bought hydrogen tanks from AGL Welding Supply in New Jersey. In the wreckage under the World Trade Center, three tanks marked "AGL Welding" were found. In addition, the terrorists had sent a letter to the *New York Times* claiming responsibility for the blast. Portions of this letter were found on computer disks taken from a suspect's office. Finally, DNA analysis of saliva on the envelope matched that of the suspect. The wealth of evidence resulted in easy convictions, and each of the men was sentenced to 240 years in prison (911 Memorial & Museum).

Ste. Elisabeth de Warwick, Quebec, Canada
June 27, 1993 Propane Explosion

On the morning of June 27th, at 09:02 a.m. Warwick Volunteer Fire Department received a call for a barn fire. They arrived about 10 minutes later they found a large cattle barn on fire. A quick size up revealed a 1,055 gallon propane tank right next to the barn. Relief valve was fully operational on the tank with flames shooting 16 feet in the air. Firefighters set up and began applying water to the exposed propane tank to cool it.

Suddenly, a BLEVE occurred involving the tank splitting the tank into two large parts. The force of the explosion sent one part of the tank 150 feet into a field where it hit a fire truck and then continued another 754 feet where it struck a vehicle on the road pinning the occupant inside. Three firefighters were killed when the second part struck the engine where they were donning SCBA and preparing hose lines. A fourth firefighter was killed when he was thrown 150 feet when the tank part slammed into the engine.

The blast also injured three firefighters as well as four civilians, including an occupant in the vehicle on the road (Ste. Elisabeth de Warwick, Quebec, Canada, Fire Department).

Firefighters That Made The Supreme Sacrifice

Pompier Rene Desharnais

Pompier Martin Desrochers

Pompier Raynald Dion

Assistant Director Raymond Michaud

Hazmatology Point: *NFPA advises that LPG tanks will BLEVE approximately 8 to 30 minutes with 58% occurring in less than 15 minutes after flame impingement begins on the vapor space. This is based on studies of previous incidents where a BLEVE has occurred. Small tanks can be as dangerous as large tanks even an LPG tank on a barbeque grill can BLEVE.*

Orrtanna, PA December 6, 1993
Ammonia Leak Knouse Foods

Two volunteer fire fighters were killed in an industrial accident at the Knouse Foods cold storage building (Figure 1.112). Both were maintenance employees at the plant that uses anhydrous ammonia as a refrigerant. They were doing routine maintenance on a valve using a fork life to reach the valve. A leak developed and they were sprayed with liquid ammonia.

Figure 1.112 Two volunteer fire fighters were killed in an industrial accident at the Knouse Foods cold storage building.

They did not have any protective equipment. Emergency responders who answered the 911 call for help realized their brother firefighters were the victims and tried to make a risky rescue. Several were injured because they did not have protective equipment to protect them from the ammonia.

Harold D. Miller, Jr. of Orrtanna, PA, was a volunteer firefighter with the Cashtown Community Fire Department, Adams County Company 4. Jack Kauffman was a firefighter at the South Mountain Fire Department. Miller and Kauffman were working draining refuse oil from the liquid ammonia refrigeration system near the ceiling at Knouse Foods, a company that processes fruits from the nearby orchards of Adams County. According to the OSHA investigation, the valve was opened without a proper hose in place. This routine maintenance procedure had been performed in the plant for 30 years without a problem. Anhydrous ammonia is used as a refrigerant to keep the fruit cold until it can be processed. Both men were working from a scissor lift and the valve was opened they were sprayed with liquid ammonia. They could not readily escape from the lift and received lethal doses of ammonia gas. Twelve other employees who tried to rescue the victims were also injured. Two hundred employees were evacuated from the plant along with about 70 Orrtanna residents who lived nearby.

Firefighters responded from the Cashtown department and were disheartened to find one of their own was a victim of the incident. Firefighters made entry into the building to rescue their comrade without wearing the proper protective equipment to protect themselves. As a result several of them were injured by the ammonia and take by ambulance to the

Gettysburg Hospital. OSHA fined Knouse $90,450 dollars for 3 willful, 10 serious and 5 other than serious violations.

> **Author's Note**: *Harold Miller was killed on his last day of work at the Knouse Plant, he had turned in his resignation to go take another job. On the day of his funeral as the department was leaving the firehouse, they were ironically toned out for another call at the Knouse Plant for a diesel spill.*

Plant Workers/Firefighters Who Lost Their Lives in the Incident

Harold D. Miller, Jr., Cashtown Community Fire Company

Jack Kauffman, South Mountain Volunteer Fire Department

> **Hazmatology Point:** *Anhydrous ammonia is commonly used for cold storage plants in addition to its agricultural use as a fertilizer. Ammonia is toxic and without water. Anhydrous is a term that means without water. Ammonia is also corrosive to metals and skin. An additional hidden hazard is ammonia flammable under certain circumstances, particularly in confined spaces and inside of buildings. Fortunately, in this instance, the ammonia did not ignite. Anyone entering an atmosphere of ammonia requires Level A Chemical Protective Clothing. Ammonia, because it is without water, seeks it out from the environment. It looks for moisture from the skin, eyes, and lungs. SCBA will protect the eyes and lung but not the skin.*

Hanover-Adams The Evening Sun, Thursday May 26, 1994.

Port Neal, IA December 13, 1994 Ammonium Nitrate Explosion

The Terra Industries ammonium nitrate plant in Sergeant Bluff, south of Sioux City, exploded after an equipment malfunction on Dec. 13, 1994, killing four and injuring 18. Some 3,000 residents of Iowa and Nebraska were evacuated. Ammonia gas wafted off the site for six days (Figure 1.113). Better safety protocols and design changes are now in place, Iowa OSHA Administrator Stephen Slater said Thursday.

"All kinds of technologies have had huge improvements," Slater said. "And we haven't had any bad experiences at the plants in the 20 years since Terra. I'm knocking on wood." And, Slater noted, large fertilizer plants are also now subject to an extra set of safety regulations that include meticulous requirements on which equipment is used and how it is replaced.

Figure 1.113 Some 3,000 residents of Iowa and Nebraska were evacuated. Ammonia gas wafted off the site for six days. (From NTSB Investigation Report.)

Such plants often use ammonium nitrate as a fertilizer because of its high nitrogen content, which promotes plant growth. But the chemical can also explode under certain conditions, and was the explosive used in the Oklahoma City bombing that killed 168 in 1995. The potential for explosions has raised concerns in Iowa. Already operating in the state are the reconstructed former Terra plant, now owned by CF Industries, and a Koch Nitrogen Co. facility in Fort Dodge.

In addition, plans have been announced to the expand the rebuilt Terra plant in the Port Neal complex near Sioux City and to build a second large fertilizer plant in southeastern Iowa near the Lee County town of Wever. Mitch Doherty, who lives less than a mile from the planned Lee County plant, worries that it could present problems that his community hadn't needed to be concerned about previously. "Not only do you have the facility, but you've got the trains and trucks hauling chemicals all the time," Doherty said. "You'd hope that this plant would be more modern, safer." But Doherty said Lee County supervisors and representatives of Iowa Fertilizer Co., the company planning the project, have offered only vague reassurances when asked about safety issues.

"That little town in Texas, it's about the same as us, Doherty said. "You look at something like that, and think 'holy cow,'" Doherty said. A spokeswoman for the project's parent company, Orascom Construction Industries of Egypt, declined to comment on the Texas explosion or safety protocols it intends to have in place at the Lee County plant. The Deerfield, Ill., company plans to invest $1.7 billion to expand the former Terra plant.

Chronicles of Incidents and Response   255

There also has been talk of a possible third fertilizer plant to be built in Mitchell County in northeast Iowa near the Minnesota border. Cronus Chemical LLC has reportedly been in discussions with Iowa development officials who would provide it with $35 million in tax incentives to proceed with the project.

The U.S. Environmental Protection Agency collects "worst case scenario" documents from such plants and monitors their emergency preparedness plans. A spokesman for the office in Lenexa, Kan., Kris Lancaster, said because of the varying condition at different plants, his staff can't say how wide an area would be evacuated if one of the Iowa plants exploded, or how large a crater might be left. If another explosion were to occur, the Iowa Department of Natural Resources would decide how big an area to evacuate, based on air temperature, humidity and the amount of gas released. The area would cover miles.

Barbara Lynch, who has worked in the DNR's environmental protection offices for 35 years, said she knows of only one large-scale fertilizer plant explosion in Iowa, the Terra plant.

Rodney Tucker, a member of the DNR emergency response team, said all three of the large plant sites in Iowa are within coverage areas served by hazardous materials squads.

Labor Commissioner Michael Mauro said the Texas tragedy forces decision makers at the state and federal level to refocus on safety issues (Chemical Safety Board Investigation Report).

Oklahoma City, OK April 19, 1995 Terrorist Bombing

Bombing of the Murrah Federal Building in Oklahoma City April 19, 1995 was the largest loss of life in a terrorist bombing in the United States. Occurring at 9:02 a.m. the blast from a 4,800 pound mixture of ammonium nitrate, nitro methane and fuel oil destroyed 1/3 of the nine story building (Figure 1.114). The blast was heard 16 miles away and registered 3.2 on the Richter Scale 16.1 miles away. Destruction and property damage occurred to 324 buildings in a 16 block radius around the Murrah building and shattered glass in another 258 buildings.

Eighty-six cars in the area were burned or destroyed by the blast pressure. The bomb was placed in a Rider rental truck and parked in front of the federal building in downtown OKC. Timothy McVeigh and Terry Nichols planned, gathered materials, built the bomb and brought the truck to OKC. They had previously surveyed the government complex and placed a getaway car a few blocks from the federal building. They thought of themselves as revolutionary's and planned to fire the first shot in a new American Revolution.

Figure 1.114 Occurring at 9:02 a.m. the blast from a 4,800 pound mixture of ammonium nitrate, nitro methane and fuel oil destroyed 1/3 of the nine story building. (From Federal Emergency Management Agency.)

Response to the Bombing

> **Author's Note**: *While teaching a one week Hazmat Incident Command Course for the National Fire Academy for the 63rd Civil Support Team, there was an officer from the Oklahoma City Fire Department in the class. Turns out he was one of the first chief officers on the scene of the bombing. He was kind enough to give me a walking tour of the bombing site that had been turned into the National Memorial. He provided an overview of his and OKCFD's actions during the response to the incident. That personal touch made my tour of the site much more meaningful and informational. One of the things I learned was the Federal Building was not the intended target of McVeigh. When McVeigh and Nichols scouted the site and they chose the Federal Courthouse across from the Federal Building. However, when they arrived to place the truck bomb, street work was being conducted in front of the courthouse and the truck could not park there. The federal building became a target of opportunity.*

Calls to the 911 center began at 9:03 a.m. when over 1,800 calls for help were received. Many responders were already in route when they heard the initial explosion. Within 23 minutes the State Emergency Operations Center had been activated. Within the "Golden Hour" 50 people had been rescued from the rubble and sent to hospitals throughout the area. During the response phase there was so much help coming from citizens and emergency personnel, no security zone could be created. A local

Chronicles of Incidents and Response   257

television station put out a broadcast without being told to, that all available doctors and nurses respond to the scene. This created more problems with accountability. Everyone was focused on saving lives, but because of the circumstances managing the scene became difficult at best. One nurse that responded to the call for help was killed while searching the site for victims when she was hit on the head by debris. During the search someone thought they found another bomb and the site was evacuated along with a 4 block area around the building.

This gave incident commanders a chance to gain control and secure the site. ID tags were issued to those who had a legal responsibility to be operating at the site. The last victim was rescued from the building debris by 7:00 p.m. Eleven FEMA Urban Search and Rescue Teams were dispatched to the site. Communications was another problem during the incident. Groups of responders were not able to talk to other groups. As a result many did not know what others were doing. People evacuating the scene clogged roads and delayed emergency vehicle response to the scene. At the time of the bombing, Incident Command was new to the OKC department and they were in a learning process. IC involved a culture change and not all officers were using it. Chief Marrs got on the radio during the bombing response and told the officers to set-up command and start using ICS. After that ICS was used by the department all the time.

Oklahoma City Fire Department did not have an Urban Search and Rescue Team (USAR) at the time of the bombing. In the years following several members received USAR training. FEMA was adding a team and OKC hoped they would be the one. However, Missouri was selected instead. Eventually, the OK Department of Homeland Security funded a state USAR team. Currently the team is composed of an OKC metro team and a Tulsa metro team. When combined they are a Type 1 USAR Team. Oklahoma Task Force 1 (OKTF1) also has a K-9 team with 8 dogs in OKC and 10 in Tulsa. Firefighter Jeff Hanlon trains OKC USAR search dogs; a dog named Willy is currently in training. Both teams have a swift water rescue component and a Helicopter Search and Rescue Team. Oklahoma Air National Guard flies the helicopters and OK-TF1 rides the hoist.

Oklahoma City Bombing National Memorial

On May 23, 1995 the remains of the Murrah Federal Building were demolished. Planning for a fitting memorial erected in its place carried on for two years. Today on the site there is memorial and museum honoring the significance of that tragic day. According to Kari Watkins, the executive director of the Oklahoma City National Memorial & Museum, "The memorial was really built to remember those who were killed and those who survived and those who were changed forever." Two of the most outstanding features of the memorial are the "Gates of Time". One gate is located at each end of a reflecting pond. The concrete gates are covered with a "Naval" and yellow bronze. On one of the gates is the time 9:01, representing the last moment of peace and on the other gate 9:03 representing the first moments of recovery.

> **Author's Note:** For me, the most moving part of the memorial was the 168 bronze chairs, each engraved with the name of one of the victims. Large chairs represented the adults and small chairs the children. Also on the memorial site is a 70 year old American elm, "The Survivor Tree," tilted by the force of the explosion, but still standing and living despite the explosion. I have been to the memorial twice. During my second visit to OKC I brought my 5 year old granddaughter Abby. My wife and I explained what the memorial was all about. Following our explanation of the chairs, Abby immediately ran to a chair and hugged it (Figure 1.115). I believe she understood the purpose of the memorial.

Investigation

McVeigh was captured just 90 minutes after the explosion by an Oklahoma Highway Patrol officer Charlie Hanger about 62 miles North of OKC. McVeigh's car was missing the rear license plate and was stopped by Officer Hanger. McVeigh was arrested for carrying a concealed weapon. Officer Hanger had no idea McVeigh was the OKC bomber. McVeigh was taken to the Noble County Jail and all of his clothing was collected and

Figure 1.115 During my second visit to OKC I brought my 5 year old granddaughter Abby. My wife and I explained what the memorial was all about. Following our explanation of the chairs, Abby immediately ran to a chair and hugged it.

Chronicles of Incidents and Response

put in paper bags. As the FBI began their investigation evidence quickly lead to McVeigh. Evidence collected at the bombing scene lead to a vehicle identification of a Ryder Truck rented in Kansas. On April 20, the FBI released a sketch of McVeigh obtained from the Ryder dealer's description.

A motel owner in Junction City, Kansas recognized the person in the sketch as a guest who registered at the motel as Timothy McVeigh. The FBI ran arrest records check on McVeigh and found he was in custody at the Noble County Jail in Perry, Ok. Additional investigation led to army buddy Terry Nichols and he was arrested as well. Nichols was tired, convicted and sentenced to life in prison. McVeigh was found guilty and sentenced to death. He was executed June 11, 2001. FBI Agent Barry Black said of the investigation "It went as it was supposed to." The system worked as it should.

Oklahoma City Bombing

April 19, 2020 was the 25th anniversary of the bombing of the Mariah Federal Building in Oklahoma City (OKC) which killed 168 and injured over 600. At the time the loss of life was the largest in an act of terrorism in the United States. This was an act of domestic terrorism carried out by Timothy McVeigh and Terry Nichols in retaliation for the sieges that occurred at Ruby Ridge, Idaho. Federal agents in a standoff with Randy Weaver resulted in the death of his 14 year old son and wife. Waco, Texas another standoff at the Branch Dividian compound April 19, 1993 resulted in 76 deaths including women, children and leader David Koresh. April 19, 1775 is also Patriots Day, the anniversary of the rebellion against the British authority at Lexington, MA (Oklahoma Historical Society).

Delaware County, PA August 6, 1995 Ammonia Leak

August 6, 1995, was a very hot and humid day with temperatures in the upper 90's with humidity not far behind. Around 8:00 a.m. in the town of Yeadon an ammonia leak occurred while contractors were removing ammonia from lines of in the old A&P warehouse (Figure 1.116). A fire department engine company had been standing by in case they were needed. EPA Region 3 response team was summoned along with a fire department hazardous materials team. Anywhere between 2-5 gallons had escaped containment. There was no danger to residents or workers, though one firefighter was taken to Fitzgerald Mercy Hospital and examined as a precautionary measure. Upon reading this article this seems like a pretty cut and dry event, which handled quickly and effectively.

Figure 1.116 Around 8:00 a.m. in the town of Yeadon an ammonia leak occurred while contractors were removing ammonia from lines of in the old A&P warehouse.

Author's Note: *Paul Harvey a radio broadcaster of years past would tell a story and then after a commercial break, he would tell "The Rest of the Story." A&P warehouse has a small anhydrous ammonia leak. Without a commercial break, here is the "Rest of the Story." How do I know the rest of the story? I was there for several hours and have over two hundred photos to back up my story. Tom Micozzie a former student of chemistry of mine at the National Fire Academy Chemistry for Emergency Response taught in Pennsylvania. We became friends and I visited him on occasion in Upper Darby, PA, a suburb of Philadelphia. He was the Hazardous Materials Coordinator for the County. One day he invited me to an exercise at the Sun Oil Refinery, so I accepted and made the trip from Maryland. About a half hour or so into the exercise he was toned out for an ammonia leak. He asked me if I wanted to go, so I did. The incident had occurred at an abandoned A&P cold storage building. The entire first alarm assignment had contacted the ammonia and were having burning on their skin. It was a very hot day everyone was sweating and ammonia loves water.*

Hazmat was called and the firefighters deconed. EPA Region 3, Regional Response Team was right in Philadelphia and was requested. EMS was called to treat exposed firefighters, police to secure the scene, and a command post was sent from emergency management. ICS was set up and this was one of the best managed incidents I have ever personally experienced. Everyone knew their job and the locals did their job. However, things kind of went south when the EPA team arrived. They had two personnel in a van and a supervisor who arrived in a car. They only had Level B PPE and ammonia required Level A. But they went in anyway. At first they were only sending in one man because that is all the PPE they had. Fortunately, the IC knew better and objected, but was overruled by the EPA Team. Fortunately, it was a small leak from piping workers were removing from the building and were remediating the waste ammonia. They had a tanker on site to haul the ammonia away. Firefighters were not seriously hurt and the ammonia dispersed by itself. That was the only experience I have had with a Regional Response Team, but that was 24 years ago and hazmat response has come a long way since then.

Weyauwega, Wisconsin March 4, 1996 Train Derailment and Fire

Just before dawn, the early morning quiet of March 4, 1996 was interrupted by the sounds of crashing metal and burning propane in this East-Central Wisconsin farming community. At the North end of town, near an industrial area, 37 cars of a Wisconsin Central Railroad train derailed. Among the 37 derailed cars were 7 tank cars of highly flammable liquefied petroleum gas (LPG), 7 tank cars of highly flammable liquefied propane gas, and 2 tank cars of sodium hydroxide, which is a non-flammable corrosive material. Initially, as a result of the derailment, three of the tank cars opened up and the propane and LPG immediately caught fire (Figure 1.117).

Figure 1.117 Initially, as a result of the derailment, three of the tank cars opened up and the propane and LPG immediately caught fire. (Courtesy: Weyauwega Fire Department.)

262   *Hazmatology: The Science of Hazardous Materials*

The characteristics of this incident are similar in many respects to previous incidents that occurred in Waverly, Tennessee, and Crescent City, Illinois, the outcome in Weyauwega however, was much different. In the Waverly incident a propane tank car exploded from a boiling liquid expanding vapor explosion (BLEVE). It was cold when the train derailed in Waverly, just as it was in Weyauwega, however, two days into the Waverly incident the temperature started warming up. This caused an increase in pressure in the tank car which had been damaged in the derailment. The BLEVE occurred when the tank could no longer withstand the increasing pressure. Crescent City experienced a similar incident to Weyauwega in that several propane tank cars derailed in the center of town. Resulting fires impinging on the tanks caused the propane tanks to BLEVE resulting in injuries to response personnel, bystanders and wide spread damage to the community.

The derailment in Weyauwega occurred at approximately 5:55 a.m. the wind was calm and the temperature was about 30° F. Twenty-Four volunteer firefighters from the Weyauwega Fire Department were on the scene within five minutes led by Assistant Chief Jim Baehnman. (Fire Chief Gary Hecker had left town on vacation just a few days before the incident occurred. He returned the second week of the incident but left Assistant Chief Baehnman in command of the incident). The Weyauwega fire department has two pumpers, engine 917 with a 1250 g.p.m. pump and engine 913 with a 750 g.p.m. front mounted pump.

The department also has two tankers, one with 1200, and the other with 3500 gallons of water. Additionally they have a brush rig and a rescue/equipment van that was used as the command post. Several days into the incident, the Pierce Fire Apparatus company in Appleton sent a brand new 1500 gallon pumper to assist in the firefighting efforts. Upon arrival, Chief Baehnman took command and immediately established the incident command system. The derailment had blocked the rail crossing on Mill Street and Chief Baehnman had a difficult time determining what was happening on the North side of the tracks. Fires were burning, with large fireballs 200 to 300 feet high and visible at times 13 miles away. It appeared that there were three separate fires.

The first was a large fire fed by a damaged tank car at the intersection of Mill Street and the railroad crossing. The second fire involved the feed mill some 75 yards to the east of the crossing. The third fire was approximately 75 yards to the west of the crossing and involved a storage building. It is believed that the third fire resulted from fallen power lines caused by the derailment. Many of the fires burned for most of the 18 days however, unlike Crescent City, Illinois and Waverly, Tennessee, no explosions, (BLEVEs) occurred involving the propane or LPG tanks. It is believed that the cold temperatures and the snow on the ground may have contributed to the lack of explosions by helping to keep the tanks

Chronicles of Incidents and Response 263

cool, although the exact reason is still a mystery. The lack of explosions accounted for the relatively small amount of damage that occurred within the town. Damage though extensive, was confined to the property of the feed mill, which was destroyed along with a storage shed. A residence to the north of the derailment site and one business experienced some heat damage. The estimated loss to the feed mill and surrounding property is 25 million dollars.

Upon sizing up the situation Chief Baehnman realized that the incident was beyond local capabilities and notified dispatch to request assistance from neighboring communities, the state of Wisconsin, and Region V EPA. Eleven neighboring fire departments from Navarino, Iola, Scandinauia, Manawa, Clintonville, Mukwa, Fremont, Waupaca, Poysippi, Tustin, and New London, responded to Weyauwega's call for assistance bringing 22 pieces of fire apparatus and over 200 firefighters. Radio communications quickly became a problem because of the volume of traffic from all of the responding agencies. Additionally agencies from the private sector as well as the state and federal governments responded including, Region V EPA from Chicago, Wisconsin Department of Natural Resources (DNR), State Emergency Management, the governor's office, Wisconsin National Guard, State EPA, the Federal Railroad Administration, Red Cross, and Salvation Army.

The first responding firefighters from Weyauwega found a tangled mass of railcars, broken rails, and a large volume of fire. The derailment occurred near a feed mill and the spreading fire was already impinging on feed mill structures and several grain and propane vehicles. Initial efforts were focused on fighting the fire in the feed mill and protecting exposures. The firefighters were lucky to the extent that a building across from the feed mill had not received its spring shipment of fertilizers and pesticides. Had the fire reached this building it could have added additional hazardous materials complications to the incident. A cheese factory just one block south of the burning tank cars had anhydrous ammonia pipes on the roof of the building. If a BLEVE had occurred the pipes could have been damaged by flying tank parts causing an ammonia leak.

The responding firefighter's view of the incident scene was initially obscured by the large volume of fire and darkness. They were unaware of the tank cars of hazardous materials that were involved in the derailment and subsequent fire. Reports from first responding firefighters are conflicting as to the exact time it took to recognize that the burning tank cars contained propane and LPG. Time estimates range from 10 minutes to 1 hour. Wisconsin Central Railroad personnel advised the firefighters that the construction of the tank cars would allow them to withstand fire for approximately 1 1/2 hours.

When this information was received it was already one hour into the incident. Once firefighters realized that there were propane and LPG

264 *Hazmatology: The Science of Hazardous Materials*

tanks involved in the derailment, and they were on fire, a decision was made to pull back. Firefighters abandoned hose lines in the streets when they realized the seriousness of the situation. The abandoned hose lines were later damaged when they froze and were driven over by clean-up and restoration personnel during the incident. Temperatures during the incident ranged from daytime highs of +15° F to +30° F to nighttime temperatures of -5° F to +15° F. As the incident progressed the weather forecast was for warmer temperatures that would cause the pressure to increase inside the derailed tanks. Much like the increased pressure that occurred in Waverly, Tennessee that caused the BLEVE of one of the rail tank cars there. This forecast of warmer temperatures caused concern for the crews and helped confirm the need to hot tap the remaining cars before this dangerous pressure increase could occur.

Firefighters were fortunate that no explosions (BLEVEs) occurred involving the propane and LPG tanks before the decision was made to move to a safer location. The National Fire Protection Association (NFPA) says that BLEVE times range from 8 to 30 minutes, with the average being less than 15 minutes in 58% of the incidents. The initial evacuation of personnel was two blocks for the first hour, this initial distance was then expanded to 7 blocks and finally to 1 1/4 miles. The Department of Transportation Emergency Response Guidebook (ERG) under guide 22 for propane and LPG recommends an evacuation distance of a minimum of 1/2 mile if propane or LPG tanks are on fire. In addition to pulling back response personnel, Chief Baehnman ordered the evacuation of the entire city.

Next to the Centex Gasoline Station, less than 100 feet from the derailment, a tractor trailer truck driver slept through the entire derailment and initial emergency response efforts. An attendant at the gas station had to awaken the truck driver to tell him to evacuate. Fire, police, and EMS personnel made notifications of the evacuation by going door to door while the local radio station provided additional announcements. As each residence was evacuated it was marked with a yellow tag so that other emergency responders would know the building had been cleared. Approximately 1700 people were evacuated from Weyauwega and another 600 from surrounding rural areas. Sirens and the Emergency Broadcast System (EBS) were not used for the evacuation. Most people, once notified, evacuated using personal vehicles. Two nursing homes with over 200 total patients were evacuated and taken to a hospital and other nursing facilities in nearby communities.

Evacuation of the nursing homes was conducted by the staff of the homes with the assistance of EMS emergency response personnel. The evacuation was facilitated using buses from the local school district for ambulatory residents and ambulances for those unable to walk on their own. The evacuation of the nursing homes was completed in approximately

Chronicles of Incidents and Response

265

two hours. Once the city was completely evacuated Chief Baehnman made the decision that no emergency response personnel would enter the city to fight any fires that may break out from the derailment or other causes. Residents of Weyauwega that were displaced by the derailment and fires were directed to an evacuation center that was set up at a former gym in Waupaca 7 miles away. However, many of the evacuees opted to stay with relatives or in hotels, some as far as 30 miles away, which were provided by the railroad. Pets were not considered in the initial evacuation. As the incident progressed more and more citizens became concerned about their pets that were left behind. Chief Baehnman did not feel that rescuing pets was worth the risk to personnel and the public. So the decision was made that no pet rescues would be conducted.

Chief Baehnman was later overruled by the governor and the pet rescue was conducted on a limited basis over the objections of the incident commander. After one week the National Guard was activated. Once on the scene, one of their initial efforts was to conduct a rescue of pets in a small portion of the city farthest from the incident. The pet rescue created additional problems as business owners then wanted to go in to check on their property. Eighteen days after the derailment occurred, the evacuated residents were allowed to return to their homes, after all danger from the burning propane and LPG tanks had diminished. There were no injuries to emergency responders or residents directly resulting from the derailment. One resident suffered an elbow injury from a fall during the initial evacuation however, she was able to continue evacuating on her own. Another resident had a heart attack upon returning home, her home was not damaged and the heart attack was not attributed to the incident. Assistant Chief Jim Baehnman said "from the start of the incident, the tone of the incident would be driven by safety and not time". This approach and the fact that no BLEVEs occurred very likely accounted for the lack of injuries to response personnel, civilians, and the minimal loss of property.

The railroad did not have a hazardous materials team but immediately sent an operations supervisor to coordinate railroad activities with the incident commander. Five hazardous materials teams responded to Weyauwega from Oshkosh, Waupaca County, Appleton, Stevens Point, and Brown County. The first team arrived on the scene within one hour after the derailment occurred. Because of the potential dangers of the burning propane and LPG tanks the hazmat teams were assigned an advisory role in the beginning until the contract companies hired by the railroads arrived on the scene. Security and traffic control were provided by the local police and the Waupaca County Sheriff's Office. Once the evacuation had been completed, law enforcement personnel were stationed at all roads into the city and did not allow anyone back in without the knowledge and consent of the incident commander. Police patrols made periodic entries into the evacuated area to ensure that no one had gone back in and that the homes and businesses were secure.

Fresh snowfall on the second day of the incident provided a means for law enforcement personnel to detect breaches of security of the evacuated area. The air space within 10 miles of the derailment was restricted to reduce curiosity flights and to control the news media, this distance was reduced to 5 miles the fourth day of the incident. Superior Environmental and the national guard provided helicopters for aerial surveillance to locate spot fires and determine the extent and effectiveness of the evacuation. For the first few days after the incident, aerial photographs were the only means of viewing the incident scene.

The railroad had conducted previous exercises with other communities on this type of incident using the incident command system. The railroad brought in hazardous materials clean-up experts from EM Tech and Hulcher Professional Services from Denton, Texas, and Superior Environmental from Appleton, Wisconsin. When it was determined that the last remaining propane tank would have to be vented and burned, an explosive expert from Louisiana was brought in also. Upon reaching the derailment site EM Tech and all other personnel took two days to study the site and determine the safest actions to take based upon the remaining tanks that were involved. Aerial photographs were used to make tactical decisions after EM-Tech arrived on the scene. Many of the tank cars were severely damaged in the derailment. Two of the cars were upside down and buried with only 10% of one and 20% of the other showing above the ground.

The actual "emergency" response phase of the incident only lasted until approximately 10:45 on the day of the derailment. Evacuation of the city had been completed by this time along with the relocation of response personnel to a position past the command post, which was now 1.25 miles from the derailment scene. The firefighters were also displaced just as other people evacuated from the town. The Fremont Fire Department shared their quarters with Weyauwega firefighters until they could return to their own quarters. At this point, the focus turned to stabilization, clean-up and restoration of the incident scene, which took the next 17 days to accomplish. The emergency response personnel assumed a role of support for EM Tech and other contractors during the process of stabilizing the burning and damaged tanks. Unmanned monitors were placed into service on the 3rd day of the incident as the pressure fires subsided and 6 or 7 of the propane and LPG tanks still had minor fires burning. Off loading of the tank cars was not undertaken because it was unknown what the extent of damage was to the cars. After carefully evaluating the derailment site, the decision was made to "hot tap" some of the propane and LPG cars to transfer the remaining fuel to a burn pit where it was allowed to burn off. The outer cars were tapped first and tapping moved

Chronicles of Incidents and Response

inward until the last car was reached which could not be hot tapped due to its position.

Hot tapping is a procedure that has been used successfully for many years in the chemical and petroleum industry but has limited applications in tank car accident situations. The process involves the welding or securing of a tapping to a vessel (such as a tank car), while it still contains gases or liquids. The tank is then drilled through a valve attached to the tapping fitting. A special drilling machine is used which threads the valve and reduces leakage. Once the drill bit penetrates the vessel wall the drilling machine is removed from the valve. Product can then be pumped or drained from the tank car through the valve. Hot tapping should only be attempted by trained and experienced contract personnel this is NOT typically an emergency response function.

Because of the potential danger to personnel, tanks containing certain commodities should not be hot tapped. They include: acetylene, ethylene, ethylene oxide, halogens, elemental sulfur, hydrocarbons in stainless steel tanks, cryogenics, hydrogen, acids, oxygen, and tank cars operating at below atmospheric pressure. Once the tank was tapped, nitrogen was injected into the tank forcing the liquid through the tap valve to a flexible hose connected to a pump and then into a burn pit through a hard pipe. The liquid was burned off until the tank is empty. It is the process of hot tapping and "flaring" that took the most time in the stabilization process. Fire department personnel stood-by with charged hose lines and rescue equipment to protect the contract personnel performing the hot tapping and flaring operations.

Personnel from EM Tech said "they had never experienced an incident with circumstances that were as difficult or as large as Weyauwega". Assistant Fire Chief Jim Baehnman, the incident commander, indicated that "the total amount of liquefied gases involved was over 1 million pounds, which is reported to be the largest incident ever in the United States in Terms of volume". Because all but 10% of the last remaining tank car was showing above ground, the decision was made to vent and burn the tank car. Charges were placed on each end of the tank opening it up to allow the remaining propane and LPG to burn freely. Venting a tank involves the placing of shaped explosive charges on the high end and low end of a damaged tank car. The resulting explosion opens up the tank car and allows the product to drain out and burn. Resources were not a problem during the incident and those needed were obtained promptly with everyone cooperating; "everyone knew just what to do". One of the main problems encountered initially was obtaining the necessary number of telephones to complete the calls to deal with the incident. The local phone company brought in 40 lines by noon on the first day of the incident. Initially power was cut to about 15% to 20% of the homes in the area to control ignition sources

268   Hazmatology: The Science of Hazardous Materials

immediately around the derailment. The natural gas service was also cut to the community and as a result up to 95% of the homes did not have heat which caused water pipes to freeze in many homes and businesses. The freezing was not as severe as first feared and only about a dozen homes were severely damaged. The main natural gas feeder line into the community ran directly under the derailment site next to Mill Street. Responders were concerned that the line may have been damaged as a result of the derailment which prompted the decision to shut down the line.

The news media was briefed each day between 0800 and 0900. As the days passed the residents were constantly wanting to know when they would be allowed to return to their homes. Noon briefings were held each day in three locations to update the evacuated citizens on the progress of the incident. Evacuees were notified by the news media that the briefings would be conducted in Appleton, Waupaca, and Stevens Point. The incident commander used examples of the Waverly, Tennessee and Crescent City, Illinois derailments to illustrate to the people what can happen when propane and LPG tanks BLEVE. This type of information helped people to understand why they couldn't go back to their homes before it was safe to do so.

During and after the incident maintenance problems were experienced with fire apparatus that had been constantly idling for almost three weeks without being shut off. The primary breakdowns involved motors and electrical generators on the motors (*Firehouse Magazine*).

Burnside, Illinois October 2, 1997
LPG Tank Explosion

Thursday October 2, 1997 at 4:39 p.m. CST the Carthage, IL volunteer fire department was alerted for a dryer fire on the Brown Family farm just north of Burnside IL. Witnesses located about.2 of a mile south of the farm, reported seeing what looked like a "giant Bic Lighter" shooting up in the air. A short time later they said they saw the fireball that looked like pictures of an atomic bomb blast, but did not hear the explosion. First arriving firefighters found not only the grain dryer fire, but 30-40 foot intermittent fire blowing from the safety relief valves on two 1,000 gallon LPG tanks, and a fully involved field tractor. The LPG fueled grain dryer was fed by the LPG tanks. Deciding on a direct attack, the first in officer positioned the engine and firefighters behind a large grain silo 100 feet away from the burning tanks. However, the engines tailboard extended beyond the silo's vertical edge and nearly in line with the end of one of the tanks. Three firefighters were advancing around the rear of the tailboard when an explosion occurred in one of the LPG tanks and a large tank section struck them. Two Carthage firefighters were killed when the LPG tank exploded.

Reports from the county coroner indicated he thought the two died from the concussion from the blast. He said they were hit by the fireball but they were not burned. The report from the State Fire Marshal indicated the two firefighters were killed by the "rocketing tank" while preparing to advance pre-connected hose lines from their high pressure fog pumper (Figure 1.118). Two other Carthage firefighters were injured, Chief Scott Carle, 41 and Firefighter Jason Livingston, 22. Livingston was airlifted by helicopter to Blessing Hospital and admitted to intensive care. Chief Carle was taken to Memorial Hospital. Both were released on Saturday (Chemical Safety Board).

Hazmatology Point: Enough LPG tank fires and explosions have occurred in the past that a particular pattern can be concluded from those incidents. When LPG tanks are impinged by fire, pressure builds up inside the tanks. When relief valves function and the gas catches fire, the burning gas shoots 30 feet or higher into the air like a flaming torch. Sounds coming from the relief valves has been described as sounding like a jet engine. The higher the flames are shooting and the louder the noise, the greater the pressure is within the tank; BLEVE is imminent. Fire

Figure 1.118 The report from the State Fire Marshal indicated the two firefighters were killed by the "rocketing tank" while preparing to advance pre-connected hose lines. (From Chemical Safety Board CSB.)

270 *Hazmatology: The Science of Hazardous Materials*

officers visualizing and hearing these sounds should evacuate the area in anticipation of a tank failure rather than placing firefighters in position to try and prevent the failure. NFPA says BLEVE will occur within 8-30 minutes of the begging of fame impingement on the vapor space of a tank. Further, if cooling the tank is determined to be a viable option, at a minimum of 500 g.p.m. uninterrupted is required to be successful in cooling tanks. LPG tank fires are historically, killers. Evacuation according to the distances in the Emergency Response Guide Book and letting them burn is the safest option.

Firefighters That Made the Supreme Sacrifice:

Mike Mapes, 35

Doug Buckert, 23

Albert City, Iowa April 9, 1998 Propane Explosion

On April 9, 1998, an 18,000-gallon propane tank exploded at the Herrig Brothers farm in Albert City, Iowa. Firefighters on scene reported that the relief valve was functioning properly and the sound coming from the valve sounded like a jet engine. Firefighters were attempting to protect the surrounding structures and let the propane burn off when the explosion occurred. Seven others, including a deputy sheriff, were injured in the explosion. Several buildings were also damaged by the blast A pipe connected to a large propane storage tank was broken by children riding four-wheel all-terrain vehicles. The leaking propane found an ignition source and ignited. Eventually, a BLEVE (boiling liquid expanding vapor explosion) occurred in the propane tank as a result of direct flame impingement from the burning propane.

The farm raised turkeys, which were housed in seven barns. Space heaters and furnaces provided heat for these turkey barns. Fuel for these space heaters and furnaces was supplied by a propane storage and handling system that included the propane tank that exploded.

On the evening of the incident, eight high-school-aged teens gathered at the farm for a party.

According to one of the co-owners of the farm, the youths had attended similar social gatherings at the farm on other dates, but with neither the knowledge nor the consent of the owners. Neither owner lived at the farm. At approximately 11:00 pm, one of the youths began driving an all-terrain vehicle (ATV) around the farm. Then the driver of the ATV picked up a passenger and continued his ride.

The ATV was heading east between the propane tank and a turkey barn when it struck two aboveground propane pipes (liquid and vapor

Chronicles of Incidents and Response   271

lines) that ran parallel to one another from the propane tank to direct-fired vaporizers approximately 37 feet to the north of the tank. (The direct-fired vaporizers were components of the system that used heat to transform liquid from the tank into a gas that was piped to space heaters and furnaces on the farm.) The ATV damaged both the liquid and vapor lines. The liquid line (which measured approximately ¾-inch inside diameter) was completely severed from the tank at the location where it was connected to a manual shut-off valve directly beneath the tank. An excess flow valve protecting the liquid line failed to function, and propane leaked out of the tank at the point of the break. As the liquid propane sprayed out of the tank, it rapidly changed to vapor.

Propane vapor may have also leaked from the damaged vapor line. Within a few minutes, propane from the damaged lines ignited, most likely when it reached one of the direct-fired vaporizers approximately 37 feet away. A fire, fed by the broken liquid line, began burning vigorously under the tank. Two of the teenagers drove to the home of a neighbor, approximately ½ mile from the farm, to report what had happened. At 11:10 pm, the neighbor called the 911 operator to report the fire.

Twenty members of the Albert City Volunteer Fire Department and two Buena Vista County Sheriff Deputies were the first responders to reach the farm. Upon arrival at about 11:21 pm, firefighters observed flames originating from two primary locations: from under the west end of the tank and from the pressure relief valve pipes located on the top of the tank. One firefighter reported that the "west end of the tank [near the broken liquid line] was *engulfed* in flames". Another stated that "the propane tank was fully engulfed and flames were 70-100 yards in the air." Fire fighters stated that the noise from the pressure relief valves was "like standing next to a jet plane with its engines at full throttle." At approximately 11:28 pm, as fire-fighting equipment was being moved into position, the tank exploded, scattering metal tank fragments in all directions. One large piece of the tank traveled in a northwesterly direction, striking and killing two volunteer firemen (Figure 1.119). Seven other emergency personnel sustained injuries as a result of the explosion.

Key Findings

The explosion that occurred at the farm is known as a Boiling Liquid Expanding Vapor Explosion or BLEVE. A BLEVE can occur when a pressure vessel containing a flammable liquid, like a propane tank, is exposed to fire. The book, *Loss Prevention in the Process Industries*, provides the following description of a BLEVE:

When a vessel containing a liquid under pressure is exposed to fire, the liquid heats up and the vapor pressure rises, increasing the pressure

Figure 1.119 One large piece of the tank traveled in a northwesterly direction, striking and killing two volunteer firemen. (From Chemical Safety Board CSB.)

in the vessel. When this pressure reaches the set pressure of the pressure relief valve, the valve operates. The liquid level in the vessel falls as the vapor is released to the atmosphere. The liquid is effective in cooling that part of the vessel wall which is in contact with it, but the vapor is not. The proportion of the vessel wall which has the benefit of liquid cooling falls as the liquid vaporizes. After a time, metal which is not cooled by liquid becomes exposed to the fire; the metal becomes hot and then may rupture.

- In this incident, the tank was engulfed in flames due to a leak of propane under the tank.
- These flames created the conditions that produced the BLEVE.
- Neither the propane tank nor its aboveground piping were protected by a fence or any other physical barrier designed to prevent damage from vehicles.
- The propane tank was equipped with an excess flow valve to protect the tank's liquid line that leads to the vaporizers. In the event of a complete break in the liquid line downstream from the valve, it was designed to close and greatly reduce the flow of propane from the broken pipe. (Even when an excess flow valve is activated, a small amount of fluid bleeds through a tiny hole in the valve).

Consequently, installation of a shut-off valve immediately downstream from the excess flow valve is required to stop all flow.) When the ATV severed the liquid line at this installation, however, the excess flow valve failed to close because the flow capacity of the outlet piping system

Chronicles of Incidents and Response 273

downstream of the valve was less than the closing rating of the excess flow valve installed in the tank. Fire fighters were positioned too close to the burning propane storage tank when it exploded. They believed that they would be protected from an explosion if they avoided the ends of the tank.

The propane storage and handling system was installed at the farm in 1988. When the tank system was installed, Iowa law provided that the 1979 edition of the National Fire Protection Association's *Standard for the Storage and Handling of Liquefied Petroleum Gases* (NFPA58) governed the installation. Under NFPA 58 and other relevant Iowa law the State Fire Marshal should have received a plan of the farm's propane tank storage and handling system.

This should have happened before it was installed. Iowa law, however, did not specifically designate which party the owner or the installer of a large propane storage facility -- was required to notify the State Fire Marshal. The CSB's investigation revealed that the State Fire Marshal had no record of the system and that it was not installed in compliance with all NFPA 58 requirements adopted as Iowa law (Chemical Safety Board).

Firefighters That Made The Supreme Sacrifice:

Tom Archer, 47, 14 Year Firefighting Veteran

Larry Walsh, 45, 16 Year Firefighting Veteran

Springer, OK September 1998 Liquid Nitrogen Asphyxiation

In early September 1998, an incident occurred involving liquid nitrogen, a cryogenic liquid that presents an asphyxiation hazard in addition to being very cold. The accident occurred while two workers were working on an oil pipeline in Springer, OK. They had been using liquid nitrogen in a pit to pressure check the pipeline. When discovered, both men were frozen by the very cold nitrogen, which has a liquid boiling point of 320 degrees below zero Fahrenheit; the vapors being released by the liquid would also have been very cold (Tulsa Oklahoma Fire Department).

> **Hazmatology Point:** *Nitrogen makes up 78% of the air we breathe. It is non-toxic, non-reactive and non-flammable. However, it can displace oxygen in a confined space. By displacing oxygen below 16% life cannot be supported. Cryogenic liquids will freeze anything they come in contact with.*

274 *Hazmatology: The Science of Hazardous Materials*

2000s
Baltimore, MD July 18, 2001
Howard Street Tunnel Fire

The train, en route with 31 loaded and 29 empty cars from West Baltimore, Maryland, to Philadelphia, Pennsylvania, departed the CSX West Baltimore Yard, about 6 miles west of the derailment site, at 2:37 p.m. on July 18, 2001. Within its consist the train had eight fully loaded tank cars containing hazardous materials regulated by the U.S. Department of Transportation (DOT). On the day of the accident, 11 trains went through the Howard Street Tunnel before the accident train. The crewmembers of the accident train reported that their train entered the west end of the tunnel at a speed of 23 mph.

The locomotive event recorders indicated that the locomotives were in throttle position 5 and that the train's speed was less than the time-table speed of 25 mph as the train entered the tunnel. About 1,343 ft into the 8,700-ft-long tunnel, the track grade changes from a slight descending grade to a slight ascending grade. The event recorders showed that at the dip, the train's speed was 24 mph in throttle position 4. As the train passed through the dip, the engineer gradually increased the throttle to position 8 as the train started the ascending grade. While the train was moving about 21 mph, the locomotive tractive effort increased and the train slowed to 18 mph. At 3:08 p.m., an uncommanded emergency air brake application was recorded, and the lead locomotives stopped in the tunnel about 1,850 ft from the east portal. Unknown to the crew at the time, the train had derailed. The emergency application of the train air brakes had occurred when the train became uncoupled ahead of the first car to derail, causing the train air brake line to separate (Figure 1.120). The derailment also resulted in the puncturing of a derailed tank car carrying tri-propylene and the subsequent ignition of this product. The puncture was a 2-in. diameter hole located near the bottom of the tank on the B-end (the leading end), left side, and on line with the interior end of the stub sill. Post accident inspection of the tank car indicated that a braking system linkage bar had disconnected and that the disconnected end of the linkage bar, when lifted upward, aligned with the hole in the tank. The fire spread to cargo in adjacent cars, which included paper and wood products, and generated heavy smoke and fumes that quickly filled the tunnel. Additionally, 2,554 gallons of hydrochloric acid were released from another derailed tank car.

The CSX chief dispatcher contacted the Baltimore trainmaster to advise him of the situation and ask if the train was transporting hazardous materials. He was told that the train did include hazardous materials cars. About 3:40 p.m., the CSX chief dispatcher determined that the

Figure 1.120 Unknown to the crew at the time, the train had derailed. The emergency application of the train air brakes had occurred when the train became uncoupled ahead of the first car to derail, causing the train air brake line to separate. (From NTSB.)

train likely had a serious problem and had possibly derailed. At 3:51 p.m., the CSX director of network operations issued a request for assistance to the railroad's hazardous materials team. About 4:00 p.m., Baltimore 911 received a call reporting smoke coming from a sewer near the Howard and Lombard Street intersection. Fire department responders were dispatched, and they traced the smoke to the west tunnel portal. Also about 4:00 p.m., the CSX chief dispatcher telephoned the CSX police communications center to ask that the Baltimore City Fire Department be notified and that emergency response personnel be dispatched to the tunnel.

At 4:04 p.m., the CSX police communications center notified the Baltimore 911 operator, who notified the Baltimore City Fire Department personnel responded to the site (Mt. Royal Station) about 4:10 p.m., but they could not enter the tunnel because of the fire and smoke. The train crew provided the *train consist* to the emergency responders. About 5:07 p.m., the incident commander, after deliberating with the responding technical experts, concluded that the derailment did not pose an immediate threat of a catastrophic explosion or a dangerous vapor release that would require an evacuation of the area. The incident commander thus did not believe a mass evacuation was necessary and instead decided to employ a "shelter-in-place" strategy for the several blocks on either side of the tunnel path along the principal length of the tunnel.

Other precautionary measures included evacuating the Camden Yards baseball stadium, activating the public alert siren system, and employing local television and radio outlets for public notifications. About 6:15 p.m., the water elevation began dropping at the city of Baltimore's Montebello II treatment plant. About 6:19 p.m., the water elevation at the Montebello water treatment plant also began dropping. At Druid Park Lake, the water flow rate abruptly increased from about 8.5 million to 9 million gallons per day (mg/d) to about 18 mg/d between about 6:15 p.m. and 6:30 p.m. A time-stamped security camera (taking a picture every 48 seconds) showed that water had broken through to the street surface at the intersection of Howard and Lombard Streets at 6:19:38 p.m.

Water flooded the intersection and flowed south on Howard Street. Water also flowed into the Howard Street Tunnel, which was below the street. The train consist shows the make-up of the train, including the placement and contents of all cars. If cars containing hazardous materials are part of the train, documentation is attached to the consist list that details emergency response information for those materials. Flow rate is a measurement of the volume of water leaving the reservoir and entering into the water system over a given period of time.

At 3:26 p.m., the director of security at a hotel above and adjacent to the derailment site in the tunnel called 911 and reported an unusual disturbance near his facility. The security director then called the Baltimore Department of Public Works to report the disturbance. About 3:34 p.m., he called the CSX communications center to advise them of a strong "rumbling" that had occurred at his building. He told the communications center that he suspected the rumbling had originated in the Howard Street railroad tunnel. About 3:36 p.m., the communications center operator forwarded the call directly to the CSX chief dispatcher.

According to city records, notification was received of a water leak about 6:19 p.m., and the city sent a crew to investigate. The crew determined that a failure had occurred in the 40-in.-diameter cast iron water main that passes directly above the Howard Street Tunnel at station 63+15. The crew closed a 40-in. valve at the intersection of Lombard and Paca Streets. They also closed valves on an interconnected 20-in.-diameter water line. A 40-in. valve located 1 block to the east (as well as numerous interconnecting lines) was also closed to isolate the area of the break. The line was shut down by 11:59 p.m. about 5h 40min after the appearance of water at street level.

The city of Baltimore estimated that about 14 million gallons of water were lost from the water main between the time of the break and the time the line was shut down. For the next 2 days, several groups of firefighters and railroad employees equipped with self-contained breathing apparatus ventured into the tunnel to determine the extent of the derailment and the status of burning equipment and cargo. Inside the tunnel, the first

Chronicles of Incidents and Response   277

45 railcars in the train consist had not derailed and had been pulled. The remaining four railcars had not derailed. The fire lasted for about 5 days as smoke emanated from both ends of the tunnel and several manholes at the Howard Street level. On Monday, July 23, at 7:42 a.m., the incident commander declared the scene officially under control. Later that morning, he authorized entry into the tunnel without self-contained breathing apparatus for qualified personnel.

Minot, ND January 18, 2002 Train Derailment & Anhydrous Ammonia Release

At approximately 1:37 a.m. on January 18, 2002, eastbound Canadian Pacific Railway (CPR) freight train 292-16, traveling about 41 mph, derailed 31 of its 112 cars about 1/2 mile west of the city limits of Minot, North Dakota. Five tank cars carrying anhydrous ammonia, a liquefied compressed gas, catastrophically ruptured, and a vapor plume covered the derailment site and surrounding area. The conductor and engineer were taken to the hospital for observation after they complained of breathing difficulties. About 11,600 people occupied the area affected by the vapor plume. One resident was fatally injured, and 60 to 65 residents of the neighborhood nearest the derailment site were rescued. As a result of the accident, 11 people sustained serious injuries, and 322 people, including the 2 train crewmembers, sustained minor injuries. Damages exceeded $2 million, and more than $8 million has been spent for environmental remediation.

On January 14 and 15, 2002, Canadian Fertilizers Limited loaded between 29,000 and 29,800 gallons of anhydrous ammonia into each of 15 tank cars in Medicine Hat, Alberta, Canada. At Medicine Hat, the 15 loaded cars were added to train 292-16, which had departed South Edmonton on January 16, 2002, bound for St. Paul, Minnesota. On Thursday, January 17, 2002, at 9:15 p.m., a train crew consisting of an engineer and conductor went on duty at Portal, North Dakota, to take train 292-16 to Harvey, North Dakota. The train consisted of 2 locomotives, 86 loads, and 26 empties. Its gross weight was 12,342 tons, and it was 7,138 ft long. The train consist included 39 tank cars containing hazardous materials, as regulated and defined by the U.S. Department of Transportation (DOT), including the 15 car loads of anhydrous ammonia, 10 car loads of liquid petroleum gas, 11 car loads of styrene monomer, and 3 empty tank cars that contained residue of a DOT-regulated hazardous material.

The conductor said he told the engineer to "bring her to a controllable stop." The engineer said he reached for the handle to apply the brakes lightly, and as he began to manipulate the controls, the train's emergency brakes automatically applied. Immediately after the emergency stop, the train crew discovered that there had been a significant derailment beginning with the fourth car behind the locomotives. The conductor told investigators, "… I had watched the explosions and the arcs from our train and the plumes of smoke that came up with the explosions. I knew there were explosions because I felt the concussion and I heard it." Additionally, the derailing equipment had knocked down power lines, disrupting electrical power to 2,820 residences and businesses in the nearby area. It would later be determined that 31 cars had derailed, including all 15 tank cars of anhydrous ammonia. The remaining hazardous materials cars were farther back in the train and were not involved in the derailment. The conductor stated that immediately after the derailment, he repeatedly called out "emergency" on the radio, as required by the operating rules. The conductor also radioed the CPR dispatcher in Minneapolis, Minnesota.

While awaiting a response from the dispatcher, he called 911 in Minot on his personal cell phone at about 1:37 a.m. and reported his train's location and the fact that the train had derailed with an explosion and hazardous materials release. The engineer used his personal cell phone to call the CPR yard office in Harvey, North Dakota to report the same information to the railroad. When the dispatcher in Minneapolis contacted the crew, they told him that the train had derailed and they could see "vapors or something." The engineer and conductor decided to evacuate the area using the train's locomotives.

The crew asked for and received permission from the dispatcher to detach the locomotives and pull away from the train. According to the dispatcher's Record of Movement of Trains, at 1:43 a.m., a crewmember told the dispatcher that something "… smells like anhydrous ammonia there at the head end at milepost 471." The conductor then walked to the rear locomotive and uncoupled it from the train. He stated that, at that time, he was in the middle of a white ammonia cloud. The conductor then went back to the lead locomotive and told the engineer to continue east toward Minot. The crew departed the area using the locomotives.

During the derailment, five anhydrous ammonia tank cars sustained catastrophic shell fractures that resulted in the separation of the tank shells and the complete and instantaneous loss of the contents. When the tanks violently ruptured, sections of the fractured tanks were propelled as far as 1,200 ft from the tracks. About 146,700 gallons of anhydrous ammonia were released from the five cars, and a cloud of hydrolyzed ammonia formed almost immediately. This plume rose an estimated 300 ft and gradually expanded 5 miles downwind of the accident site and over a population of about 11,600 people.

Over the next 5 days, another 74,000 gallons of anhydrous ammonia were released from six other anhydrous ammonia tank cars. Upon receiving the 1:37 a.m. call from the conductor, the Ward County 911 dispatcher

Chronicles of Incidents and Response 279

immediately paged the Minot Rural Fire Department. The fire department chief responded directly to the scene from his house (approximately 2 miles away); the assistant chief responded directly to the Minot Rural Fire Department fire hall.

Six fire department units responded from the fire hall to the scene (approximately 6 miles away). At 1:44 a.m., the Minot Rural Fire Department requested mutual aid from the Minot City and Burlington Fire Departments. At 1:47 a.m., the chief of the Minot Rural Fire Department arrived on-scene at the West 83 Bypass at the intersection of 4th Avenue NW (approximately 1/2 mile east and 1/2 mile north of the train derailment site). He immediately assumed incident command and performed an initial site and accident assessment. At approximately 1:50 a.m., the chief established a field incident command post along the West 83 Bypass near the intersection of 19th Avenue NW.

Meanwhile, the crew of CPR train 292-16 was traveling east, away from the derailment site, with the two locomotives. At 1:47 a.m., near the Arrowhead grade crossing (at 16th Street and approximately 2nd Avenue SW), the crew met a Minot City Fire Department battalion chief waiting at the crossing. The battalion chief was responding to a different call at the time, but when the train crew approached him on foot and told him about the derailment, he notified the Minot City Fire Department and went toward the derailment location.

The conductor and engineer remained at the crossing and prevented entry to the area by private vehicles. They were later relieved by law enforcement personnel, and they were transported by car to Minot City Fire Station Number 1. At the fire station, the crew provided all of the train paperwork, including information about the train's hazardous cargo. They also described the ammonia fog at the derailment site. The two crewmembers were then transported to Minot Trinity Hospital for observation and treatment.

According to interviews and 911 records, immediately after the accident, two residents of Tierracita Vallejo, the neighborhood closest to the derailment, went outside their homes, became disoriented, and were unable to get back to their homes for some time. When one of these residents did return to his home, he and his wife drove their car away from the neighborhood. Another couple attempting to flee their home in their truck crashed the vehicle into a house diagonally across the street. The occupants of the house were able to assist the female passenger into the house, but the male driver collapsed in their yard, and they were unable to move him. At approximately 2:06 a.m., one of the occupants of the house called Ward County 911 to report the man on the ground outside the house. The 911 operator told the resident that emergency responders were in the area, but in the meantime, residents must take precautions.

280 *Hazmatology: The Science of Hazardous Materials*

At 2:09 a.m., an initial staging area was set up at the West 83 Bypass near 21st Avenue NW. Because of the vapor plume, responding units were directed to travel around the city of Minot to reach the north side of the accident. At 2:13 a.m., the Minot Rural Fire Department requested that the Burlington Fire Department come to Behm's Truck Stop just west of the 83 Bypass along Highways 2 and 52 (southwest of the derailment location).

At 2:23 a.m., State Radio paged Des Lacs and Berthold Fire Departments to request mutual aid assistance.

At 2:37 a.m., the emergency operations center was opened at the Minot City Fire Station Number 1. At that time, Minot Rural Fire Department engine 214 was assigned as the mobile command unit, a Minot Rural Fire Department assistant chief was assigned as the on-scene incident commander, and the Minot Rural Fire Department chief maintained command at the emergency operations center.

At 2:39 a.m., two firefighters who were driving Minot Rural Fire Department tanker 212 drove through the vapor cloud when the wind shifted. The firefighters reported their eyes watering a minute later. At 2:40 a.m., Minot Rural Fire Department engines 214 and 216, staged at 21st Avenue NW, and Minot Rural Fire Department unit 218 reported that all the civilians that had been encountered on the local roads were assembled inside Behm's Truck Stop. By 2:42 a.m., the vapor cloud was reported to cover the Highway (Routes) 2/5/83 Bypass completely. At 2:43 a.m., the mobile command post was repositioned on a hill farther south of the derailment.

At 2:45 a.m., a decision was made to evacuate the people at Behm's Truck Stop. At 2:52 a.m., a city bus was sent to Behm's Truck Stop to take the people outside the affected area. As early as 1:41 a.m., 911 operators were telling residents to stay in their homes and close their windows. By the time the emergency operations center was opened at 2:37 a.m., emergency responders, because of the ammonia vapor cloud and the dangers it posed to the residents of both the Tierracita Vallejo neighborhood close to the derailment and to the city of Minot, had decided not to evacuate residents.

This response, called "sheltering in place," differs from an evacuation in that people who shelter-in-place take precautions but remain within the "hot zone." The emergency responders then issued additional guidelines and implemented the public notification procedures by contacting the local media and sounding the outdoor warning system. At approximately 3:40 a.m., Edison Elementary School was opened as an emergency shelter and triage area for residents of Minot.

At 4:29 a.m., the Minot Rural Fire Department relocated the staging area for rescue operations to Behm's Truck Stop where the levels of ammonia had diminished. This located the staging area near the affected neighborhood. At approximately 4:39 a.m., the resident who had called

Chronicles of Incidents and Response

911 about the man on the ground outside his house called a second time to report that the man was still outside and that the man's wife, who had been outside in the cloud, was in poor condition. The resident explained that there was no cloud around the house at the time.

At 4:47 a.m., Minot Rural Fire Department unit 219 went into the Tierracita Vallejo neighborhood to rescue the residents. This first unit into the area found the man, but in attempting to recover him, the firefighters exited their unit without first donning their self-contained breathing apparatus (SCBA). They were unable to recover the injured man and had to leave the scene and regroup at the staging area. At approximately 5:07 a.m., the residents of the house and the wife of the injured man had gone to Behm's Truck Stop, and emergency responders returned to pick up the injured man. At approximately 5:15 a.m., the assistant chief of the Minot Rural Fire Department, wearing an SCBA, found the man lying on the driveway. Ten minutes later, the Burlington Fire Department transported the man to Behm's Truck Stop, where he was assessed by responders from Community Ambulance and found to be unresponsive.

At about 5:30 a.m., firefighters entered the Tierracita Vallejo neighborhood and went door-to-door removing residents from their homes and putting them on Minot City buses. The residents were then transported to a triage area near Behm's Truck Stop. By this time, some of them were able to leave the area in their own vehicles. At 6:41 a.m., the firefighters continued their rescue efforts as the ammonia odor continued to permeate. By 8:21 a.m., after a second check of all the houses to ensure no one was left behind, the rescue operation in the neighborhood was complete. The chief estimated that between 60 and 65 residents of Tierracita Vallejo were rescued.

In the afternoon of January 18, 2002, the shelter and triage area at Edison Elementary School was closed, as was Minot Rural Fire Department's field command post. At 10:00 a.m. on January 20, 2002, the Minot Rural Fire Department chief relocated the emergency operations center to the Minot Municipal Auditorium. The fire department remained on scene until 2:00 a.m. on January 22, 2002, assisting the environmental cleanup being performed by Earthmovers, Inc. The emergency operations center remained open on a limited basis until March 19, 2002. Some residents of Tierracita Vallejo were not able to return to their homes until the second week of March 2002.

After the accident, the Minot Police Department made emergency notifications to the public that included cable television interrupts, radio broadcasts, and outdoor warning sirens. However, many residents did not hear the emergency broadcasts because their homes had lost power as a result of the derailment. Additionally, residents of the houses in the neighborhood closest to the derailment did not hear the outdoor warning sirens because the sirens are positioned to be heard within the city limits of Minot.

The Minot Police Department attempted to contact the designated local emergency broadcast radio and television stations. At the time of the accident, only one person was working at the designated local emergency broadcast radio station (KCJB-AM), and the police department's calls to the station went unanswered. The designated local emergency broadcast television station (KMOT) did not have an overnight crew at the station. To arrange emergency broadcasts, the police department had to contact the KMOT news director at his home. Of the 122 firefighters who responded to the accident, 7 sustained minor injuries.

The injuries to six Minot Rural Fire Department firefighters and one Burlington Fire Department deputy chief were headaches, sore throats, eye irritation, and/or chest pain. An additional 11 Minot Police Department officers sustained minor injuries while blocking and directing traffic around the perimeter of the accident scene. Their injuries were eye irritation, chest discomfort, respiratory distress, and/or headaches. One Ward County Sheriff's Department lieutenant sustained minor injuries as a result of the accident. The lieutenant had stationed his vehicle south of 4th Avenue on the West Bypass to prevent traffic from entering the area. Soon afterwards, a chemical cloud engulfed his vehicle, and he became disoriented. While attempting to exit the area, he drove his car into a ditch and remained inside his vehicle for approximately 45 min until rescuers arrived. He was then taken to Trinity Hospital and released after being treated for toxic effects of anhydrous ammonia.

The driver of the truck that crashed into a house in the Tierracita Vallejo neighborhood while attempting to flee the area, a 38-year-old male, sustained fatal injuries. The Ward County coroner determined that the cause of death was prolonged exposure to anhydrous ammonia. Three residents of the Tierracita Vallejo neighborhood sustained serious injuries as a result of the accident and were admitted to Trinity Hospital. Their injuries included chemical burns to the face and the feet, respiratory failure, and erythema of the eyes and the nose.

Eight other residents of Minot sustained serious injuries as a result of the movement of the ammonia cloud over parts of the city of Minot. The injuries, which included shortness of breath, difficulty breathing, and/or burning of the eyes, were determined to be have been complicated by preexisting health problems such as asthma and heart conditions. A total of 301 other persons sustained minor injuries as a result of the accident. Of these, 11 were admitted to Trinity Hospital for less than 48 h. The remaining 290 individuals were treated and released at either Trinity Hospital, the triage center established at Edison Elementary School in Minot, the Minot Air Base Health Clinic, Kenmare Community Hospital, and/or St. Alexius Medical Center.

National Transportation Safety Board Report and Minot Fire Department.

Macdona, TX June 28, 2004 Train Derailment Chlorine Leak

About 5:03 a.m., central daylight time, on Monday, June 28, 2004, a westbound Union Pacific Railroad (UP) freight train traveling on the same main line track as an eastbound BNSF Railway Company (BNSF) freight train struck the midpoint of the 123-car BNSF train as the eastbound train was leaving the main line to enter a parallel siding. The accident occurred at the west end of the rail siding at Macdona, Texas, on the UP's San Antonio Service Unit. The collision derailed the 4 locomotive units and the first 19 cars of the UP train, as well as 17 cars of the BNSF train. As a result of the derailment and pileup of railcars, the 16th car of the UP train, a pressure tank car loaded with liquefied chlorine, was punctured.

Chlorine escaping from the punctured car immediately vaporized into a cloud of chlorine gas that engulfed the accident area to a radius of at least 700 ft before drifting away from the site (Figure 1.121). Three persons, including the conductor of the UP train and two local residents, died as a result of chlorine gas inhalation. The UP train engineer, 23 civilians, and 6 emergency responders were treated for respiratory distress or other injuries related to the collision and derailment. Damages to rolling stock, track, and signal equipment were estimated at $5.7 million, with environmental cleanup costs estimated at $150,000.

The National Transportation Safety Board determined that the probable cause of the June 28, 2004 collision at Macdona, Texas was Union Pacific Railroad train crew fatigue that resulted in the failure of the engineer and conductor to appropriately respond to wayside signals governing the movement of their train. Contributing to the

Figure 1.121 Chlorine escaping from the punctured car immediately vaporized into a cloud of chlorine gas that engulfed the accident area. (From NTSB.)

284   *Hazmatology: The Science of Hazardous Materials*

crewmembers' fatigue was their failure to obtain sufficient restorative rest prior to reporting for duty because of their ineffective use of off-duty time and Union Pacific Railroad train crew scheduling practices, which inverted the crewmembers work/rest periods. Contributing to the accident was the lack of a positive train control system in the accident location. Contributing to the severity of the accident was the puncture of a tank car and the subsequent release of poisonous liquefied chlorine gas.

The initial notification to local emergency response authorities came via a 911 call placed at 5:06 a.m. from a residence on Nelson Road to the Bexar County 911 Emergency Call Center. The caller reported difficulty breathing and the presence of white smoke outside the residence. The caller also, in what could be described as a weak voice, referred to a train derailment. The 911 operator heard the word smoke and understood that the caller was experiencing breathing difficulty but apparently did not recognize the words train derailment, and the caller was transferred to a fire department dispatcher. The caller again reported train derailment and smoke, but the fire dispatcher also did not recognize that the incident involved a train derailment.

The response was thus processed as a difficulty breathing and smoke in the residence response action. The principal agency responsible for responding to emergencies, including hazardous materials incidents, within the Macdona District of Bexar County, where the accident occurred, was the Southwest Volunteer Fire Department. Fire department emergency responders were dispatched to the Nelson Road residence at 5:08 a.m. followed shortly thereafter by Bexar County Sheriff's Office patrol units that had been dispatched for support. None of the responders were yet aware that they were responding to a train accident, and that their path to the Nelson Road residence was blocked by derailed equipment. When fire department responders approached the accident site in darkness about 5:15 a.m., they began to have difficulty breathing as they became exposed to the vapor cloud of chlorine from the punctured railcar. They immediately withdrew from the scene and requested mutual aid from other agencies. Some of the firefighters obtained protective clothing and self-contained breathing apparatus, and commencing about 5:40 a.m., then reentered the scene to conduct a search for survivors. The initially requested mutual aid resources began to respond to the scene shortly thereafter, which included the Bexar County Office of Emergency Management. About 6:10 a.m., the Bexar County Office of Emergency Management established the unified (incident) command system, activated the Bexar County Emergency Operations Center, and initiated the Bexar County emergency management plan.

Additional mutual aid resources were also being dispatched to the scene, including the San Antonio Fire Department. About 6:15 a.m., Southwest Volunteer Fire Department officers, who were advancing west along Nelson Road, came upon an individual who would later be identified as the UP train engineer stumbling along the roadway about 240 ft east of the grade crossing. He was in respiratory distress, and was transported from the scene for medical attention. A short time later, responders determined that the derailment wreckage at the grade crossing prevented access to residences at the west end of Nelson Road, one of which was their dispatch destination. The obstructed grade crossing also prevented the immediate rescue of three individuals who were reported to be trapped in their residence by the vapor cloud several hundred feet to the south of the emergency dispatch destination.

Responders told investigators that early in the response effort they considered, then rejected, a plan to use a helicopter to access Nelson Road south of the grade crossing and evacuate the two occupied residences there. They decided that such a plan was ill-advised until the gas plume had reduced or stabilized because of the vulnerability of the helicopter equipment and crew and the possibility that rotor wash could spread the gas. Commencing about 6:33 a.m., hazardous materials response contractors retained by the UP began to arrive and conduct a technical assessment of the chlorine release.

Further access to the accident site through the wreckage pileup at the grade crossing area was restricted until the assessment could be made and appropriate personal protective equipment (principally level-A hazardous materials suits) could be donned by firefighters and hazardous materials personnel. Sunrise occurred at 6:37 a.m., and the wind was moderate but steady toward the northwest.

An evacuation zone was established around the accident site with a radius of about 2 miles. Other than assisting with the evacuation of residents within the evacuation zone and assisting hazardous materials release mitigation responders with the technical assessment of the chlorine gas release, no further direct rescue attempt activity of the responding firefighters and mutual aid responders for the three individuals who were reported to be trapped in their residence by the vapor cloud was documented for about the next 3 h. During this time, pursuant to 911 call center instructions, the three residents were attempting, without success, to flee and find a safe shelter from the chlorine fumes that had engulfed their residence.

Also during this time, according to post accident interviews, the principals of the primary responding emergency services agencies (the San Antonio Fire Department, the Bexar County Office of Emergency Management, and the Southwest Volunteer Fire Department) were involved in what was described as a certain amount of discordant debate regarding jurisdictional boundaries and incident command authority.

About 9:45 a.m., with a preliminary technical assessment of the chlorine gas release having been completed, the first of three firefighter entry teams entered the accident area to attempt a rescue of three trapped persons on Nelson Road who were unable to escape the chlorine vapor cloud that had enveloped their residence. This first entry team, however, became disoriented while attempting to advance through the wreckage pileup, and inadvertently diverted down the wrong roadway (actually a long driveway leading to another residence) and away from their objective. Along that roadway, the team encountered the body of a person who was later identified as the UP train conductor. Shortly thereafter, one of the entry team firefighters showed signs of dehydration, prompting the dispatch of a second entry team to come to the aid of the first. About 10:12 a.m., a third entry team, consisting of two firefighters and one UP employee, was dispatched to carry on with the rescue mission that had been aborted by the first entry team. This team successfully advanced through the wreckage pileup and, about 10:55 a.m., reached the three persons who had been trapped at their residence by the gas cloud. All three were found to be in considerable respiratory distress.

About 11:46 a.m., after the responders had revived and stabilized them, the three individuals were transported by helicopter to a local hospital for medical attention. About 11:55 a.m., the entry team entered another Nelson Road residence and found two persons who had sustained fatal injuries.

National Transportation Safety Board Report.

Graniteville, SC March 24, 2005 Chlorine Disaster

January 6, 2005, a train derailed including three cars of chlorine, which killed 9 and injured 554 (Figure 1.122). About 2:39 a.m. eastern standard time on January 6, 2005, northbound Norfolk Southern Railway Company (NS) freight train 192, while traveling about 47 mph through Graniteville, South Carolina, encountered an improperly lined switch that diverted the train from the main line onto an industry track, where it struck an unoccupied, parked train (NS train P22). The collision derailed both locomotives and 16 of the 42 freight cars of train 192, as well as the locomotive and 1 of the 2 cars of train P22. Among the derailed cars from train 192 were three tank cars containing chlorine, one of which was breached, releasing chlorine gas. The train engineer and eight other persons died as a result of chlorine gas inhalation.

About 554 people complaining of respiratory difficulties were taken to local hospitals. Of these, 75 were admitted for treatment. Because of the chlorine release, about 5,400 people within a 1-mile radius of the derailment site were evacuated for several days. Total damages exceeded $6.9 million. When train 192 struck train P22, both locomotives and the first 16 cars of train 192 Derailed. The ninth car from the locomotive units containing 90 tons of chlorine was punctured during the derailment and

Figure 1.122 January 6, 2005 a train derailed including three cars of chlorine, which killed 9 and injured 554. (From NTSB.)

released chlorine gas. The sudden release and expansion of the escaping gas caused the product remaining in the tank to auto-refrigerate and remain in the liquid state, slowing the release of additional gas.

Immediately after the collision, at 2:39:43 a.m. on January 6, a female employee on duty at one of the Avondale Mills facilities (about 200 ft from the collision site) placed a 911 call to the Aiken County Sheriff's Office 911 Emergency Call Center. The caller identified herself and reported, "I think there's been a train wreck at Graniteville at Hickman Mills." She said that she was alone and that when she went outside to investigate, she could see smoke but could not discern exactly what had happened. She further indicated that the accident was at the Hickman Street railroad grade crossing. Near the end of the 48 second call, the caller appeared to become increasingly agitated, saying, "I smell smoke." The caller then exclaimed, "I got to get out of here" which point the call abruptly ended.

Over the next 10 min, about a dozen additional calls were made to 911, with callers reporting that there had been a train wreck. Some callers reported a low-lying yellow haze that smelled like bleach. Within about an hour after the accident, more than 100 additional 911 calls were received. By 6:00 a.m., more than 200 calls had been received. At 2:40:40 a.m., resources of the Graniteville, Vaucluse, and Warrenville Volunteer Fire Department were dispatched to the scene, with the first responding unit reported to be en route less than a minute later. At 2:42 a.m., upon hearing a report from the dispatcher that a smell of chemicals was reported in the area, the initially responding fire department senior officer (the fire

288   *Hazmatology: The Science of Hazardous Materials*

chief) advised further responding fire department personnel to stand by at their locations away from the scene until the situation could be further assessed.

At 2:45 a.m., emergency dispatch advised the fire department that it had confirmed the possibility that two trains had collided head on. Additional confirmation came from the NS about 3 min later. The fire chief, upon approaching the scene, smelled an intense chemical odor and experienced difficulty breathing.

At 2:46 a.m., a hazardous materials team was requested. At 2:48 a.m., the fire chief advised dispatch that he could not breathe and was withdrawing from the area. At 2:49 a.m., the fire department asked dispatch to initiate the Aiken County Reverse 911 Emergency Notification System, with instructions for residents to shelter indoors. Commencing about 2:50 a.m., additional resources:

- Ambulances, hazardous materials personnel, and equipment,
- Aiken city and other mutual aid services were asked to respond. At 2:52 a.m., dispatch advised responders that three persons were trapped inside the Hickman plant,
- At 2:54 a.m., fire department personnel asked dispatch to advise Aiken Hospital that persons overcome by chemical fumes were en route from the scene,
- At 2:57 a.m., the fire department asked that approach roads be blocked (which effectively initiated a 1-mile-radius buffer around the accident site) and reiterated the earlier reverse 911 request to shelter indoors,
- At 3:05 a.m., while awaiting delivery of train consist information that had been faxed from NS, the fire chief directed that an incident command center be established; the fire chief would become incident commander for the search and rescue effort. About this time, firefighters also asked dispatch to obtain wind direction information (from Bush Field in Augusta), which was received about 3 min later,
- At 3:06 a.m., firefighters were informed that the sixth through ninth cars on the train contained chlorine and sodium hydroxide and that additional information would be forthcoming,
- At 3:08 a.m., the fire department began staging equipment and personnel at the incident command center, which had been set up at a nearby car dealership. At 3:10 a.m., firefighters asked that all Aiken County emergency medical services ambulances be placed on standby,
- At 3:13 a.m., an Avondale Mills employee told firefighters that workers on duty at the Stevens steam plant could not be contacted. The employee expressed concern about a possible explosion if the

Chronicles of Incidents and Response 289

workers had departed the plant without properly shutting down the boilers,

- At 3:21 a.m., personnel from the Aiken County hazardous materials team arrived at the incident command center. At 3:24 a.m., a copy of the faxed train 192 consist list was delivered, and fire department authorities advised all responding personnel to report with their equipment to the command center. At 3:30 a.m., the fire department requested all available self-contained breathing apparatus,
- At 3:33 a.m., a report was received of a steady stream of individuals departing the Ascauga Lake Road area. At 3:35 a.m., authorities established the first of four decontamination stations to treat individuals exposed to the chlorine gas,
- At 3:37 a.m., dispatch advised firefighters that it had received reports of people down inside the Avondale Mill's facility. A Firefighter and mill supervisor entered the steam plant to prevent a possible boiler explosion. A decision was made to shut down and evacuate the plant,
- At 3:38 a.m., a second decontamination station was set up,
- At 3:39 a.m., the incident commander, concerned that the incident command center was too close to the accident site, directed that it be moved from the car dealership to a location about a mile away. Some emergency response apparatus remained at the initial site as a forward command location,
- At 3:40 a.m., a firefighter wearing personal protective equipment got close enough to the accident site to note the number on the tank car that had been breached. He also encountered an individual suffering from gas inhalation and discovered another individual trapped in an automobile beneath a fallen tree near the derailment site. (Both these individuals were later successfully rescued.),
- 3:50 a.m., several entry teams, consisting of firefighters wearing personal protective equipment and riding in privately owned pickup trucks, were organized and dispatched to the accident site for search and rescue operations. Upon locating individuals or groups affected by the chlorine gas, the teams transported them to one of the decontamination sites. The entry teams then returned to the site to repeat the search and rescue cycle for several hours,
- At 3:53 a.m., a third decontamination station was established at a local high school,
- At 4:10 a.m., an entry team reported a downed electrical power line near the wreckage site, and a request was made to the utility company to respond to the scene to disconnect the power feed. Also at 4:10 a.m., an entry team reported no visible fire in the derailment wreckage. This report was revised at 4:24 a.m., when responders reported seeing bright orange smoke emanating from one railcar and green smoke from another,

290 *Hazmatology: The Science of Hazardous Materials*

- At 4:55 a.m., an entry team entered the steam plant to shut down the facility. Moments later, the mission was revised when five or six individuals were reported to be trapped in a room at the plant. One person was found and rescued, after which the team completed the shutdown of the steam plant,
- At 5:07 a.m. the entry team performed a final sweep of the plant. Several additional entry teams began missions to assess the condition of the railroad equipment. The immediate area around the accident site remained relatively stable until 1:00 p.m., when a fire was reported at the steam plant. A fire department entry team entered the plant and found that a fire had ignited in coal chutes feeding several of the boilers. A pumper truck supplied water to an unmanned waterline-fed monitor nozzle that discharged a spray on all the coal feeders. At this point, the fire was under control but not extinguished. The discharging monitor nozzles were left in place while workers evacuated the area. The scene remained relatively stable for the balance of the day. For the next several days, fire department entry teams monitored and contained the fire in the steam plant, while a clean-up of the railroad wreckage continued,
- At 8:00 a.m. on January 14, with the report that the hazardous materials had been removed from the un breached railcars, support operations were concluded and hazardous materials response personnel were released. At about noon, fire department personnel returned to the steam plant to extinguish the remaining fire in the coal feeders. The fire was reported extinguished at about 4:00 p.m., which concluded the fire departments operations for the incident,
- About 11:00 p.m. on January 6, responders inserted a temporary polymer patch in the opening of the punctured ninth tank car,
- At 7:00 p.m. on January 8, responders began unloading sodium hydroxide from the eighth tank car. At 8:50 p.m., the temporary polymer patch in the ninth tank car failed, releasing chlorine vapors and causing the unloading of the eighth tank car to be temporarily discontinued,
- By 8:37 a.m. on January 9, responders had inserted a second polymer patch in the opening in the punctured ninth tank car. Chlorine vapor was then drawn from the car to create a vacuum that would reduce the amount of chlorine gas escaping. The chlorine vapor removed from the tank was transferred into a sodium hydroxide solution to neutralize it. Following these measures, the unloading of the eighth tank car was resumed and completed by 3:30 p.m. While the eighth car was being unloaded, construction began on a permanent lead patch for the punctured ninth car,
- At 9:30 a.m. on January 10, the punctured chlorine tank car was rotated so the puncture was at the highest elevation on the tank car.

This rotation disturbed the liquid chlorine in the tank and caused a delay in efforts to unload other tank cars,

- At 12:10 p.m. on January 10, responders began unloading the chlorine from the derailed sixth car in the train. On the morning of January 11, responders rejected the plan for a lead patch on the punctured ninth tank car and decided to use a steel patch instead. Unloading of the sixth car was completed by 2:00 p.m. on January 11,
- About 1:10 a.m. on January 12, responders began unloading chlorine from the derailed seventh car,
- By 9:30 a.m., the steel patch was in place on the punctured tank car, and the unloading was started. Because the punctured tank car had extensive damage, the remaining chlorine could not be removed as it had been from the un breached cars. The chlorine in this car had to be vaporized and transferred from the tank as a gas, after which it was bubbled through a sodium hydroxide solution held in a separate tank,

  This process, which converted the chlorine into a relatively safe and easily transportable liquid bleach and salt solution, required several days to complete morning hours of January 13, the two unbreached chlorine tank cars had been unloaded, placed on railroad flat cars, and moved from the site. By midnight on January 18, the unloading of the punctured tank car was complete,
- On January 19, by 9:00 a.m., the tank car was cleaned and purged on site. It was then loaded on a flat car and moved to the Augusta Yard the following morning, January 20.

National Transportation Safety Board Report

A Tale of Two Hurricanes

New Orleans, LA August 29, 2005 Hurricane Katrina

News media personnel and others have said you couldn't appreciate the devastation in New Orleans from the news video footage and photographs presented in the media. What an understatement! I had the opportunity to visit New Orleans on December 23, 2005 and talk with hazmat team members. Under ordinary circumstances, a visit to the New Orleans Hazmat Team would begin at Station 7 located at 1441 Saint Peter Street. There HM 1 and HM 2 are headquartered along with Rescue 1, Engine 7, a mass decontamination trailer and other hazmat support equipment. However, Station 7 became one of the casualties of Hurricane Katrina with 4 ft of water inside that remained there for 10 days.

Currently, Station 7 is uninhabitable due to mold, bacteria, and structural damage. The building housing part of Station 7 was built around 1900 as a bar with living quarters above. Upstairs rooms were later changed to accommodate a brothel. The building fell into disrepair, and a fire occurred in the attic at some point. Charred wood framing members still remain today. Eventually, the city purchased the building and converted it into a firehouse, adding a three bay apparatus floor, an office area, and a day room. It is currently unknown if the station will be restored or demolished and a new station built.

My day began by following directions from Chief Woodridge to the site where HM 1 was located. I proceeded to an area along the Mississippi River behind the Hilton Hotel on the river side of the flood gates next to the Cruise Ship Sensation. There I found HM 1 parked next to a security fence in front of several bright yellow tents which turned out to be the headquarters for the hazmat team (Figure 1.123). New Orleans HM 1 is a 2000 custom built American LaFrance. They also respond with a Chevy Suburban purchased with federal grant funds for assessment and control. Both units have satellite communications capability that had ironically been installed just 1 week before Katrina hit. New Orleans Hazmat normally responds for Hazmat and weapons of mass destruction (WMD) calls to the parishes of Orleans, St. Bernard, Jefferson, and Plaquimines.

Captain Don Birou was the officer in charge when I arrived. We sat down at a table in his tent quarters, and I began to hear the most amazing stories of victim rescue, firefighter survival, and hazmat response following Hurricane Katrina. Having spent over 26 years in the emergency services, I never dreamed I would hear about fire department or hazmat personnel having to operate under such desperate conditions within the United States. Some of the conditions I was about to hear sounded like something right out of Hollywood. Firefighters in New Orleans who

Figure 1.123 I found HM 1 parked next to a security fence in front of several bright yellow tents which turned out to be the headquarters for the hazmat team.

Chronicles of Incidents and Response 293

remained there to deal with the aftermath of Hurricane Katrina are truly heroes in the truest sense of the words. Following the hurricane, they survived, operated, and rescued thousands of victims under impossible conditions. This would be a day burned into my mind for as long as I live.

Awaiting the arrival of Hurricane Katrina, New Orleans hazmat personnel were staged in the city's convention center to ride out the storm. Once the storm subsided, personnel emerged to survey the aftermath of the Category 3 storm. Surprisingly, there wasn't the catastrophic damage many had expected. It looked as though New Orleans had dodged the bullet once again. Power was out but back-up systems designed for 24–48 h of operation in the emergency communications center and other facilities in most cases were in full operation. Reports began to filter in on Monday, August 29th that water was flowing over some levees and some were beginning to fail. The city rapidly began to fill up with water in low lying areas as Lake Pontchartrain swollen by the storm surge of the hurricane began to flow into New Orleans. The real disaster was just beginning. Ultimately, almost 80% of the city would be under water including the 911 and communications centers and many of the city's fire and police stations. Nearly, 70% of New Orleans fire stations were under water or damaged. Some had been looted. Following the storm nearly 75% of firefighter families are living outside of New Orleans and most have lost their homes and belongings.

Prior to Hurricane Katrina the New Orleans Fire Department had no boats. Their only fire boat had been decommissioned a few months earlier and would have been useless any way because of its size and location (on the Mississippi River). Firefighters were asked and many brought their own boats to work with them the day of the hurricane. Thousands of people were rescued by fire fighters using their own personal boats. The city was isolated by the floods. There was no way for mutual aid and other resources to reach the city by road, everything was flooded. On August 31, 2005 backup batteries and generators for the department's communications network went dead. There was no way to replenish the fuel or batteries needed to keep the communications network on line. Firefighters who were on duty the day the levees broke were stranded in the city and on their own once radio and telephone communications systems went down. They were without food or water and in many cases without shelter. All cellular and regular phone service was disabled by the hurricane.

Firefighters were forced to use written messages delivered by runner to exchange information. Satellite phones were the only means of communications and only two of them were available. Captain Don Birou was able to get through to his sister at Phillips/Conoco Emergency Operation Center in Houston. He expressed the dire communications situation in New Orleans and asked if there was any help that could be provided. Captain Birou's sister made some contacts and within 2 days there were 10

294   *Hazmatology: The Science of Hazardous Materials*

hand held satellite phones at FedEx in Baton Rouge. According to Captain Birou, once received these "10 handheld phones became the most important items on the fire department." After about 5 days the direct connection on their Nextel phones began working. Nine hazmat specialists on duty the day of the hurricane ultimately worked 24/7 until December 18th except for a mandatory rehab September 612! Personnel from the New York City Fire Department were assigned to New Orleans Hazmat during the rehab period. As New Orleans slipped into anarchy and police protection was limited or non-existent the Superintendent of the Fire Department told firefighters who had personal weapons to carry them and they became their own police force for protecting themselves and other personnel.

Each new day began with hazmat team members foraging for boats, fuel, flashlights, tire repair kits, food and water to accomplish the rescue mission before them and for survival themselves. The first 2½ weeks their time was spent on rescue operations. One of the rescue operations at the Louisiana State University Dental School turned out to be 100 plus firefighters, police officers and New Orleans Health Department Paramedics who were using the structure as a safe haven from hurricane Katrina. Flood waters from the broken levees surrounded the Dental School trapping the personnel for 4 days. When not conducting rescue operations the hazmat team members spent time evaluating a list of 93 Tier II Facilities with Extremely Hazardous Substances (EHS). They also searched through 430 facilities without EHS substances looking for railcars and other hazmat containers with leaks or damage. They also searched for "orphaned" containers carried away from facilities by the flood waters. On a New Orleans City map a grid system was drawn to identify areas of the city that needed to be searched. As of December 7, 2005 over 10,817 containers of hazardous materials had been located and evaluated.

All containers found were checked and marked. Five major Level A entries were conducted for atmospheric testing, evaluation and discovery of hazards in the following locations:

- New Orleans Medical Center of Louisiana (Charity Hospital)
- University Hospital
- LSU Dental School
- Dennis Sheen Transfer Company (Following fire and explosions)
- 2321 Timoleon Street (unknown release at debris removal site)

The Louisiana National Guard, 62nd Civil Support Team, and the EPA on-scene coordinator and START teams eventually reached the city as waters started to recede and were an excellent resource during hazardous materials operations. They provided personnel and hazmat equipment and helped the firefighters survive the first couple of weeks (Figure 1.124).

Figure 1.124 The Louisiana National Guard 62nd Civil Support Team and the EPA on-scene coordinator and START teams eventually reached the city as waters started to recede and were an excellent resource during hazardous materials operations. (Courtesy: New Orleans Fire Department.)

Civil Support Teams and a small EPA group were assigned under the command of New Orleans Hazmat for operations within the city. Hazmat personnel from Illinois Departments of Carpentersville, Chicago, Chicago Heights, and Decatur assisted New Orleans Hazmat in addition to Gonzales, Louisiana. Approximately 2,000 firefighters from around the country supplemented the entire New Orleans Fire Department in the weeks following Katrina.

There were over 3,000 railcars in Orleans Parish during Hurricane Katrina. Of those, 640 contained hazardous materials and 105 of those were derailed by the flood waters (Figure 1.124A). Two releases occurred from railcars, one at the CSX rail yard releasing ethylene oxide and another at Air Products Company releasing hydrogen chloride. The CG Railway Company is a Mexican operation where railcars are shipped from Mexico by specially designed barges to offload in New Orleans. Four of the tracks in the CG Rail Yard were under water with five derailed cars. Consists provided by CG Railway Company showed there were 14 railcars containing hazardous materials at their New Orleans facility. Materials in the railcars included methylamine, anhydrous ammonia, potassium hydroxide solution, and chlorine. Strong winds and storm surge also caused several cars to derail in the CSX rail yard.

One of the derailed cars contained fuming sulfuric acid. Other nearby cars contained methyl acrylate monomer, heptanes, and combustible liquids. Several barges were also washed over the levees by the storm surge. One such barge was about 2/3 full of benzene. The barge traveled

Figure 1.124A There were over 3,000 railcars in Orleans Parish during Hurricane Katrina. Of those, 640 contained hazardous materials and 105 of those were derailed by the flood waters. (Courtesy: New Orleans Fire Department.)

approximately 100 ft from the levee across an industrial yard, a four-lane highway, through power poles and through a swamp. No damage occurred to the barge and there were no leaks. The barge still rests where it landed today awaiting removal of the benzene by the owners.

Hazmat team members responded to one explosion with fire at 3200 Chartres Street with phosphorus, arsenic, calcium, and tin metal involved. They also responded to the Mandeville Street Wharf fire that destroyed a large portion of the massive warehouse. It is estimated that over 1,000 natural gas leaks occurred, a total that was likely much higher. Other missions included Hazmat Response for Presidential visits on October 10 and 11. Ammonia releases occurred from two separate companies. Two leaks occurred at New Orleans Cold Storage and one at Browns Velvet Dairy.

As time went on and rescues were concluded and the waters receded hazmat team members focused on identifying hazardous materials in buildings and homes. As more people returned to the city call volume increased for hazardous materials discovered in debris piles. Refrigerators and other appliances with Freon had to be identified so that contractors could remove the Freon before disposal. Fumigation contractor procedures were assessed by hazmat personnel and permits issued for fumigation operations to insure they were conducted safely.

Chronicles of Incidents and Response

EMS personnel from the New Orleans Department of Health, about 60% of Fire personnel including the hazmat team and other City Hall employees were living on the cruise ship Sensation, including the Superintendent and Assistant Superintendent of the fire department. Police personnel were living aboard the other cruise ship at the same dock. Most of the hazmat team members spent Christmas 2005 away from their families. Life aboard the cruise ships includes room, meals, and a nightly movie. There are no phones or television. It is reported that the cruise ships were provided by FEMA until March 2006. No timetable has been determined for the renovation or rebuilding of damaged fire stations, including Hazmat Station Personnel are just taking things 1 day at a time and hoping that it won't be too long before things begin to return to normal in New Orleans.

Houston, TX August 24, 2017 Hurricane Harvey

On August 24, 2017, Hurricane Harvey, a Category 4 hurricane, made landfall in southeast Texas. Over the next several days, the storm produced unprecedented amounts of rainfall over southeast Texas and southwest Louisiana, causing significant flooding. Hurricane Harvey turned out to be the most significant tropical cyclone rainfall event and the second most costly hurricane in U.S. history, after Hurricane Katrina. Over the course of the storm, Hurricane Harvey killed 68 people and flooded over 300,000 structures, forcing roughly 40,000 people to evacuate their homes. Hurricane Harvey required a large emergency response effort supported by local, state, and Federal officials.

Much was learned from what New Orleans experienced during Hurricane Katrina. During Hurricane Harvey assets including FEMA Urban Search and Rescue Teams (USAR) were pre-positioned so that they can respond to the aftermath of the storm. Hurricane Harvey was the only time that all 28 USAR Teams were activated for the same incident and on scene at the same time. FEMA also pre-positioned supplies for victims and personnel to assist victims following storm. Some of the USAR teams brought rescue boats and specialize in flood rescues. Lincoln Nebraska Task Force One is one such team.

Author's Note: My friend Captain Mark Majors from Lincoln Fire and Rescue went to Houston along with Nebraska Task Force 1 (USAR) during Harvey (Figure 1.124B). They were deployed to an area of SW Harris County near George W. Bush Park. Their mission was the rescue of residents from the area and moving them to higher ground (Figure 1.124C). During the fall of 2018 I went to Houston and Mark asked me to return to the area to take some after the flood photos. It is primarily residential with supporting commercial facilities.

Figure 1.124B Lincoln Nebraska Task Force One is one of the water rescue teams. My friend Captain Mark Majors from Lincoln Fire and Rescue went to Houston along with Nebraska Task Force 1 (USAR) during Harvey. (From FEMA.)

Figure 1.124C Nebraska Task Force 1 was deployed to an area of SW Harris County near George W. Bush Park. Their mission was the rescue of residents from the area and moving them to higher ground. (From FEMA.)

George Bush Park is a county park in Houston, Texas, United States, located on the far west side of the city. Situated entirely within Barker Reservoir, a large flood control structure, the park covers 7,800 acres (32 km^2), most of which is undeveloped forest used for the storage of floodwater (Figure 1.124D). A variety of public recreation facilities are located along Westheimer Parkway, which bisects the park, including soccer and

Figure 1.124D George W. Bush Park is Situated entirely within Barker Reservoir, a large flood control structure, the park covers 7,800 acres (32 km^2), most of which is undeveloped forest used for the storage of floodwater.

baseball field complexes, a shooting range, and a dog park. Named in honor of former Houston-area U.S. Representative and President George H.W. Bush, the park was known as Cullen–Barker Park until 1997.

The large park, located on the far west side of Houston, serves as an attraction and nature reserve for the Buffalo Bayou, a major water source in the park. Since the park is entirely located within the normally dry Barker Reservoir (Figure 1.124E), it is subject to routine flooding and closure during times of high rain fall. However, Hurricane Harvey was more than the reservoir was designed to accommodate. Most of the attractions are located on Westheimer Parkway (not to be mistaken with Westheimer Road/FM 1093), a major thoroughfare in the park. The park hosts a large group of soccer fields, baseball/softball fields, a shooting range, model aircraft flying fields, and numerous pavilions, playgrounds, ponds, and jogging trails. The park is also a popular geocaching destination.

Hurricane Katrina affected almost all of New Orleans, which is a small area compared to Houston. Hurricane Harvey largely affected residential areas. There were some hazmat issues, but residential rescue was the most significant mission of rescue personnel. One major hazmat emergency occurred from the flooding. Arkema Inc. a chemical plant that specializes in organic peroxides was flooded and a part of the mandatory evacuation areas.

Figure 1.124E The large park, located on the far west side of Houston, serves as an attraction and nature reserve for the Buffalo Bayou, a major water source in the park. Since the park is entirely located within the normally dry Barker Reservoir.

Crosby, TX August 29, 2017 Chemical Plant Explosion & Fire During Hurricane Harvey

On August 29, 2017, flooding from Hurricane Harvey disabled the refrigeration system at the Arkema plant in Crosby, TX, which manufactures organic peroxides. The following day people within a 1.5 mile radius were evacuated. August 31 as temperatures increased inside the trailers the peroxides spontaneously combusted. Officials ignited the remaining trailers, on Sunday, September 3, 2017. The evacuation zone was lifted on September 4, 2017. The Arkema Crosby facility is located within the 100-year and 500-year flood plain. Extensive flooding caused by heavy rainfall from Hurricane Harvey exceeded the equipment design elevations and caused the plant to lose power, backup power, and critical organic peroxide refrigeration systems.

Consequently, Arkema used its standby refrigerated trailers to keep the organic peroxide products cool. This flooding also eventually forced all Arkema's employees to evacuate from the facility. A Unified Command established a 1.5-mile evacuation zone around the Arkema facility and helped transport residents out of this zone for their safety. On August 31, 2017, organic peroxide products stored inside a refrigerated trailer decomposed, causing the peroxides and the trailer to burn (Figure 1.125). Twenty-one people sought medical attention from exposure to fumes generated by the decomposing products when the vapor travelled across a public highway adjacent to the plant. Emergency response officials initially decided to keep this highway open, despite the fact that it ran through the established evacuation zone around the Arkema facility, because this road served as an important route for hurricane recovery efforts.

Over the next several days, a second fire and a controlled burn conducted by the Unified Command consumed eight more trailers holding Arkema's remaining organic peroxide products that required low-temperature storage. Over the course of the three fires, in excess of 350,000 lb of organic peroxide combusted. As a result, more than 200 residents living within 1.5 miles of the facility who had evacuated the area could not return home for a week provides a timeline for the incident events and activities that unfolded at the Arkema Crosby facility as Hurricane Harvey moved through southeast Texas and flooded the site.

While it was becoming apparent that a reactive chemical incident would occur at the Arkema Crosby facility and while emergency

Figure 1.125 On August 31, 2017, organic peroxide products stored inside a refrigerated trailer decomposed, causing the peroxides and the trailer to burn. (From Chemical Safety Board CSB.)

302   *Hazmatology: The Science of Hazardous Materials*

responders were dealing with that eventuality the massive emergency response to Hurricane Harvey remained underway. As the hurricane and subsequent flooding advanced from the Houston area eastward into the Beaumont region, emergency responders needed to move as well. Rainfall from the storm flooded portions of Interstate 10, leaving eastbound Highway 90, which cut through the middle of the evacuation zone for the Arkema Crosby incident, as the best route for transporting hurricane relief and rescue resources.

Due to this constraint and the importance of transporting personnel and equipment to where they were needed, Harris County officials kept eastbound Highway 90 open to traffic even while enforcing the remainder of the evacuation zone. Emergency responders were also staged to block the road if the contents of one of the refrigerated trailers began to combust. On Wednesday, August 30, 2017, just before midnight, two of the police officers assigned to monitor the exclusion zone perimeter reported that they drove through a cloud of white smoke coming from the Arkema Crosby facility as they drove west on Highway 90 to respond to a call related to flooding in the area. After they reported the white smoke cloud, the Unified Command closed Highway 90.

The police officers who drove through the white smoke cloud reported to other police officers that their vehicle's dash cam recorded the white cloud and that some type of release was occurring at the facility. Three other police officers then drove their vehicles east on Highway 90 to check on the officers and to review the dash cam footage. As these officers passed by the Arkema facility, they also reported driving through a white cloud of smoke coming from the facility. Shortly thereafter, all five police officers recognized they might have been exposed to chemicals by driving through the cloud of white smoke coming from the Arkema Crosby facility. The combination of their symptoms, and their desire to get prompt medical attention led the police officers to drive west on Highway 90 toward the command post and medical aid to obtain treatment.

In all, five police officers in four police vehicles drove west down Highway 90 toward the command post. The officers called to request medical assistance when they reached the southwestern end of the exclusion zone. During this trip, the officers reported being exposed to a black smoke cloud coming from the Arkema Crosby facility. By the time they reached the southwestern end of the exclusion zone, the officers reported experiencing nausea, headaches, sore throats, and itchy watering eyes. Other emergency responders examined the police officers, who were complaining of exposure to the smoke, and flushed the officers' exposed skin with water. When the officers reached the forward command post they told the emergency responders what they had experienced. Harris County emergency response officials then shut down travel in both directions on Highway 90. Highway 90 was not reopened until the all-clear

Chronicles of Incidents and Response

was sounded days later. As the night progressed, organic peroxide products in one of the nine refrigerated trailers continued to decompose and then caught on fire.

The next day, Friday, September 1, 2017, at about 5:00 p.m., two more refrigerated trailers ignited and burned. At this point, the three refrigerated trailers that could not be moved to the high ground in the lay down area due to the high waters had all burned and the remaining six refrigerated trailers located on higher ground had still not combusted. It is likely that the refrigeration units on these trailers were still operating, but with no way to check on the contents inside of the trailers and only relying on telemetry data, the Unified Command could not safely remove the organic peroxide contents.

Organic peroxides are reactive chemicals that are inherently unstable. Because of this instability, these reactive chemicals require special storage and handling precautions to prevent the organic peroxides from decomposing and producing heat and byproducts. Organic peroxides continually decompose at a rate based on the temperature of the product. An important organic peroxide safety property is the Self-Accelerating Decomposition Temperature or SADT. All of the organic peroxides produced by Arkema must be stored at a temperature below their SADT to limit the rate of decomposition to a safe level. The first three refrigerated trailers burned in two separate fires at the Arkema Crosby facility when the trailers lost refrigeration and the organic peroxide products inside, which required low-temperature storage, decomposed and combusted. Companies need to ensure they have sufficient safeguards in place to maintain organic peroxide products below their SADTs (United States Chemical Safety Board (CSB)).

St. Louis, MO June 24, 2005 Flammable Gas Cylinder Fire & Explosion

On June 24, 2005, fire swept through thousands of flammable gas cylinders at the Praxair gas repackaging plant in St. Louis, Missouri. Dozens of exploding cylinders were launched into the surrounding community and struck nearby homes, buildings, and cars, causing extensive damage and several small fires (United States Chemical Safety Board (CSB)).

Shepherdsville, KY January 16, 2007 Crude Oil Train Derailment & Fire

Executive Summary

On January 16, 2007, about 8:43 a.m., E.S.T northbound CSX Transportation (CSX) freight train Q502-15, traveling about 47 mph through a curve, derailed 26 of its 80 cars near Shepherdsville, Kentucky. Twelve of the derailed cars contained hazardous materials. Three of those cars breached and released significant amounts of flammable hazardous liquids, which ignited and burned (Figure 1.126). About 500 people were evacuated from the area near the accident. No one was injured during the derailment; however, 50 people and 2 emergency responders were treated at local hospitals for minor injuries related to the hazardous materials release and fire. CSX estimated the total costs associated with this accident at $22.4 million. The weather was dry and cloudy, although recent rains had left the soil well saturated. The temperature was 28°F with 14 mph winds.

Probable Cause

The National Transportation Safety Board determines that the probable cause of the accident was the failure of the 18th rail car to properly negotiate a curve because of the inadequate side bearing clearance of the B-end truck assembly, likely due to a broken side bearing wedge plate attachment bolt, which caused a wheel to climb the rail, which derailed the car. Contributing to the derailment was (1) the undesirable contact of the truck bolster bowl rim with the car body center plate and (2) the hollow worn wheels on the 18th car, which further diminished the steering ability of the truck assembly. After notification from the Bullitt County E911 Emergency Call Center, the ZFD responded to the scene about 8:47

Figure 1.126 Twelve of the derailed cars contained hazardous materials. Three of those cars breached and released significant amounts of flammable hazardous liquids, which ignited and burned. (From NTSB.)

a.m. The ZFD firefighters were supported by mutual aid from neighboring communities. ZFD responders arrived on scene in about 4 min, and soon thereafter, the additional mutual aid resources arrived.

Upon arriving on scene, firefighters reported an intense fire encompassing an area of about 35,000 ft^2. As other responders arrived on scene, they observed that several of the derailed cars within the blaze were tank cars displaying hazardous materials cargo labeling. Firefighters confirmed the contents of the railcars with the CSX conductor and initiated efforts to control and suppress the fire. A voluntary evacuation order was issued for residents within 1 mile of the accident. Because of the dense, drifting black smoke from the accident, an 8-mile section of Interstate 65 was closed from about 9:11 a.m. to about 7:54 p.m. on January 16, 2007, and vehicular traffic was detoured. Fire suppression and hazardous material cleanup activities concluded on January 20. On January 21, the evacuation order was lifted, residents returned to their homes, and repairs to the tracks were finished. The on-scene emergency response activities were discontinued on Monday, January 22, 2007, about 5:05 p.m. (National Transportation Safety Board (NTSB)).

Ghent, WV January 30, 2007
Propane Explosion and Fire

Four people were killed and five others were seriously injured when propane vapors from a storage tank ignited and exploded at the Little General convenience store and gas station in Ghent, West Virginia

Figure 1.127 Four people were killed and five others were seriously injured when propane vapors from a storage tank ignited and exploded at the Little General convenience store and gas station. (From Chemical Safety Board.)

(Figure 1.127). Propane was used as fuel inside the building, which was completely destroyed. On the day of the incident, a junior propane service technician employed by Appalachian Heating was preparing to transfer liquid propane from an existing tank, owned by Ferrellgas, to a newly installed replacement tank. The existing tank was installed in 1994 directly next to the store's exterior back wall in violation of West Virginia and U.S. Occupational Safety and Health Administration regulations. When the technician removed a plug from the existing tank's liquid withdrawal valve, liquid propane unexpectedly released. For guidance, he called his supervisor, a lead technician, who was offsite delivering propane. During this time propane continued releasing, forming a vapor cloud behind the store. The tank's placement next to the exterior wall and beneath the open roof overhang provided a direct path for the propane to enter the store.

About 15 min after the release began, the junior technician called 911. A captain from the Ghent Volunteer Fire Department subsequently arrived and ordered the business to close. Little General employees closed the store but remained inside. Additional emergency responders and the lead technician also arrived at the scene. Witnesses reported seeing two responders and the two technicians in the area of the tank, likely inside the propane vapor cloud, minutes before the explosion. Minutes after the emergency responders and lead technician arrived, the propane inside the building ignited. The resulting explosion killed the propane service technicians and two emergency responders who were near the tank. The blast also injured four store employees inside the building as well as two other emergency responders outside the store.

The IC arrived at the Little General store shortly after the initial dispatch call. In the approximately 5 min from his arrival until the explosion at 10:53 a.m., the IC took several actions. He:

- Assessed the frostbite injury to the junior technician;
- Ordered the business to close;
- Directed the EMTs to the rear of the building to treat the junior technician's frostbite;
- Ordered the EMTs to ensure that the business was closed, that no one was smoking, and that no gasoline was being pumped; and
- Ordered the firefighter to ensure that everyone was out of the building.

Fire departments from the neighboring communities of Beckley, Beaver, and Princeton responded to the explosion. Later that day a team from the West Virginia Office of the State Fire Marshal arrived to investigate, assisted by an agent from the U.S. Bureau of Alcohol, Tobacco, Firearms, and Explosives (ATF). United States Chemical Safety Board (CSB)

Emergency Responders Who Made the Supreme Sacrifice

Captain, Lawrence Dorsey, 24

EMT, Jeffrey Lee Treadway, 31

Oneida, NY Monday, March 12, 2007 Tank Car Explodes

On Monday, March 12, 2007, about 6:58 a.m., CSX Transportation (CSX) train No. Q39010, a mixed freight train, derailed near Oneida, New York. The train was en route from Buffalo, New York, to Selkirk, New York. At the time of the derailment, the train was traveling about 47 mph. The train consisted of 3 locomotives and 78 cars. Twenty-nine cars derailed. Six tank cars were breached, including four carrying liquefied petroleum gas, one carrying toluene, and one carrying ferric chloride. An explosion and fire followed that led local emergency response officials to close two elementary schools and evacuate a 1-mile area around the derailment site (Figure 1.128). Four firefighters were taken to a hospital for observation as a precaution because they had stepped in a pool of ferric chloride. There were no fatalities. Estimated damages and environmental cleanup costs were $6.73 million.

The engineer said that he saw a fire near the middle of the train. The crew contacted the train dispatcher and reported that the train was on fire. After coming to a stop, they exited the locomotive and walked ahead of the train. The conductor made contact with the arriving emergency responders and gave them the written consist of the train.

Figure 1.128 An explosion and fire followed that led local emergency response officials to close two elementary schools and evacuate a 1-mile area around the derailment site. (From NTSB.)

308   *Hazmatology: The Science of Hazardous Materials*

Emergency Response On the morning of March 12, 2007, the Oneida Fire Department chief was driving north on Broad Street to his station located at 109 North Main Street. He noticed a large fire in the distance and began driving in that direction. After he arrived on West Elm Street, he drove along a dirt access road between West Elm Street and the railroad. From this road, he saw a large fire and rail cars near the track. He went back to West Elm Street and established a command post there. He radioed his station and requested that one engine (Engine 2) respond to the north side of the railroad. A deputy fire chief responded with this engine. Two engines (Engines 1 and 3) responded to the south side of the railroad at West Elm Street. The fire chief also requested a tanker from the Canastota Fire Department, a tanker from the Wampsville Fire Department, and the Oneida County Hazmat Team. An ambulance was also dispatched to the scene to stand by.

The Madison County Communications Center operates the 911 and emergency communications for the county. The first 911 call was received about 7:03 a.m. The caller reported that a train had exploded on the train tracks on Canal Street in Oneida. According to the fire chief, the deputy fire chief met the train crew on the north side of the track. The crew gave the deputy fire chief the train consist. The fire chief estimated that he received this information within about 30 min of his arrival on scene. The initial evacuation area was a 1-mile radius around the accident site. After a further assessment, the evacuation area was reduced to a 1/2-mile radius that included eight houses.

The mayor of Oneida declared a state of emergency that made this evacuation mandatory. The Oneida Police Department, the Madison County Sheriff, and the New York State Police conducted the evacuation. A school bus and an ambulance were used to assist those without transportation. The Oneida City School District bus garage was used as a shelter for evacuated residents. Two elementary schools (Durhamville and North Board Street) were closed, and students were taken to Oneida High School.

Throughout the morning, local and railroad responders began to assess the accident scene and develop action plans. Based on observations from a New York State police helicopter that flew over the accident site with a CSX representative on board, it was estimated that cars 19 through 535 were involved in the derailment; two propane cars were breached; one propane car was burning; and the toluene car was burning and venting. Because the toluene car was venting, responders planned defensive actions. No fire suppression was conducted for the tank car fires; however, unmanned water monitors were used to cool the cars. During the afternoon and evening, local railroad responders began to assess the damage to the derailed cars and tested the pressure of the derailed cars. Based on these assessments, they made plans for transferring the contents of the damaged cars to either tanker trucks or empty rail tank cars brought from other locations. They also decided which cars had limited damage and could be re-railed. The state of emergency was lifted at 3:00 p.m. on March 15, 2007, 3 days after the derailment, and evacuated residents were allowed to return to their homes (National Transportation Safety Board).

Dallas, TX July 25, 2007 Bottled Gas Plant Fire and Explosions

A series of explosions at Southwest Industrial Gases, Inc. sent flaming debris raining onto highways and buildings near downtown Dallas on Wednesday and seriously injured at least three people (Figure 1.129). Authorities evacuated a 1/2-mile (0.8 km) area surrounding the Southwest Industrial Gases, Inc. facility and shut down parts nearby Interstates 30 and 35. Video footage showed numerous small fires burning in the area as stacks of gas cylinders caught fire and exploded.

The canisters held acetylene and propane gas, said Texas Commission on Environmental Quality spokeswoman Andrea Morrow. It was not immediately clear what caused them begin exploding around 9:30 a.m. By noon, fire crews were hosing down the charred metal wreckage to extinguish any lingering flames. Earlier, about a dozen cars burned in a nearby parking lot and a grassy areas of a highway median. "I thought it was artillery. It was just coming just boom, boom, boom," said witness Tony Love, a former Army soldier.

Figure 1.129 A series of explosions at Southwest Industrial Gases, Inc. sent flaming debris raining onto highways and buildings near downtown Dallas on Wednesday and seriously injured at least three people. (Chemical Safety Board.)

Parkland Hospital spokesman Robert Behrens said two people injured by the explosions had been brought to his hospital in serious condition. A third man was taken to Methodist Dallas Medical Center, hospital spokeswoman Sandra Minatra said. She did not give out his condition.

According to the industry Web site gasworld.com, Southwest Industrial is a distributor that carries a range of gases, including acetylene, helium and hydrogen, as well as welding equipment. Calls to a phone listing for the company were not answered. At the edge of the evacuation zone is Dallas County's main jail and criminal courts building, but operations continued their uninterrupted, said Deputy Michael Ortiz of the Dallas County Sheriff's Department. Carol Peters, a spokeswoman for Oncor Electric Delivery, said about 30 buildings near the blasts were without power and would stay shut off until fire crews extinguished the blaze (United States Chemical Safety Board).

Jacksonville, FL December 19, 2007 Chemical Plant Explosion

Four people were killed and several injured after an explosion Wednesday of a chemical plant sent a thick plume of smoke over a section of Jacksonville, authorities said. "Literally, it's a hellish inferno (Figure 1.130). There is no other way to describe it," said Fire Department spokesman Tom Francis. Fourteen people were hospitalized after the blast at the T2 Lab of Faye Road, in an industrial area on the waterfront in north Jacksonville, Francis

Figure 1.130 Four people were killed and several injured after an explosion Wednesday of a chemical plant sent a thick plume of smoke over a section of Jacksonville, authorities said. "Literally, it's a hellish inferno. (Courtesy: Jacksonville Fire Department.)

said. Officials initially ordered an evacuation of nearby businesses, but by 4 p.m. the order had been lifted after tests of the air found no toxicity, Francis said. Firefighters were still battling hot spots, and the effort will be going on for "quite some time," he said. Six of those injured were transported to Shands Hospital in Jacksonville, hospital spokeswoman Kelly Brockmeier said. A Shands official said the hospital incident command system had been activated - something done to put the staff in high alert in anticipation of trauma patients.

A woman who answered the T2 Lab's 24-h facility emergency phone said the plant manufactures ecotane, a gasoline additive that reduces tailpipe emissions, according to the laboratory's Web site. The billowing black smoke could be seen from the city's downtown, said Florida Times-Union reporter Bridget Murphy. Murphy said she talked to several witnesses as they walked out of the area, and they were "shaken to the core." "They described a hissing noise and then a sound wave," she said. Antonio Padrigan was trying to get in touch with his son, who works in a plant in the area, but was having no luck reaching him on his cell phone.

"He was shook up when he called me, but I can't get through to him anymore," Padrigan said. "I don't know if he's in the hospital or what." CNN I-Reporters Jonathan Payne and his son Calvin, 16, shot pictures of the explosion. They felt the blast shake their home, about 15 min away and went to see what was going on. Carlton Higginbotham, 63, was working at home on Townsend Boulevard in Jacksonville when a loud boom shook his house, he said. "It was a gunshot-type explosion; it wasn't a rumble," he said (United States Chemical Safety Board).

> "The cloud that come out of it was white, some would say mushroom-shaped," Higginbotham said. "It was followed by dark, dark smoke."
>
> CNN News Story *2007-12-19*

Cherry Hill, IL June 19, 2009 Ethanol Train Derailment and Fire

About 8:36 p.m. C.D.T on Friday, June 19, 2009, eastbound Canadian National Railway Company freight train U70691-18, traveling at 36 mph, derailed at a highway/rail grade crossing in Cherry Valley, Illinois. The train consisted of 2 locomotives and 114 cars, 19 of which derailed. All of the derailed cars were tank cars carrying denatured fuel ethanol, a flammable liquid. Thirteen of the derailed tank cars were breached or lost product and caught fire. At the time of the derailment, several motor vehicles were stopped on either side of the grade crossing waiting for the train to pass. As a result of the fire that erupted after the derailment, a passenger in one of the stopped cars was fatally injured, two passengers in the same car received serious injuries, and five occupants of other cars waiting at the highway/rail crossing were injured (Figure 1.131). Two responding firefighters also sustained minor injuries. The release of ethanol and the resulting fire prompted a mandatory evacuation of about 600 residences within a 1/2-mile radius of the accident site. Monetary damages were estimated to total $7.9 million.

Probable Cause

The National Transportation Safety Board determines that the probable cause of the accident was the washout of the track structure that was discovered about 1 h before the train's arrival, and the Canadian National Railway Company's (CN) failure to notify the train crew of the known washout in time to stop the train because of the inadequacy of the CN's emergency communication procedures. Contributing to the accident was the CN's failure to work with Winnebago County to develop a comprehensive storm water management design to address the previous washouts in 2006 and 2007. Contributing to the severity of the accident was the CN's failure to issue the flash flood warning to the train crew and the inadequate design of the DOT-111 tank cars, which made the cars subject to damage and catastrophic loss of hazardous materials during the derailment.

At 8:36 p.m., callers began contacting local emergency response authorities to report the incident. At 8:36 p.m. a cell phone call reported a

Figure 1.131 As a result of the fire that erupted after the derailment, a passenger in one of the stopped cars was fatally injured, two passengers in the same car received serious injuries, and five occupants of other cars waiting at the highway/rail crossing were injured.

Chronicles of Incidents and Response

bus/truck fire at the crossing. One minute later, at 8:37 p.m., several other callers reported the derailment. The Winnebago County 911 Emergency Call Center received a total of 19 calls reporting the accident. The information in this section is based on incident response data and communications information supplied by the emergency service agencies involved, as well as on individual interviews with key personnel of the emergency services agencies and the railroad.

A follow-up dispatch can occur when an event location is at or close to the boundary with another emergency response jurisdiction. The operational protocol prescribes that the agency closest to the physical location of the event should be expediently dispatched to the scene.

Characteristics of a pressure fire include a focused flame indicating that the substance fueling the fire is under pressure as it is ignited. Tank cars are equipped with valves designed to open automatically to relieve pressure and prevent a tank rupture in the event that excess heat or other factors cause pressure within the tank to rise to a dangerous level. Pressure tank cars are designed to carry chemicals or petroleum products under pressures usually exceeding 100 psi. The ethanol being transported by train U70691-18 was not under pressure during regular transit. It is permissible to transport denatured ethanol in both general service and pressure tank cars.

At 8:38 p.m., Rockford 911 Dispatch began dispatching resources of the Rockford Fire Department (RFD). About a minute later, Rockford 911 issued a secondary, or follow-up, dispatch of resources of the Cherry Valley Fire Protection District (CVFPD), which immediately responded. The first responding CVFPD resource arrived at the scene at 8:46 p.m., followed shortly by additional responding CVFPD units and mutual aid resources from neighboring communities. The CVFPD chief subsequently determined that the incident had actually occurred within the jurisdiction of Cherry Valley rather than Rockford. With his arrival on scene, the CVFPD chief assumed the role of incident commander. He observed from a distance obvious major fire involvement over a relatively widespread area. The characteristics of discharging flame suggested a pressure fire. Flames were impinging on several non burning tank cars in proximity to the burning cars, and the chief stated that he heard numerous pressure relief valves opening.

By this time, the flames extended several hundred feet into the air. It was not known if the pressure relief valves on the derailed and overturned tank cars were either damaged or inoperable. It was not known if any pressure tank cars were involved in the wreckage pileup. At that point, responders did not know the contents of the burning cars because the placards indicating contents were not readily visible. They also did not know if any locomotives or train crew might be involved. From his vantage point on the north side of the site, the CVFPD chief could see several

314   *Hazmatology: The Science of Hazardous Materials*

motor vehicles in the roadway on the north side of the tracks near the grade crossing; however, no determination could be made as to whether any of the vehicles had been damaged by the derailing railcars or the subsequent fire. The chief said he assumed that a similar situation (vehicles stopped in a queue that may have been directly affected by the derailment and fire) existed on the south side of the grade crossing.

The nearest dwelling was about 600 ft from the derailed equipment. The closest building to the site was a commercial facility on the south side of the tracks about 300 ft from the crossing. The business was closed at the time of the accident, and the building was unoccupied. The CVFPD chief requested a mutual aid response to the scene, including an RFD hazardous materials response team with a decontamination unit. He also established an initial incident command (IC) at the intersection of Mulford Road and Abbington Court, about 1,400 ft north of the fire perimeter. By radio, he advised the next responding fire department chief, who was an RFD responder, to set up an initial south sector command at the intersection of Mulford Road and Sandy Hollow Road, about 900 ft south of the fire perimeter.

A situation report from the south sector command to the CVFPD chief indicated major fire involvement with multiple railroad tank cars burning. South of the crossing, several vehicles were in the roadway near the derailed equipment, with one vehicle on fire. An emergency medical services (EMS) crew was attending to several injured persons. At this time, responders had not been able to identify the contents of the tank cars or to determine whether pressurized rail tank cars were involved. Based on the volume of fire involvement and the observation of flame discharge from the pressure relief devices, the CVFPD chief consulted the *2008 Emergency Response Guidebook* and determined that the area needed to be evacuated. At 9:02 p.m., the CVFPD chief contacted RFD Dispatch to implement a mandatory evacuation within a radius of about 1/2 mile from the fire perimeter. The evacuation was to be executed by the local law enforcement personnel.

At 9:09 p.m., RFD Dispatch advised the CVFPD chief that the CN had informed the RFD that the tank cars contained ethanol. Based on that information, the CVFPD chief requested that RFD Dispatch locate quantities of fire-suppressing foam. A few minutes later, at 9:12 p.m., RFD Dispatch notified the IC that the railroad indicated that only one tank car contained ethanol. In order to make a positive identification of the contents of the burning rail cars, the IC initiated a search for the train crew. The crew would have a train consist (a listing of all the cars and their order within the train) as well as the shipping papers for the hazardous materials being carried. Shortly before 9:50 p.m., a firefighter notified the IC that the crew had been located and that the two crewmembers were on their way to the IC with the train's shipping papers.

Chronicles of Incidents and Response 315

At 10:10 p.m., law enforcement officers requested fire department help in implementing the mandatory evacuation. Fire companies and an ambulance were dispatched to assist in the effort. About 10:30 p.m., two CN dangerous goods officers arrived on scene and became hazardous materials liaisons to the fire departments. The officers provided additional guidance for the incident response. About 10:20 p.m., the two CN crewmembers arrived at the IC post and presented the train consist and the hazardous materials shipping papers documenting that the tank cars contained ethanol. Upon receipt of the shipping papers, the IC recognized that the car positions shown on the printed train consist were not correct based on the visual identification of the cars in the pileup. It was not until 1:22 a.m. the next morning that the IC received an e-mail from the CN containing the correct consist. (Errors in the train consist are discussed in detail in the Train Consist Inaccuracy section of this report.)

Having learned that the source of the fire was ethanol, the incident commander considered the following factors:

- the volume of burning cargo
- the fact that extinguishing the fire would require large quantities of foam, which were not available on short notice
- the overall topography of the widespread wreckage pileup, the fact that the fire no longer was an immediate hazard to life or property
- the fact that the fire would consume the cargo content, which would help reduce the effects of a hazardous materials product release into the environment

Based on these considerations, the incident commander decided to allow the cargo content to burn itself out; the CN dangerous goods officers concurred with this decision.

After about 1/2 hour at the IC, the two train crewmembers were asked by IC and CN personnel to return to their locomotive and move the train forward a short distance to provide a greater separation between the fire and the remaining (nonburning) railcars of the train. The two crewmembers returned to the train, and the train was subsequently pulled forward a short distance.

About 11:00 p.m., two CVFPD firefighters, using an off-road vehicle, were assigned to check the area southeast of the accident site for fire and to determine the location and condition of the remaining railcars of the train. During that mission, the firefighters were apparently exposed to toxic fumes, with one of them experiencing disorientation. The two firefighters were subsequently transported, as a precautionary measure, to a local medical facility for evaluation. The IC principals had seen yellow fiberglass warning markers in the vicinity of Mulford Road near the

316 *Hazmatology: The Science of Hazardous Materials*

accident site indicating the presence of an underground natural gas pipeline in the area. They were not aware of the specifics of the pipeline, such as its exact location, its size and product pressure, and how deep it was buried.

Because the incident commander believed underground natural gas pipelines are typically buried deep enough to be protected from the impact of a heavy surface vehicle making forceful contact with the soil, he was not immediately concerned that the integrity of the pipeline might be threatened by the derailment and wreckage pileup. The IC was also in possession of map documentation received from a pipeline training contractor that did not indicate the presence of a pipeline in that area. About midnight, as a precautionary measure, one of the IC officers made an inquiry to Nicor Gas, the local pipeline operator, for information about the pipeline. Nicor Gas personnel reported that no gas pipelines were located in that area and that the closest gas main pipeline was about 0.7 mile south of the accident site.

By midday on June 20, 2009, the ethanol was substantially burned off, and it was expected that the fire would self-extinguish within a few hours. At 5:00 p.m. on June 20, the IC declared that all fires were extinguished. At 5:30 p.m., the IC suspended the mandatory evacuation, and about 8:00 p.m., released mutual aid resources. The CVFPD firefighting resources stood by, monitoring the situation throughout the night and through the following day in case the fire rekindled. At 5:00 p.m. on June 21, after conferring with principals of the CN, the shipper, and other responding organizations about the status of the event, the incident commander terminated the on-scene operations. A total of 35 separate fire department entities with 250 personnel and about 80 vehicles responded or were part of the mutual aid response. Additionally, the Winnebago County Sheriff's Department responded with 20 units (officers), which were supported by resources from 31 other law enforcement agency (National Transportation Safety Board).

Norfolk, NE December 10, 2009 Propane Storage Tank Fire Protient Propane Fire

December 10, 2009 at 07:03 p.m. Norfolk Fire and Rescue was challenged with a once in a career type of incident when an alarm came in for a possible explosion at 704 Omaha Avenue in Norfolk. Initial information indicated a large building fire with flames and heavy smoke showing. However, upon arrival responders encountered a 30,000 gallon fixed propane tank on fire (Figure 1.132). First order of business was to evacuate 78,000 people in harm's way. Norfolk Police and Firefighting staff began evacuating a 1/2 mile radius of homes and businesses. Mutual aid

Figure 1.132 Initial information indicated a large building fire with flames and heavy smoke showing however upon arrival responders encountered a 30,000 gallon fixed propane tank on fire. (Courtesy: Norfolk Fire Department.)

was summoned to cover city fire stations and the Nebraska State Patrol (NSP) provided a helicopter for aerial surveillance of the incident scene. Command staff reported that this was the game changer in the successful handling the incident, which was concluded without any injuries to the public or emergency responders. The incident was brought under control by 12:00 h (Nebraska State Fire Marshal).

Tiskilwa, IL October 7, 2011 Train Derailment Ethanol Fire

Train derailment at 2:14 a.m. causes 800 Tiskilwa residents to be evacuated. The lead locomotive experienced a severe impact under the wheels and the train went into emergency braking. When the conductor looked back he saw that the cars near the locomotive were on fire. The railcars had uncoupled from the locomotive. The engineer stopped the locomotives about 20 car lengths beyond where the first car of the train derailed. The 131 car eastbound train lost 26 cars in the derailment, 12 of which carried flammable liquid Ethanol.

Those cars were damaged by the derailment and leaked and caught fire. Tiskilwa firefighters responded quickly at 02:15 h. After ordering an evacuation of 1/2 mile in all directions, as recommended by the DOT Emergency Response Guide Book, he requested a hazmat team from LaSalle Fire Department. Five additional fire departments were called for water supply and operations began to cool the tanks. Two other ethanol plants provided firefighting foam to crews battling the fires. Cooling of the tanks continued until approximately 11:00 a.m. and the fire was declared under control. There were no injuries reported (National Transportation Safety Board).

West, Texas April 17, 2013
Ammonium Nitrate Explosion

On April 17, 2013 a fire and subsequent explosion occurred at the West Fertilizer Company in West, Texas. Firefighters from the West Volunteer Fire Department were fighting a fire at the facility when the explosion occurred, approximately 20 min after they started fighting the fire. Ammonium nitrate was located in a bin inside a seed and fertilizer building on the property. The explosion registered 2.1 on a seismograph reading from Hockley, Texas, 142 miles away (Figure 1.133). Fifteen people, mostly emergency responders, were killed, over 260 were injured and 150 buildings sustained damage. Victims included 5 volunteers with the West Volunteer Fire Department, 4 volunteers from neighboring volunteer departments that were attending an (EMS) class in West and went to the scent to help, one career captain off duty went to the scene to offer help, 2 civilians that went to help the West volunteers and 3 residents who lived around the facility. Investigators confirmed that ammonium nitrate was the source of the explosion. According to the United States Environmental Protection Agency (EPA) there was a report of 240 tons of ammonium nitrate on the site in 2012. According to the Department of Homeland Security the company had not disclosed to them their ammonium nitrate

Figure 1.133 Ammonium nitrate was located in a bin inside a seed and fertilizer building on the property. The explosion registered 2.1 on a seismograph reading from Hockley, Texas, 142 miles away. (From Chemical Safety Board.)

Chronicles of Incidents and Response

stock. Federal law requires that the DHS be notified whenever anyone has more than 1 ton of ammonium nitrate on hand, or 400 lb if the ammonium nitrate is combined with combustible material.

The fire and explosion at West, Texas was investigated by the U.S. Chemical Safety Board. Listed below are some of the observations and preliminary findings following the initial investigation. The explosion at West Fertilizer resulted from an intense fire in a wooden warehouse building that led to the detonation of approximately 30 tons of ammonium nitrate stored inside the wooden bins. Not only were the warehouse and bins combustible, but the building also contained significant amounts of combustible seeds, which likely contributed to the intensity of the fire. The building lacked a sprinkler system or other systems to automatically detect or suppress fire, especially when the building was unoccupied after hours. By the time firefighters were able to reach the site, the fire was intense and out of control. The detonation occurred just 20 min after the first notification to the West Volunteer Fire Department.

Although some U.S. distributors have constructed fire-resistant concrete structures for storing ammonium nitrate, fertilizer industry officials have reported to the CSB that wooden buildings are still the norm for the distribution of ammonium nitrate fertilizer across the U.S. No federal, state, or local standards have been identified that restrict the sitting of ammonium nitrate storage facilities in the vicinity of homes, schools, businesses, and health care facilities. In West, Texas, there were hundreds of such buildings within a mile radius, which were exposed to serious or life-threatening hazards when the explosion occurred on April 17. West volunteer firefighters were not made aware of the explosion hazard from ammonium nitrate stored at West Fertilizer, and were caught in harm's way when the blast occurred. NFPA recommends that firefighters evacuate from ammonium nitrate fires of "massive and uncontrollable proportions." Federal DOT guidance contained in the Emergency Response Guidebook, which is widely used by firefighters, suggests fighting even large ammonium nitrate fertilizer fires by "flooding the area with water from a distance." However, the response guidance appears to be vague since terms such as "massive," "uncontrollable," "large," and "distance" are not clearly defined. All of these provisions should be reviewed and harmonized in light of the West disaster to ensure that firefighters are adequately protected and are not put into danger protecting property alone (United States Chemical Safety Board).

Hazmatology Point: More firefighters have been killed fighting fires with ammonium nitrate than any other hazardous material. Firefighters should be aware of the potential for ammonium nitrate to be present at agricultural fertilizer facilities. They should not attempt to fight fires where the ammonium nitrate is burning. We have enough historical

320   *Hazmatology: The Science of Hazardous Materials*

*information about ammonium nitrate disasters that tell us these fires are
"losers".*

Journal of Hazardous Materials

Explosions of Ammonium Nitrate Fertilizer in Storage or Transportation

Vitenis Babrauskas   Ammonium nitrate (AN) is a detonable substance
which has led to numerous disasters throughout the 20th century and
until the present day. Needed safety lesson have not been learned, since
typically each accident was viewed as a great surprise and investigations focused on finding some unique reason for the accident, rather than
examining what is common among the accidents. A review is made of
accidents which involved AN for fertilizer purposes, and excluding incidents involving ANFO or additional explosives apart from AN. It is found
that, for explosions in storage or transportation, 100% of these disasters
had a single causative factor—an uncontrollable fire. Thus, such disasters
can be eliminated by eliminating the potential for uncontrolled fire. Two
actions are required to achieve this:

1. adoption of fertilizer formulations which reduce the potential for
   uncontrolled fire and for detonation; and
2. adoption of building safety measures which provide assurance
   against uncontrolled fires.

Technical means are available for achieving both these required measures. These measures have been known for a long time and the only reason that disasters continue to occur is that these safety measures are not
implemented. The problem can be solved unilaterally by product manufacturers or by government authorities, but preferably both should take
necessary steps.

West Memorial

The upright stone monument stands at the corner of the firehouse with
the words "Last Alarm" and the blast date etched above the 12 names of
the deceased firefighters and first responders (Figure 1.134). Of the 15 people who died in 2013, five worked for the West Volunteer Fire Department,
and the remaining seven volunteered from the area. The memorial was
donated by Phipps Memorial, Nors said. A stone bench also was installed
for visitors to have a place to sit, Nors said. "This is a good place to remember," he said. West Mayor Tommy Muska said the fire station memorial
is appropriate and hopes to add another marker across town at Parker's
Park with all 15 names on it. West City Council plans to discuss installing

Chronicles of Incidents and Response

Figure 1.134 The upright stone monument stands at the corner of the firehouse with the words "Last Alarm" and the blast date etched above the 12 names of the deceased firefighters and first responders. (Courtesy: West Volunteer Fire Department.)

bathrooms and rebuilding the park's pavilion at Tuesday's meeting, Muska said. When those projects are underway, Muska said he plans to begin the process of installing a city memorial. The final spot hasn't been decided, but it will probably go in the park since it is closest to where the blast occurred, Muska said. But he isn't in a hurry to begin the process. "I want it to be perfect and I want everybody to be really proud of it," he said.

Emergency Responders that Made The Supreme Sacrifice

Morris Bridges, 41, West VFD

Perry Calvin, 37, Mertens VFD and Navarro Mills VFD

(Student at Hill College Fire Academy)

Jerry Dane Chapman, 26, Abbott VFD

Cody Frank Dragoo, 50, West VFD

(Foreman at the West Fertilizer Co.)

Kenneth "Lucky", Harris, Jr.

Career Captain Dallas, TX FD

James Matus, 52, West VFD

Joseph Pustejovsky, 29, West VFD

Cyrus Adam Reed, 29, Abbott VFD

Kevin William Sanders, 33, EMT

Douglas Snokhous, 48, West VFD

Robert Snokhous, 50, West VFD

William "Buck" Uptmor, Jr., 45, Abbott VFD

Citizen Fatalities Not Firefighters

Adolph Lander, 96, Retired Farmer

Mariano C. Saldivar, 57, Retired

Judith Monroe, 65

Chemical Safety Board (CSB) Investigation

West Volunteer Fire Department (WVFD)

Emergency dispatchers paged the WVFD, and firefighters responded to the scene with two fire engines, two initial attack apparatus or brush trucks, and a water tender truck at various times. Dispatchers also paged mutual aid personnel from neighboring counties, including Abbott, which responded. Many of the firefighters also responded by using their personally owned vehicles (POVs). According to eyewitness accounts, the fire intensified very quickly and was described as a rolling fire that moved from the northeast end of the fertilizer building (in the seed storage area north of the office) toward the southern end of the building.

Five firefighters arrived on scene in two fire engines at different times. The first fire engine arrived on scene and staged east of the burning structure while one of the brush trucks staged to the north of the first fire engine. Four other firefighters directed water (using two 1.5-in. hoses) from the first fire engine's internal tank onto the fire through the northeast doorway of the bagged fertilizer room, where fire was present. Once the second fire engine arrived on scene, the two firefighters from that fire engine began laying 1,000 ft of 4-in. hose line from the fire hydrant near the high school (1,600 ft away) toward the fertilizer facility. After laying all of the hose lines from the second fire engine, they discovered that the hose was approximately 700 ft short of the length needed to effectively fight the fire.

After assessing the situation, one firefighter arranged to take the first fire engine, which had a better pump with greater pressure capabilities and additional hose that would allow him to continue to reverse-lay the lines. However, rather than resuming where the first fire engine ran out of hose, the firefighter went back to the fire hydrant near the high school to connect the first engine to the hydrant without laying the additional length of hose needed to supplement the hose that had already been laid from the second engine.

He saw flames (40 to 50 ft high) coming out of the cupola atop the fertilizer storage building and out of the door on the northeast corner of the building. Before the firefighter could make his way back to the end of the hose run, the explosion occurred. Before the explosion, the WVFD assistant chief arrived at the WFC facility, spoke with the police officer on scene, and advised him to begin evacuating nearby homes. He also made a radio request to the dispatch center, asking for a ladder truck to set up at the West Terrace Apartments in case a fire started there, but a ladder truck was not available. The WVFD chief and assistant fire chief were assessing the situation just before the explosion and were considering a total evacuation, even though neither believed that the FGAN would explode.

On the basis of interviews that CSB conducted after the incident, the WVFD came to understand that it did not have enough water to effectively fight the fire. Accordingly, the WVFD was considering the appropriate course of action possibly standing down, letting the structure burn, and focusing on evacuation.

Abbott, Bruceville-Eddy, Mertens, and Navarro Mills Volunteer Fire Departments

On the evening of the incident, a group of volunteer firefighters from neighboring city fire departments (including Bruceville-Eddy, Mertens, and Navarro Mills), who were taking an Emergency Medical Technician (EMT)–Basic class at the West Emergency Medical Services (EMS) building, responded to the fire. The West EMS facility is located a few blocks west of the WFC facility. When these volunteer firefighters heard the sirens activated in the city, they immediately made their way to the site. In addition, an ambulance responded with two EMTs and a volunteer firefighter. According to interviews that CSB conducted with emergency responders, radio and cell phone capabilities at the scene were limited after the explosion. Following the explosion, officials established two different staging areas. The first staging area, at the high school football field about 0.25 miles from the blast site, was used as a triage area for injured residents. Injured personnel and residents were relocated from the football field to the second staging area, at the community center about 1 mile away. After the explosion at approximately 8:15 p.m., additional volunteer firefighters from the neighboring cities of Abbott, Bruceville-Eddy, Mertens, and Navarro Mills responded to the WFC facility.

During the investigation, CSB noted two potential scenarios that could have led to more severe consequences. First, if the fire had started during the middle of a normal school day instead of the evening and if all other conditions remained unchanged (specifically, if onsite WFC employees

324 *Hazmatology: The Science of Hazardous Materials*

were unable to extinguish the fire), students would have been present at the intermediate and high schools. Had the schools evacuated, students likely would have assembled in areas such as the gymnasium and multi-purpose rooms within the schools (and in other pre-designated areas outside of the schools) before the evacuation to conduct a head count. Given the short time that elapsed before the explosion, many students and staff members might have been injured in the 20 min from the first discovery of a fire until the explosion. Second, a railcar loaded with more than 100 tons of FGAN toppled during the explosion but did not detonate. If the contents of the railcar had detonated, the damage, injuries, and fatalities would have been significantly worse. These scenarios are important to consider because throughout the United States, there are many facilities that, like WFC, are located near public structures such as schools.

The two 12,000-gallon anhydrous ammonia pressure vessels were approximately 30% full of ammonia at the time of the explosion. The anhydrous ammonia pressure vessels were south of the crater. The pressure relief valves (PRVs) on the northern anhydrous ammonia pressure vessel still had their weather caps on and consequently did not relieve the pressure. The weather caps were missing on the PRVs in the middle of the southern anhydrous ammonia pressure vessel. Two additional liquid fertilizer storage tanks sat parallel to the railroad track southwest of the anhydrous ammonia pressure vessels. The blast of the explosion also damaged the tracks on the railroad between the WFC property and the park. The blast was sufficiently powerful to shift the tracks more than 2 ft to one side, creating a prominent curve in the tracks.

Ammonium Nitrate Hazards

Under normal conditions, pure solid ammonium nitrate is a stable material; it usually is not sensitive to mild shock or other typical sources of detonation (such as sparks or friction). However, AN exhibits three main hazards in fire situations:

- Uncontrollable fire
- Decomposition with the formation of toxic gases
- Violently explosive

Key Contributing Factors to Emergency Responders' Fatality

CSB identified the following seven key factors that contributed to the fatalities of firefighters and other emergency responders in West:

1. Lack of incident command system.
2. Lack of established incident management system.

Chronicles of Incidents and Response

3. Lack of hazardous materials (HAZMAT) and dangerous goods training.
4. Lack of knowledge and understanding of the detonation hazards of FGAN.
5. Lack of situational awareness and risk assessment knowledge on the scene of an FGAN-related fire.
6. Lack of pre-incident planning at the WFC facility.
7. Limited and conflicting technical guidance on AN.

Lack of Incident Command System

CSB found that none of the responding emergency response personnel trained and certified in the National Incident Management System (NIMS) process formally assumed the position of Incident Commander (IC) who would have been responsible for conducting and coordinating an incident command system (ICS). Senior emergency response personnel at the WVFD arrived at the scene of the WFC incident at different times and did not delegate an IC to be in charge of the incident. Also, there was no record that arriving firefighters conducted an initial incident size-up or risk assessment to determine initial actions (offensive or defensive) that would be most suitable in responding to the incident based on the situation and available resources without putting emergency personnel at risk.

Despite multiple responders having ICS training, none of them reportedly established command or took control of the fire ground. On the basis of a review of radio communications and interviews with surviving firefighters, CSB found no clear messaging or discussion among the responding firefighters on who should assume the role of the designated IC.

Without a delegated IC officially taking control of the fire ground operations, no ICS was established. Consequently, no senior emergency response personnel or IC was responsible for coordinating the various response activities carried out by individual firefighters on the scene. The West fire chief arrived on scene at about 7:41 p.m. and did not critically assess the conditions on the ground before the explosion 10 min later, at about 7:51 p.m. The fire chief and assistant chief provided support and advisory functions but did not actively engage in fire ground function or take control of the fire ground; no record indicated that the West fire chief took command of the incident upon his arrival. Without direction to the contrary, the firefighters immediately took offensive action against the flames coming from the doors on north end of the east side of the structure.

CSB interviews with surviving firefighters indicated that before the arrival of the fire chief, the other senior firefighters who had reached the incident scene about 6 min earlier had not delegated senior personnel with the training and expertise needed to formally assume responsibility as the IC. The firefighters had not reached a conclusion about how to

326 *Hazmatology: The Science of Hazardous Materials*

establish a best approach and how to respond to the fire when the explosion occurred. Despite being trained for the ICS and NIMS process, none of the certified firefighters had prior practical experience in establishing incident command or coordinating and maintaining control of any previous emergency that merited the same approach as an FGAN-related fire scene.

Lack of Established Incident Management System

CSB found that the emergency response personnel who responded to the WFC incident did not take time to set up, implement, and coordinate an effective incident management system plan that would have ensured evacuation of the nearby residents. Because no formal IC was in charge of the incident, none of the firefighters took responsibility for formally establishing and coordinating an effective incident management system.

Firefighter Training

Firefighters must cope with extraordinary situations and circumstances that threaten their personal safety. To improve execution and reduce the threat of injury or loss of life, it is vital for both volunteer and career firefighters to receive thorough training and information supporting effective decision making. CSB's investigation of the WFC incident revealed that no standardized training requirement applies to volunteer firefighters across the nation.

Nationally, CSB found that the curriculum used for HAZMAT training does not fully address the hazards and severity of FGAN-related fires and explosions. A review of the U.S. Fire Administration (USFA) National Fire Academy HAZMAT field course outlines confirmed that they place little emphasis on emergency response to storage sites containing dangerous reactive chemicals and oxidizers such as FGAN.

CSB concludes that the current training resources at the local, state, and federal levels do not provide sufficient information for firefighters to understand the hazards of FGAN. It is therefore essential for firefighter and emergency response training institutions to collaborate with fire departments to develop and implement a realistic process for ensuring that hazard response knowledge, once attained, does not become unused and obsolete.

Although many firefighter training courses provide overviews of initial fire scene size-up, assessment, incident planning, and execution, CSB found that none of the firefighter HAZMAT field training courses provide sufficient information on firefighter situational awareness and risk assessment that could help them make informed decisions while at the fire scene. The firefighters who initially responded to WFC did not have the tools to effectively perform the situational awareness and risk assessment that would have enabled them to make an informed decision to not fight the fire. Situational awareness in firefighting involves the capability to "read" the scene of a fire or emergency, including changes in the behavior of a fire. Effective situational awareness supports prompt decision making to either evacuate the scene of a fire or continue fighting the fire by taking a defensive or offensive stance. Chapter 4 of NFPA 472 (*Standard for Competence of Responders to Hazardous Materials/Weapons of Mass Destruction Incidents*, 2013 Edition) provides guidance on situational awareness competencies for responder-level personnel.

In fires involving HAZMAT, it is not always possible for firefighters to obtain needed information before acting, but they might be able to characterize a HAZMAT incident based on initial information acquired from the emergency call center and dispatcher; emergency response manuals and guides; knowledge base on the response area; and visual, auditory, and olfactory (odorous) clues. In some cases, the fire department's standard operating procedures (SOPs) and the level of training of the emergency response crew might be insufficient to respond at the incident scene to changing events and scenarios that were not planned for or anticipated—hence, the need for effective training on situational awareness and risk assessment.

The fire department did not have a formal pre-incident planning program for FGAN at WFC. Firefighters responding to the incident were aware of the risks associated with anhydrous ammonia leaking from the tanks and that it could form a toxic flammable cloud that could leave the facility, drift into nearby homes, and potentially explode. Although some responding firefighters knew that FGAN was onsite, they did not anticipate a possible FGAN explosion. Some of the West fire department officials reported that they were aware of the chemicals routinely stored at the WFC, but there was never any formal training to prepare for a fire or chemical emergency. Effective site-specific pre-incident planning for emergency responders is essential to guide initial and subsequent actions while responders are at an emergency. Onsite pre-incident planning might have identified the possible FGAN explosion hazard. CSB did not find evidence of regularly scheduled training exercises to ensure that the WVFD conducted incident pre-planning and facility tours to address fire safety and chemicals onsite.

Most firefighting apparatuses have a copy of the ERG. After the WFC incident, NIOSH investigators found copies of the 2012 ERG in the glove boxes of some of the damaged fire equipment and apparatuses. However, CSB does not have any evidence that indicates whether the West firefighters consulted the ERG on the night of the explosion. The ERG is especially useful in situations when the relevant SDS is not readily available to firefighters. The ERG gives direction (based on DOT Hazard Classification

328   Hazmatology: The Science of Hazardous Materials

Criteria) on response to HAZMAT and dangerous goods emergencies during transportation.

Lessons Learned

On May 29, 2014, at around 5:45 p.m., a fire involving FGAN occurred at the East Texas AG Supply facility in downtown Athens, Texas. Emergency dispatchers and the Athens Police Department promptly notified firefighters from the Athens Fire Department (AFD). Emergency response units from the AFD arrived on the scene of the fire at 5:50 p.m. and found fire and smoke coming from the northwest end of the 3,500-square-foot East Texas Ag Supply facility. This facility was built with masonry bricks and combustible wooden structures, similar to construction at the WFC facility. The AFD chief arrived about 2 min after the first responding units were dispatched to the site of the incident, and he found that the fire had self-ventilated at the northwest end. On the basis of his observation of the enormous scope of the fire and the possibility of detonation of FGAN in the engulfed building, the fire chief promptly decided to let the East Texas Ag Supply facility burn to the ground instead of attempting to fight the fire. He ordered his firefighters to retreat from the scene and began an extensive evacuation of the downtown Athens, Texas, area.

Key Findings

The presence of combustible materials used for construction of the facility and the fertilizer grade ammonium nitrate (FGAN) storage bins, in addition to the West Fertilizer Company (WFC) practice of storing combustibles near the FGAN pile, contributed to the progression and intensity of the fire and likely resulted in the detonation. The WFC facility did not have a fire detection system to alert emergency responders or an automatic sprinkler system to extinguish the fire at an earlier stage of the incident.

The West Volunteer Fire Department (WVFD) did not conduct preincident planning or response training at the WFC facility to address FGAN-related incidents because was no such regulatory requirement. Thus, the firefighters who responded to the WFC fire did not have sufficient information to make an informed decision on how best to respond to the fire at the fertilizer facility.

Federal and state of Texas curriculum manuals used for hazardous materials (HAZMAT) training and certification of firefighters placed little emphasis on emergency response to storage sites containing FGAN.

Lessons learned from previous FGAN-related fires and explosions were not shared with volunteer fire departments, including the WVFD. If previous lessons learned had been applied in West, the firefighters and

Chronicles of Incidents and Response

emergency personnel who responded to the incident might have better understood the risks associated with FGAN-related fire.

CSB's analysis of the emergency response concludes that the West Volunteer Fire Department did not conduct pre-incident planning or response training at WFC, was likely unaware of the potential for FGAN detonation, did not take recommended incident response actions at the fire scene, and did not have appropriate training in hazardous materials response.

CSB found several shortcomings in federal and state regulations and standards that could reduce the risk of another incident of this type. These include the Occupational Safety and Health Administration's Explosives and Blasting Agents and Process Safety Management standards, the Environmental Protection Agency's Risk Management Program and Emergency Planning and Community Right-to-Know Act, and training provided or certified by the Texas Commission on Fire Protection and the State Firefighters' and Fire Marshals' Association of Texas.

Chemical Safety Board Investigation Report

Lac-Megantic, Quebec, CA July 6, 2013
Train Derailment Crude Oil Fire

With all the locomotives shut down, the air compressor no longer supplied air to the air brake system. As air leaked from the brake system, the main air reservoirs were slowly depleted, gradually reducing the effectiveness of the locomotive air brakes. At 00:56, the air pressure had dropped to a point at which the combination of locomotive air brakes and hand brakes could no longer hold the train, and it began to roll down hill toward Lac-Mégantic, just over seven miles away. A witness recalled watching the train moving slowly toward Lac-Mégantic without the locomotive lights on. The track was not equipped with track circuits to alert the rail traffic controller to the presence of a runaway train.

Gathering momentum on the long downhill slope, the train entered the town of Lac-Mégantic at high speed. The TSB's final report concluded that the train was travelling at 105 km/h (65 mph), more than triple the typical speed for that location. The rail line in this area is on a curve and has a speed limit for trains of 16 km/h (10 mph) as it is located at the west end of the Mégantic rail yard. Just before the derailment, witnesses recalled observing the train passing through the crossing at an excessive speed with no locomotive lights, "infernal" noise and sparks being emitted from the wheels. The unmanned train derailed in downtown Lac-Mégantic at 1:14, in an area near the grade crossing where the rail line crosses Frontenac Street, the town's main street. This location is approximately 600 m (2,000 ft) northwest of the railway bridge over the Chaudière River and is also immediately north of the town's central business district.

People on the terrace at Musi-Caféa bar located next to the centre of the explosions saw the tank cars leave the track and fled as a blanket of oil generated a ball of fire three times the height of the downtown buildings.

Between four and six explosions were reported initially as tank cars ruptured and crude oil escaped along the train's trajectory (Figure 1.135). Heat from the fires was felt as far as 2 km (1.2 mile) away. Blazing oil flowed over the ground, it entered the town's storm sewer and emerged as huge fires towering from other storm sewer drains, manholes, and even chimneys and basements of buildings in the area. The equipment that derailed included 63 of the 72 tank cars as well as the buffer car. Nine tank cars at the rear of the train remained on the track and were pulled away from the derailment site and did not explode. Almost all of the derailed tank cars were damaged, many having large breaches. About 6 million L of petroleum crude oil was quickly released; the fire began almost immediately.

Around 150 firefighters were deployed to the scene, described as looking like a "war zone". Some were called in from as far away as the city of Sherbrooke, Quebec, and as many as eight trucks carrying 30 firefighters were dispatched from Franklin County, Maine, United States (Chesterville, Eustis, Farmington, New Vineyard, Phillips, Rangeley and Strong). The fire was contained and prevented from spreading further in the early afternoon. The local hospital went to Code Orange, anticipating a high number of casualties and requesting reinforcements from other medical centers, but they received no seriously injured patients. A

Figure 1.135 Between four and six explosions were reported initially as tank cars ruptured and crude oil escaped along the train's trajectory. (Courtesy: Franklin County Maine, USA.)

Canadian Red Cross volunteer said there were "no wounded. They're all dead". Approximately 1,000 people were evacuated initially after the derailment, explosions, and fires. Another 1,000 people were evacuated later during the day because of toxic fumes. Some took refuge in an emergency shelter established by the Red Cross in a local high school.

After 20 h, the centre of the fire was still inaccessible to firefighters and five pools of fuel were still burning. A special fire-retardant foam was brought from an Ultramar refinery in Lévis, aiding progress by firefighters on the Saturday night. Five of the unexploded cars were doused with high-pressure water to prevent further explosions, and two were still burning and at risk of exploding 36 h later. The train's event recorder was recovered at around 15:00 the next day and the fire was finally extinguished in the evening, after burning for nearly 2 days.

Forty-two bodies were found and transported to Montreal to be identified. Thirty-nine of those were identified by investigators by late August 2013 and the 40th in April 2014.

Identification of additional victims became increasingly difficult after the August 1 end of the on-site search and family members were asked to provide DNA samples of those missing, as well as dental records. The bodies of five presumed victims were never found. It is possible that some of the missing people were vaporized by the explosions. At least 30 buildings were destroyed in the centre of town, including the town's library, a historic former bank, and other businesses and houses. 115 businesses were destroyed, displaced, or rendered inaccessible (Federal Railroad Administration (FRA) Canada).

Casselton, ND December 30, 2013
Crude Oil Train Derailment & Fire

On Monday, December 30, 2013, at 2:10 p.m. central standard time, a westbound BNSF Railway Company (BNSF) train with 112 cars loaded with grain derailed 13 cars while traveling on main track 1 at milepost 28.5 near Casselton, North Dakota. The first car that derailed (the 45th car) fouled the adjacent track, main track 2. At 2:11 p.m. an eastbound BNSF train with 104 tank cars loaded with petroleum crude oil (crude oil), traveling on main track 2, struck the derailed car that was fouling the track and derailed two head-end locomotives, a buffer car, and 20 cars loaded with crude oil. After the collision, about 476,000 gallons of crude oil were released and burned. On the day of the accident, the weather was cloudy with a temperature of −1°F and winds from the north at 7 mph. No injuries were reported by residents or either of the train crews. The BNSF reported damages of $13.5 million, not including lading and environmental remediation.

The National Transportation Safety Board determines that the probable cause of the collision of the oil train with the derailed grain train car was a broken axle on the 45th car of the grain train caused by an internal void that was created during axle manufacture. Contributing to the cause of the derailment were inadequate interchange rules used to locate internal material defects in second-hand-use axles. Contributing to the severity of the accident was the release and pooling of a highly flammable product that resulted in a fire and caused additional cars to fail (National Transportation Safety Board).

Lynchburg, VA April 30, 2014 Crude Oil Train Derailment & Fire

On April 30, 2014, at 1:54 p.m. eastern daylight time, 17 CSX Transportation (CSXT) tank cars on petroleum crude oil unit train K08227 derailed in Lynchburg, Virginia. Three of the derailed cars were partially submerged in the James River. One was breached and released about 29,868 gallons of crude oil into the river, some of which caught fire (Figure 1.136). No injuries to the public or crew were reported. At the time of the accident, it was cloudy and raining lightly; the temperature was 53°F. The CSXT estimated the damages at $1.2 million, not including environmental remediation. The National Transportation Safety Board determines that the probable cause of this accident was a broken rail caused by a reverse detail fracture with evidence of rolling contact fatigue (National Transportation Safety Board).

Figure 1.136 Three of the derailed cars were partially submerged in the James River. One was breached and released about 29,868 gallons of crude oil into the river, some of which caught fire. (From NTSB.)

Heimdal, ND May 6, 2015 Crude Oil Train Derailment & Fire

On May 6, 2015, at 7:21 a.m. central daylight time, a BNSF Railway (BNSF) crude oil unit train derailed six cars (81 through 86) near Heimdal, North Dakota. The train, consisting of three locomotives, two buffer cars, and 107 loaded tank cars carrying crude oil, was operating at 45 mph when the cars derailed. The train separated after a broken wheel on the 81st car struck the leading edge of the highway-rail grade crossing at milepost 149.01. A mark on the track structure at milepost 153.87 indicated that the broken wheel could not maintain its normal position on the rail at that point and the derailment sequence began. The momentum of the train pulled the 81st car and the following five cars off the track. Five of the derailed tank cars breached and released about 96,400 gallons of crude oil, which fueled a fire about 1 mile east of Heimdal (Figure 1.137). About 30 people were evacuated from Heimdal and the surrounding area due to the smoke plume that extended north. At the time of the accident, the sky was overcast and the temperature was 57°F. BNSF estimated damage at $5 million.

The National Transportation Safety Board determines that the probable cause of the collision of the oil train with the derailed grain train car was a broken axle on the 45th car of the grain train caused by an internal void that was created during axle manufacture. Contributing to the cause of the derailment were inadequate interchange rules used to locate internal material defects in secondhand-use axles. Contributing to the severity of the accident was the release and pooling of a highly flammable

Figure 1.137 Five of the derailed tank cars breached and released about 96,400 gallons of crude oil, which fueled a fire about 1 mile east of Heimdal. (From NTSB.)

product that resulted in a fire and caused additional cars to fail (National Transportation Safety Board).

Houston, TX July 16, 2017 Propane Tanker Crash
RIMS Incident

July 16, 2017, 8:15 a.m. an 18 wheeler which included an MC331 Propane Tanker went around a curve and hit a concrete embankment at 55–60 mph and skidded 200 ft on the concrete roadway. Friction between the steel tank and the concrete left scrape marks on the tanker's side (Figure 1.138). Considering the mechanism of injury (damage) to the tank, hazmat team members on arrival did a visual inspection of the tank and air monitoring to determine if there was a leak. Liquid propane is transported in a non-insulated tank so the temperature of the liquid is much the same as ambient temperature.

When the tank was loaded in Arkansas the ambient temperature was cool. As ambient temperature increases, so does the pressure inside the tank. Increases in heat of any kind when chemicals are involved are dangerous. Increase in heat from any source is the worst case scenario when dealing with a hazardous material inside a container. MC 331 tankers are equipped with temperature, pressure and liquid level gauges. These can be helpful in determining what is going on in a tank that is not leaking. If a tank is leaking or on fire, it is too dangerous to

Figure 1.138 Friction between the steel tank and the concrete left scrape marks on the tanker's side. (Courtesy: Houston Fire Department.)

Chronicles of Incidents and Response

worry about these gauges. This situation was ideal for using the gauges for important tank information. Gauges give you a visual indication there is a leak. At the time of the incident, the temperature in the container was 85°F and the tank pressure was 130. Weather reports indicated that summer temperatures in Houston would reach 110°F–115°F during the day. That temperature increase would raise the pressure inside the tank. The decision was made to keep the tank cool with a Ventura using a hose stream, which would also provide a secondary cooling effect.

This accident occurred on an elevated section of the expressway and there was no water supply. 1000 ft of 5 in hose was for a supply line was used to supply a ladder pipe that was used to hook the hose to, like an artificial standpipe. A tent was fashioned from the ever innovative firefighting tool, the salvage cover. The tank was tented and hose line placed to begin the cooling process. The entire well planned process worked as intended and kept the tank cool until it could be safely offloaded and the accident scene investigated and cleared of debris. No further damage or injury occurred (Houston, Texas Fire Department).

Cambria, WI May 31, 2017 Didion Mill Explosion

At approximately 11:00 PM on May 31, 2017, explosion(s) at the Didion Milling (Didion) facility in Cambria, Wisconsin, resulted in 5 worker deaths and an additional 14 workers injured (Figure 1.139). Because the event occurred at night, only 19 employees were working within the facility at the time of the incident. Shortly before the explosion(s) at Didion, workers saw or smelled smoke on the first floor of one of the mill buildings. In trying to find its source, workers focused on a piece of equipment called a gap mill. While inspecting the equipment, workers witnessed a filter connected to an air intake line for the mill blow off, resulting in corn dust filling the air, and flames shooting from the air intake line, followed by one or more explosions.

After the CSB's initial assessment on site, the agency mobilized additional investigators, structural and blast engineers with expertise in dust explosions, and a drone team to enhance its field capabilities. The CSB and its contractors conducted limited entries into a number of the mill buildings and completed ground and elevated surveys of the blast damage in an effort to determine the origins of the event and the possible contributing role of the dust (see call-out box). Based on evidence the CSB collected from the structural damage patterns in the mill buildings and from employee reports of corn dust and fire coming out of the gap mill air intake line just before the explosion(s), the CSB considers the Didion incident to be one or more dust explosions.

Figure 1.139 Explosion(s) at the Didion Milling (Didion) facility in Cambria, Wisconsin resulted in 5 worker deaths and an additional 14 workers injured. (From Chemical Safety Board.)

This interim Factual Investigative Update provides a multi perspective narrative of the May 31 event through detailed accounts from Didion workers present in the facility before and during the incident. The CSB's full technical analysis of the incident and its examination of dust management at Didion will be published after it is completed.

The dry corn milling process, like the one used at Didion, is inherently a dust-producing operation. Corn dust is combustible and known to be capable of generating overpressures under the right conditions. Dust fires, like all fires, occur when fuel (combustible dust) is exposed to heat (an ignition source) in the presence of oxygen (air). To become an explosion, a dust fire requires two additional elements: dispersion to create a dust cloud and confinement. Burning a dust cloud in the open generally results in a flash-fire and the generation of combustion gases. If that same fire is confined within a building, vessel, or piece of equipment, the expansion of the combustions gases can build up pressure until the enclosure bursts and releases a pressure wave capable of damaging buildings and causing serious human injuries. Dust explosions are typically classified as either *primary* or *secondary*. A primary, or initial, explosion can trigger secondary explosions when combustible dust or fines that have accumulated on floors or other surfaces are lofted into the air, or when either is thrown into the air from damaged equipment, and ignites. If a sufficient volume of dust or fines are lofted or secondary releases occur, a relatively weak primary explosion can trigger very powerful secondary dust explosions (Chemical Safety Board).

Hyndman, PA August 2, 2017 Train Derailment & Fire

On August 2, 2017, about 4:54 a.m. eastern daylight time, CSX Transportation (CSX) train Q38831, with 5 locomotives and 178 railcars (128 loaded and 50 empty), derailed railcars 33 through 65 in Hyndman Borough, Bedford County, Pennsylvania. Three hazardous material tank cars were breached and released material: one containing propane, one containing asphalt, and one containing molten sulfur. Both the propane tank car and the molten sulfur tank released caught fire (Figure 1.140). About 1,000 residents within a 1-mile radius were evacuated and several highway-railroad grade crossings were closed. There were no injuries or fatalities. On August 5,

Figure 1.140 Three hazardous material tank cars were breached and released material: one containing propane, one containing asphalt, and one containing molten sulfur. Both the propane tank car and the molten sulfur tank released caught fire. (NTSB.)

338 *Hazmatology: The Science of Hazardous Materials*

about 12:00 p.m., the evacuation was lifted. At the time of the accident, the sky was clear, visibility was 10 miles, and the temperature was 64°F.

Two train crews were involved in the movement of the train before the accident. The first crew stopped the train on a descending grade after encountering suspected air brake problems. The crew applied 58 hand brakes while inspecting and recharging the air brake system. The Conductor of the first crew found the leak on a railcar about 20 railcars from the rear of the train. A CSX mechanical employee arrived to repair the air leak. However, by the time this issue was resolved, the crew did not have enough remaining duty time to complete the trip. Therefore, CSX relieved them with a new train crew.

The second crew, thinking the train may still have air brake problems, kept all 58 hand brakes applied and unsuccessfully tried to pull the train downhill. The conductor of the second crew then released the first 25 hand brakes, leaving 33 hand brakes still applied. The engineer applied a minimum air brake application and started the train with locomotive power down the grade. The train speed varied from 20 to 30 mph. The engineer transitioned from locomotive power to dynamic braking three times before the train derailed.

National Transportation Safety Board (NTSB) investigators determined that the 35th railcar derailed one set of wheels on a curve 1.7 miles before the location of the general derailment and fire. When the derailed railcar reached a highway railroad grade crossing, the railcar moved further off the rail, initiating the derailment of the other railcars. NTSB investigators discovered that several railcar wheels east and west of the derailment cars had flat spots and build-up tread from the hand brakes not allowing the wheels to rotate, and bluing due to pad friction.

The weight of the train from the 1st railcar in the consist to the 35th was about 1,631 tons. The weight of the remaining 143 railcars trailing the 35th, was about 16,621 tons. The 35th railcar from the front of the train was an empty high sided gondola within a block of 27 empty railcars. NTSB is investigating many factors into the cause of the derailment, including the length, make-up, and operation of the train, as well as the condition of the railcars and track. Fifteen tank cars transporting hazardous materials were involved in the derailment, including the three tank cars containing propane, eight tank cars containing molten sulfur, two tank cars containing asphalt, and two tank cars containing phosphoric acid residue.

The 46th car from the front of the train, a 23,467 gallon specification U.S. Department of Transportation (DOT) 111 general service tank car, released its load of elevated temperature asphalt from a bottom outlet valve that opened during the derailment sequence. The released asphalt pooled and solidified near the railcar pileup. The 49th railcar from the front of the train, a 11,880-gallon specification DOT-111 general service tank car, released its load of molten sulfur from two tank shell tears. The molten sulfur ignited and burned for more than 48 h. The 53rd car, a 33,710-gallon specification DOT 112 pressure tank car, released its load of odorized propane from a punctured tank shell. The puncture measured about 3½ in. in length and about 1/4 inch wide. Released propane gas escaped through the void space between the jacket and tank shell and emerged burning vigorously as several locations on the tank car for more than 49 h (National Transportation Safety Board).

Oklahoma City, OK August 16, 2018 Oil Well Blow Out

Oklahoma City, Oklahoma, August 16, 2018: Today the U.S. Chemical Safety Board (CSB) released a factual update into its ongoing investigation of the January 22, 2018, blowout and fire at the Pryor Trust Gas Well located in Pittsburg County, Oklahoma, that killed five workers (Figure 1.141).

The CSB has determined the incident occurred shortly after drilling crewmembers removed the drill pipe from the well in a process known as "tripping." To date, the CSB's investigation has determined the following timeline related to the blowout and fire:

Figure 1.141 Oil well blow out near Oklahoma City, OK killed five workers.

340   *Hazmatology: The Science of Hazardous Materials*

- January 21, 2018: Crewmembers from the Patterson-UTI Drilling Company had been drilling a gas well for over a week. Activities were being overseen by the operator of the well, Red Mountain Operating, LLC (or RMO) in Pittsburg County, Oklahoma.
- At 3:36 p.m., the Patterson crew stopped drilling to remove the drill pipe from the well and change the drill bit.
- At 6:48 p.m., crewmembers began the process of removing the drill pipe from the well. The Patterson crew pumped mud into the well during the removal of the drill pipe with the intent to keep the well full of mud. That operation involved closing an isolation valve to prevent mud from flowing out of the well. By 10:30 p.m., the end of the drill pipe reached the top of the curve in the well.
- January 22, 2018: At 12:35 a.m., the crew pumped fluid also referred to as a "weighted pill" above the top of the curve to prevent gas influx into the lateral portion of the well.
- At 1:12 a.m., the crew began removing the drill pipe from the vertical section of the well. For this portion of the operation, the Patterson crew performed a "Continuous Fill" tripping method. Mud was continuously circulated in the wellbore using the Trip Tank pumps to keep the well full by replacing the volume of the drill pipe removed with drilling mud. The isolation valve was open for this operation.
- At the start of this procedure, the drill pipe started pulling "wet" – meaning the drill pipe being removed had not drained and still contained mud. The Patterson crew was aware that the pipe was plugged. At this point the Patterson crew attempted to pump a volume of mud also referred as a "slug" into the drill pipe to push the mud that remained in the drill pipe out, but this was not successful as the drill pipe was plugged. Therefore, the drill pipe in the vertical section was removed while it still contained mud.
- By 6:10 a.m., the drill pipe and drill bit were completely removed from the well. At that time, the driller closed the blind rams on the well's blowout preventer.
- At 7:57 a.m., the driller opened the blowout preventer blind rams so that a new piece of drilling equipment called a bottom hole assembly could be lowered into the well. At 8:09 a.m., mud was pumped through the bottom hole assembly to test the new equipment.
- While the rig crew tested the bottom hole assembly equipment, the mud pits gained 107 barrels of mud. Mud pit gains are an indication of a possible gas influx in the well. Data obtained by the CSB indicates that conditions existed that could have allowed a gas influx into the wellbore during the tripping operation.

Chronicles of Incidents and Response 341

- At 8:35 a.m., with testing complete, the bottom hole assembly was lifted out of the wellbore. At 8:36 a.m., mud blew upwards out of the well. The mud and gas from the well subsequently ignited causing a large fire.

In addition to the factual report, the CSB has released an animated timeline of the events leading up to the fatal blowout and fire. The animation can be found at the CSB website. The CSB investigation is ongoing. Investigators will develop a root cause analysis of this incident based on evidence collected during the investigation. A final report, including facts, analysis, conclusions, and recommendations will be issued at the end of the investigation (Chemical Safety Board).

Tilford, SD September 8, 2018
Propane Tank Explosion

Firefighters from Sturgis Fire Department received a call for a house fire in Tilford located between Rapid City and Sturgis at 4:00 p.m. for a reported structure fire. Heavy smoke was visible as firefighters approached the scene and they found a fully involved single family dwelling upon arrival. Initial attempts to battle the fire were hampered by the neighborhoods narrow gravel streets, congestion of nearby homes, sheds and garages, parked vehicles, and over growth of summer grasses and underbrush. Witnesses said there were several very large propane tanks on the property.

The property where the fire was located further hampered firefighting efforts by downed power lines and several propane tanks while fighting the fire. Fire spread from the house to several outbuildings and impinged upon a 500 gallon propane tank that sat on the south side of the structure. It was reported that the tank had been recently filled. While fighting the fires a BLEVE occurred involving a propane tank that failed due to exposure to the fire. Chief Fischer was attempting to move a Sturgis Volunteer Fire Department vehicle, a Chevrolet Suburban parked in a driveway parked north of the burning home, when the portion of the exploding tank cleared the burning home, a fire engine, and the Suburban and struck Fischer, killing him instantly. Sturgis Assistant Fire Chief David Fischer was a 22-year veteran of the fire department, and also worked for Sturgis Ambulance Service. He was also a member of the South Dakota National Guard's 82nd Civil Support Team.

Staff Sgt. Fischer was 43 years old. Chief Fischer was struck by a large portion of the propane tank when it exploded. It was determined that the fire was started by an elderly man who was in bed smoking while on oxygen. He perished in the fire. It is unclear whether he died from the fire in

342   *Hazmatology: The Science of Hazardous Materials*

the dwelling or the propane tank explosion. Reportedly the man could not get around very well on his own. Another woman in the house escaped unharmed (Rapid City Journal Newspaper, Rapid City, SD September 8, 2018).

> **Hazmatology Point:** *Propane tank explosions over the past 60 years have been the second leading cause of firefighter deaths involving hazardous materials incidents and on the scene of structural fires. Over the past 20 years, the fatalities have involved smaller tanks and have been found as a result of a structural fire response. Firefighters in Lincoln, Nebraska told me in 2018 they had a close call with a propane tank leaking on a barbeque grill. They had just turned the corner of a house when the tank exploded. Had they been standing next to it, the outcome could have been tragic.*

Deer Park, TX March 17, 2019
Petrochemical Storage Tank Fires

The U.S. Chemical Safety Board will investigate the recent fire at the Intercontinental Terminals Company (ITC) site in Deer Park, TX. CSB investigators will start interviews next week and plan to be on site for several days to document the scene and collect evidence. The massive fire, which began on March 17th, engulfed 11 above ground storage tanks containing a variety of hydrocarbons, resulting in multiple orders for community members to Shelter in Place (Chemical Safety Board).

September 16, 2019 Farmington,
ME Propane Explosion

One firefighter was killed and 6 were injured when an explosion occurred in a newly constructed 40 x 60 foot, two story building housing administrative offices for LEAP, Inc. The company provides support for people with developmental, cognitive and intellectual disabilities. Also destroyed were 11 mobile homes. The building was located at 313 Farmington Falls Road, which is also State Route 2. It was reported that a school bus with children aboard had just left the area before the explosion occurred. Firefighters responded to a report of smell of gas and were investigating when the explosion occurred without warning at 8:30 a.m. e.s.t. Injured firefighters were taken to Maine Medical Center in Portland. Deputy Chief Ross and an EMT from an ambulance company were treated and released. Five firefighters were admitted, four to ICU and one with less serious injuries. The LEAP building was completely and literally blown to pieces. State Fire Marshal Joseph Thomas said "Damage may be so severe it may be hard to pinpoint what happened." Farmington Police Chief Jack Peck commented: "We train, but I don't know that we've ever trained for something that we saw this morning, and we certainly don't expect to lose someone we've worked with for years, a loved one and member of our family."

Farmington, ME Sun/Journal Monday September 16, 2019

Firefighter that Made The Supreme Sacrifice

Captain Michael Bell, 68, a 30-year member of the Farmington Fire Rescue Department and brother of the department's chief. (Figure 1.142)

Figure 1.142 Captain Michael Bell

344 *Hazmatology: The Science of Hazardous Materials*

Firefighters Injured

Chief Terry Bell, 62

Deputy Chief Clyde Ross

Captain Timothy Hardy, 40

Captain Scott Baxter, 37

Firefighter Theodore Baxter, 64 Father of Scott Baxter

Firefighter Joseph Hastings, 24

Author's Note: This incident occurred while I was in the process of editing proofs of this volume. Information presented here was on the day the incident happened and was the best available at the time.

CSB Releases Call to Action on Combustible Dust Hazards

Washington, DC, October 24, 2018: Today, the U.S. Chemical Safety Board (CSB), as part of its investigation into the May 2017 Didion Mill explosion, issued the "Call to Action: Combustible Dust" to gather comments on the management and control of combustible dust from companies, regulators, inspectors, safety training providers, researchers, unions, and the workers affected by dust-related hazards. According to CSB Interim Executive Kristen Kulinowski, "Our dust investigations have identified the understanding of dust hazards and the ability to determine a safe dust level in the work place as common challenges." She further stated, "While there is a shared understanding of the hazards of dust, our investigations have found that efforts to manage those hazards have often failed to prevent a catastrophic explosion. To uncover why that is, we are initiating this Call to Action to gather insights and feedback from those most directly involved with combustible dust hazards."

This initiative asks for information from all individuals and entities involved in the safe conduct of work within inherently dust-producing environments at risk for dust explosions. The agency seeks input on a variety of complex issues, including recognizing and measuring "unsafe" levels of dust in the workplace, managing responsibilities and expectations that sometimes are at odds with each other (e.g., performing mechanical integrity preventative maintenance while simultaneously striving to minimize dust releases in the work environment), and the methods for communicating the low-frequency but high-consequence hazards of combustible dust in actionable terms for those working in and overseeing these environments. A full list of questions is given at the end of this section.

Chronicles of Incidents and Response 345

Comments can be emailed to combustibledust@csb.gov until November 26, 2018. The CSB will use the information to explore new opportunities for safety improvements. Dust incidents continue to impact a wide swath of industries. In 2006, it identified 281 combustible dust incidents between 1980 and 2005. One hundred and nineteen workers were fatally injured, seven hundred and eighteen were hurt, and industrial facilities were extensively damaged. These incidents occurred in 44 states, in many different industries, and involved a variety of different materials.

Since the publication of the study in 2006, the CSB has confirmed an additional 105 combustible dust incidents and has conducted in-depth investigations of five, including the recent Didion Milling dust explosion in Cambria, Wisconsin that fatally injured five workers and demolished the milling facility.

The CSB has issued four recommendations to Occupational Safety and Health Administration (OSHA) calling for the issuance of a comprehensive general industry standard for combustible dust. Combustible dust safety is on the agency's Drivers of Critical Chemical Safety Change list. To date, no general industry standards exist.

According to CSB Investigator Cheryl MacKenzie, "Our investigation of the Didion incident continues and we are analyzing evidence to understand the specifics leading up to the tragic event. However, this investigation reinforces what we are seeing across many industries—that there needs to be a more inclusive approach to creating and maintaining a safe work environment amid processes that inherently produce dust."

The CSB is an independent, non-regulatory federal agency whose mission is to drive chemical safety change through independent investigations to protect the people and the environment. The agency's board members are appointed by the president and confirmed by the senate. CSB investigations look into all aspects of chemical incidents, including physical causes such as equipment failure, as well as inadequacies in regulations, industry standards, and safety management systems. For more information, contact public@csb.gov.

The CSB asks for comment from companies, regulators, inspectors, safety training providers, researchers, unions, and workers of dust-producing operations on some very fundamental questions. Some of these questions are:

- In real-world working conditions, where dust is an inherent aspect of the operation, can a workplace be both dusty and safe?
- In such working environments—where the amount of ambient/fugitive dust cannot be wholly eliminated 100% of the time—how does an individual or organization distinguish between an acceptable or safe dust level and one that has been exceeded? How often does judgment or experience play a role in such decisions? Should it?

346   *Hazmatology: The Science of Hazardous Materials*

- How are hazards associated with combustible dust communicated and taught to workers? What systems have organizations successfully used to help their employees recognize and address dust hazards?
- What are some of the challenges you face when implementing industry guidance or standards pertaining to dust control/management?
- If companies/facilities need to use separate or different approaches in order to comply with both sanitation standards for product quality or food safety and those associated with dust explosion prevention, then how do you determine what takes priority? Is the guidance clear?
- How should the effectiveness of housekeeping be measured? What methods work best (e.g., cleaning methods, staffing, schedules)?
- As equipment is used and ages, it requires mechanical integrity to maintain safe and efficient operability. How does inspection, maintenance, and overall mechanical integrity efforts play a role in dust accumulations, and how are organizations minimizing such contributions in the workplace?
- What are some of the challenges to maintaining effective dust collection systems?
- How common are dust fires in the workplace that do not result in an explosion? Does this create a false sense of security?
- Are workers empowered to report issues when they feel something needs to change with regard to dust accumulation? What processes are in place to make these concerns known?
- How can combustible dust operators, industry standard organizations, and regulators better share information to prevent future incidents?

The CSB will review all responses submitted by November 26, 2018 and use the information to explore the conditions that influence the control and management of combustible dust to seek out a deeper understanding of the real-world challenges to preventing dust explosions and, more importantly, new opportunities for safety improvements.

To Date, the CSB Has Issued Four Recommendations

OSHA calling for the issuance of a comprehensive general industry standard for combustible dust and combustible dust safety is on the agency's Drivers of Critical Chemical Safety Change list. Yet, development of a general industry standard has not come to fruition. With this publication, the CSB aims to spearhead actionable dialogue between industry,

Chronicles of Incidents and Response

regulator, workforce, and others to achieve safety improvements in the management and control of combustible dust beyond regulatory promulgation. As mentioned earlier, in 2006, the CSB identified 281 combustible dust incidents between 1980 and 2005. Since the publication of this study in 2006, the CSB has confirmed an additional 105 combustible dust incidents, with the CSB conducting in-depth investigations of five of them, including most recently the Didion Milling dust explosion in Cambria, Wisconsin. These five incidents alone have taken the lives of 27 workers and injured 61 others. In each investigation, one or more witnesses conveyed similar messages to CSB, such as their site was "the cleanest it has ever been," that "people cleaned all the time," or that "they never thought it [the incident] would happen here." Overall, workers and management personnel from various CSB investigation sites had similar perceptions of their work environments: Dust was present, normal, and maintained at a "safe" or "manageable" level. These commonalities between companies, which differ in their dust-producing operations and their industry (e.g., sugar, corn, automotive insulation), suggest that similar real-world challenges exist for each of them regarding the safe identification and management of dust. There is value in unearthing these challenges and the extent to which they hinder dust-producing facilities from preventing the "next" dust explosion. As has been noted by others, there is a deep need after any fatal incident to understand what happened so that everything possible can be done to prevent another incidence. Prevention is not as easy as learning what people should or should not have done at a specific incident. It requires a thorough examination of the system that puts people in positions where they felt that their actions were the best option.

As such, while the CSB continues its investigation of the Didion incident to understand the specifics leading up to the tragic event, the agency aims to explore relevant topics with members of various combustible dust-producing industries, stakeholders, and technical experts to better understand the challenges to achieving a safe work environment amid processes that produce dust. This report is a springboard for those dialogues.

Perceptions about Dust Vary

As discussed in the *Didion Factual Investigative Update*, Didion employees had varying perceptions of dust accumulations within the facility. They also expressed varying sentiments regarding what they considered "clean" versus "dusty" with respect to dust accumulation. Some employees characterized the plant as "spic and span," while others reported that dust was constantly present in the work area. The perceived level of safety they each had, as it relates to the dust within their work environment, also varied. Interestingly, the CSB found strikingly similar variations in

348   *Hazmatology: The Science of Hazardous Materials*

the levels of hazard awareness and dust level perceptions between those working at Didion and at incident sites of previous CSB combustible dust investigations. The wide spectrum of perceptions that can be seen in individuals' statements regarding combustible dust call into question:

1. industry's collective understanding of the risks of combustible dust;
2. the adequacy of current efforts to manage the hazard; and
3. the effectiveness of current inspection methods for a proactive identification of "unsafe" levels of dust.

Factors Influencing Dust Hazard Perceptions

From an examination of statements made by workers and management, a number of factors appear to influence these varying perceptions of dust hazard risks and one's personal sense of safety. They include but are not limited to:

- Hazard awareness
  a. The degree to which workers and management have practical real-world understanding of combustible dust hazards will impact how they react to their environment when they observe dust. For example, do workers know how much dust is too much?
- Previous incidents and fires
  b. Observing fires or hot work activities in a combustible dust environment that did not result in an explosion could create a false sense of security. Regulatory oversight: regulatory requirements do not reinforce one another.

For example, sanitation requirements under the Food and Drug Administration (FDA) may meet food quality concerns but may not be sufficient to prevent a dust explosion.

Sanitation:
- Management and workers focus on cleaning all the time, providing a sense of vigilance; however, hazardous dust accumulation rates may exceed cleaning efforts.
- Ability to recycle material: in facilities where material can be recycled or reprocessed, there may be a greater tolerance for spills or leaks.
- Perceived difficulty in housekeeping efforts: as dust accumulates on hard-to-reach and overhead surfaces, workers perceive that those surfaces are too hard, or too dangerous, to reach for cleaning. Further exploration of these factors may yield new opportunities for accident prevention (Figure 1.143) (Chemical Safety Board).

21 Food Products

20 Motor Vehicle Manufacturing

20 Other Industrial

13 Other*

7 Chemical Manufacturing

6 Electric Services

6 Rubber and Plastic Products

5 Metal Industries

4 Lumber and Wood Products

3 Equipment Manufacturing

105 Total

Figure 1.143 Number of dust incidents by industry, 2006–2017. *Dust events in non-industrial facilities.

Dust Incidents 2006–2017

U.S. Chemical Safety Board (CSB)

1. Saft America, Inc. 3/29/2006 Cockeysville, MD **No injuries** Other Industrial
2. Leiner Health Products 4/18/2006 Fort Mill, SC **2 injured** Other Industrial
3. Package Kare, Inc. 5/2/2006 Middlesex, NJ **3 injured** Chemical Manufacturing
4. Midwest Generation, LLC 9/12/2006 Romeoville, IL **3 injured** Electric Services
5. Madison-Kipp Corporation 1/1/2007 Madison, WS **1 injured** Metal Industries
6. Risley Pellet Solutions, LLC 2/6/2007 Monticello, WS **1 injured** Lumber and Wood Products
7. CTA Acoustics, Inc. 3/14/2007 Corbin, KY **4 injured** Motor Vehicle Manufacturing
8. Imco 4/2/2007 Schulenburg, TX **1 injured** Motor Vehicle Manufacturing
9. Tires Into Recycled Energy & Supplies, Inc. 5/15/2007 Jackson, GA **1 fatality** and **1 injured** Rubber and Plastic Products

350   *Hazmatology: The Science of Hazardous Materials*

10. Maryland Cork Company, Inc. 6/12/2007 Elkton, MD **3 injured** Lumber and Wood Products
11. V.I.M. Recycling, Inc. 6/14/2007 Elkhart, IN **1 fatality** and **1 injured** Lumber and Wood Products
12. Deltic Timber Corporation 8/9/2007 Waldo, AZ **1 fatality** and **2 injured** Lumber and Wood Products
13. Rich Products 8/14/2007 Morristown, TN **2 injured** Food Products
14. International Specialty Alloys, Inc. 8/29/2007 New Castle, PA **3 injured** Metal Industries
15. Imperial Sugar Company 2/7/2008 Port Wentworth, GA **14 fatalities** and **36 injured** Food Products
16. Wells Pet Food Company 3/4/2008 Monmouth, IL **1 injured** Food Products
17. Port City Industrial Finishing 3/24/2008 Muskegon, MI **14 injured** Metal Industries
18. Advanced Environmental Recycling Technologies, Inc. 5/21/2008 Springdale, AR **1 injured** Lumber and Wood Products
19. Weiser Tent Company 6/19/2008 Monett, MO **2 injured** Rubber and Plastic Products
20. McCormick & Company, Inc. 7/21/2008 Hunt Valley, MD **1 injured** Food Products
21. New England Wood Pellet, LLC 8/10/2008 Jaffrey, NH **2 injured** Lumber and Wood Products
22. Industrial Roller Company 8/19/2008 Smithton, IL **1 injured** Rubber and Plastic Products
23. Port of Stockton District Energy Facility 9/2/2008 Stockton, CA **6 injured** Electric Services
24. Arizona Grain 12/29/2008 Maricopa, AZ **3 injured** Food Products
25. Indiana Laminates 1/9/2009 Jasper IN, **8 injured** Lumber and Wood Products
26. Eurest Services, Inc. 1/24/2009 Torrance, CA **1 injured** Other Industrial
27. WE Energies 2/3/2009 Oak Creek, WI **4 injured** Electric Services
28. Pacific Fruit Farms 2/25/2009 Walnut Grove, CA **1 injured** Food Products
29. Grand Auto Electric 3/31/2009 Monrovia, CA **1 injured** Motor Vehicle Manufacturing
30. RMD Americas 4/22/2009 Rockledge, FL **1 injured** Rubber and Plastic Products
31. The Gun Room 6/19/2009 Elk Grove, CA **1 injured** Other Industrial
32. Herdco, Inc. 6/23/2009 Bartlett, NE **2 injured** Food Products
33. Meadwestvaco Corporation 6/30/2009 Evandale, TX **1 fatality** Lumber and Wood Products

Chronicles of Incidents and Response 351

34. Ford Motor Company 7/22/2009 Lima, OH **1 injured** Motor Vehicle Manufacturing
35. Custom Alloy Scrap Sales 8/28/2009 Oakland, CA **1 injured** Metal Industries
36. Lock Joint Tube 9/29/2009 South Bend, IN **1 injured** Metal Industries
37. Berg Steel Pipe Corporation 10/5/2009 Panama City, FL **3 injured** Other Industrial
38. OPPD Power (Omaha Public Power District) 10/7/2009 Nebraska City, NE **2 injured** Electric Services
39. Quaker Oats Plant 12/15/2009 Cedar Rapids, IA **3 injured** Food Products
40. Bremer Manufacturing 12/29/2009 Elkhart Lake, WS **1 fatality** and **9 injured** Metal Industries
41. Roseburg Forest Products Company 2/12/2010 Dillard, OR **1 injured** Lumber and Wood Products
42. ConAgra 4/27/2010 Chester, IL **4 injured** Food Products
43. United Alloys & Metals, Inc. 7/13/2010 Los Angeles, CA **7 injured** Metal Industries
44. Kingsford Manufacturing Company 9/24/2010 Burnside, KY **1 fatality** and **3 injured** Lumber and Wood Products
45. Mid-America Repair Service 12/1/2010 Fairland, OK **3 injured** Other
46. AL Solutions 12/9/2010 New Cumberland, WV **3 fatalities** Metal Industries
47. Severstal North America 12/10/2010 Dearborn, MI **3 injured** Metal Industries
48. DSM Nutritional Products, Inc. 12/21/2010 Pendergrass, GA **2 injured** Other Industrial
49. Federal Mogul Corporation 12/31/2010 Blacksburg, VA **4 injured** Motor Vehicle Manufacturing
50. Hoeganaes Corporation 1/31/2011 Gallatin, TN **2 fatalities** Metal Industries
51. Yokohama Tire Corporation 3/4/2011 Salem, VA **1 injured** Rubber and Plastic Products
52. Progress Energy 3/15/2011 Wilmington, NC **1 fatality** Electric Services
53. Hoeganaes Corporation 3/29/2011 Gallatin, TN **1 injured** Metal Industries
54. Agilent Technologies 4/26/2011 Santa Rosa, CA **2 injured** Metal Industries
55. Eckart America 5/9/2011 Louisville, KY **2 injured** Metal Industries
56. Hoeganaes Corporation 5/27/2011 Gallatin, TN **3 fatalities** and **2 injured** Metal Industries

57. Universal Woods 6/23/2011 Louisville, KY **2 injured** Lumber and Wood Products
58. Delta Oil Mill 6/27/2011 Greenwood, MS **1 fatality** and **1 injured** Food Products
59. Walker Manufacturing 7/27/2011 Harrisonburg, VA **1 injured** Motor Vehicle Manufacturing
60. Newpage Wisconsin System, Inc. 8/12/2011 Biron, WI **3 injured** Lumber and Wood Products
61. Stimson Lumber Company 8/16/2011 Gaston, OR **3 injured** Lumber and Wood Products
62. Archer Daniels Midland Company 8/18/2011 Clinton, IA **1 injured** Food Products
63. Stollberg, Inc. 9/9/2011 Niagara Falls, NY **1 injured** Chemical Manufacturing
64. Sara Lee 9/15/2011 London, KY **2 injured** Food Products
65. Sunset Hutterite Colony 9/15/2011 Britton, SD **2 fatalities 1 injured** Other
66. Gilster-Mary Lee Corporation 10/6/2011 Steeleville, IL **2 injured** Food Products
67. Bartlett Grain Company, LP 10/29/2011 Atchison, KS **6 fatalities** and **2 injured** Food Products
68. Polymer Partners, LLC 12/7/2011 Henderson, KY **4 injured** Chemical Manufacturing
69. Heritage-WTI 12/17/2011 East Liverpool, OH **1 fatality** and **1 injured** Other Industrial
70. NOV Tuboscope 1/12/2012 Edmond, OK **4 injured** Metal Industries
71. Purity Zinc Metals, LLC 2/10/2012 Clarksville, TN **No injuries** Metal Industries
72. Conn-Weld Industries, Inc. 3/5/2012 Princeton, WV **1 fatality** and **2 injured** Equipment Manufacturing
73. Carolina Precision Fibers 3/20/2012 Ronda, NC **4 injured** Lumber and Wood Products
74. Deceuninck North America 5/23/2012 Monroe, OH **3 injured** Rubber and Plastic Products
75. Jordan General Contractors, Inc. 8/12/2012 Hockley, TX **2 fatalities** and **1 injured** Other
76. Scott A. Humphreys, Inc. Dba Metal Crafters 8/28/2012 Simi Valley, CA **1 injured** Metal Industries
77. US Ink Corporation 10/9/2012 East Rutherford, NJ **7 injured** Other Industrial
78. Allete Inc., Dba Minnesota Power 10/22/2012 Schroeder, MN **1 injured** Electric Services
79. Environmental Enterprises, Inc. 12/28/2012 Cincinnati, OH **1 fatality** and **1 injured** Other Industrial

Chronicles of Incidents and Response 353

80. Northstar Metal Products, Inc. 4/1/2013 Glendale Heights, IL **2 injured** Metal Industries
81. Dawson Feeders, Inc. 4/3/2013 Lexington, NE **1 injured** Food Products
82. Global Ag Inc. 5/20/2013 Otho, IA **1 injured** Equipment Manufacturing
83. Advanced Environmental Recycling Technologies, Inc. 7/17/2013 Springdale, AR **1 fatality** and **2 injured** Lumber and Wood Products
84. CB Collective, LLC 7/28/2013 Los Angeles, CA **1 injured** Other
85. Titan Metals, Inc. 10/18/2013 Glendale Heights, IL **2 injured** Metal Industries
86. International Nutrition, Inc. 1/20/2014 Omaha, NE **2 fatalities** and **10 injured** Food Products
87. Georgia-Pacific, LLC 4/26/2014 Corrigan, TX **2 fatalities** and **2 injured** Lumber and Wood Products
88. Expera Specialty Solutions LLC 9/6/2014 Kaukauna, WI **1 injured** Lumber and Wood Products
89. Nestle Purina 9/14/2014 Flagstaff, AZ **4 injured** Food Products
90. NBI Construction Services, Inc. 1/8/2015 Farmington Hills, MI **1 fatality** and **3 injured** Other Industrial
91. International Paper Company 1/23/2015 Ticonderoga, NY **1 fatality** Lumber and Wood Products
92. Corbion Caraan, LLC 3/11/2015 Grandview, MO **4 injured** Food Products
93. Noranda Aluminum, Inc. 8/3/2015 New Madrid, MO **No injuries** Metal Industries
94. Nakanishi Manufacturing Corporation 9/23/2015 Winterville, GA **1 injured** Motor Vehicle Manufacturing
95. Giles & Kendall, Inc. 11/10/2015 Huntsville, AL **1 fatality** Lumber and Wood Products
96. Massachusetts Water Resource Authority Nut Island Headworks 1/25/2016 Quincy, MA **5 injured** Other Industrial
97. JCG Farms of Alabama, LLC 2/2/2016 Rockmart, GA **1 fatality** and **4 injured** Food Products
98. Voluntary Purchasing Groups, Inc. 9/8/2016 Bonham, TX **2 injured** Chemical Manufacturing
99. Tate Lyle Grain Incorporated 9/26/2016 Francesville, IN **1 fatality** Equipment Manufacturing
100. Unilin North America, LLC 5/16/2017 Mt Gilead, NC **1 fatality** and **1 injured** Lumber and Wood Products
101. Didion Milling, Inc. 6/1/2017 Cambria, WI **5 fatalities** and **12 injured** Food Products

354   Hazmatology: The Science of Hazardous Materials

102. 3d Idapro Solutions, LLC 7/26/2017 Stanfield, OR **1 injured** Food Products
103. Northeast Agricultural Sales, Inc. 10/11/2017 Detroit, ME **1 injured** Chemical Manufacturing
104. Spectro Coating 11/2/2017 Leominster, MA **1 injured** Other Industrial
105. Desert Whale Jojoba Company/Vantage 11/9/2017 Tucson, AZ **2 fatalities** and **6 injured** Other Industrial (Chemical Safety Board).

Author's Note: Throughout history, there have been periodic instances of dust explosion incidents. For the most part, most other explosions, except for previously noted ammonium nitrate, have been mitigated. Dust is not generally recognized as a hazardous material and is not widely known to be the dangerous material it is. However, dust explosions continue to this day on a frequent basis, killing and injuring employees in the workplace. Fortunately, emergency responders have avoided death and injury from dust explosions because the explosion usually occurs before they arrive.

Bibliography

Volume 1

911 Memorial and Museum, World Trade Center Bombing Investigation, 1993. www.911memorial.org/connect/blog/1993-world-trade-center-bombing-investigation.

Abilene Reporter-News, Texas, August 16, 1948.

Adams Sentinel Gettysburg, Pennsylvania, August 30, 1841.

Alton Evening Telegraph, Illinois, December 30, 1946.

Anniston Star, Alabama, June 29, 1959.

Babrauskas, V. Explosions of Ammonium Nitrate in Storage or Transportation are Preventable. *Journal of Hazardous Materials* Volume 304, pp. 134–149, 2016.

BBC News. Lac-Megantic: The Runaway Train that Destroyed a Town, Lac-Megantic, Quebec, CA. Train Derailment Crude Oil Fire, July 6, 2013. www.bbc.com/news/world-us-canada-42548824.

Bhopal.org. www.bhopal.org/what-happened/union-carbides-disaster/

Big Spring Daily Herald, Texas, July 30, 1956.

Bridgeport Standard Telegram, Connecticut, September 15, 1919.

Bridgeport Standard Telegram, Connecticut, January 16, 1919.

Bridgeport Standard Telegram, Connecticut, October 10, 1960.

Bucks County Courier Times, Doylestown, Pennsylvania, June 5, 1971.

Buffalo News. The, Six Dead in Tragic Blast, Wednesday December 28, 1983; Buffalo NY, Propane Gas Explosion 35 Years Ago, December 27, 1983.

Burke, Robert, Jacksonville FL, First Emergency Services Hazmat Team in the United States, *Firehouse Magazine*, November 30, 2006.

Burke, Robert, Martin County Florida: First Volunteer Hazmat Team in United States, *Firehouse Magazine*, September 1, 2018.

Burke, Robert, Houston TX, Pioneer Hazmat Team, Visit to Houston Fire Department, Oral Interviews with Houston Hazmat Team Members.

Burke, Robert, HazMat Team Spotlight: Houston Fire Department, *Firehouse Magazine*, August 16, 2006.

Burke, Robert, Houston's Hazmat Team Marks 25 Years of Service, *Firehouse.com*, February 28, 2005.

Burke, Robert, Memphis TN Pioneer Hazmat Team, Visit to Memphis Fire Department, Oral Interviews with Memphis Fire Department members.

Burke, Robert, Memphis' Evolution of Hazmat to All-Hazards Rescue, *Firehouse Magazine*, April 1, 2019.

Bibliography

Burke, Robert, "The Great Disaster of 1895," Telephone Interviews and Email Exchanges with Butte, MT Fire Chief, John Miller.

Burke, Robert, The Day Texas City Lost Their Fire Department, *Firehouse Magazine*, May 1, 2007.

Burke, Robert, Kansas City KS, August 18, 1959; Southwest Boulevard Fire, *Firehouse Magazine*, November 30, 2009.

Burke, Robert, Southwest Boulevard Fire, Visit to Kansas City, Kansas and Kansas City, MO Fire Departments on Three Separate Occasions, Oral Interviews with Department Members, Survivors of the Incident and Attended the 50th Anniversary Memorial Service and Dedication of a Memorial to Those Who Died. Information also obtained from Charles Gray a KMBC-TV Reporter Who Was on Scene That Day. Ray Elder of the Kansas City, MO Fire Museum Also Provided Information and Materials.

Burke, Robert, 41 Dead, Businesses Decimated In Historic Indiana Explosions, *Firehouse Magazine*, February 28, 2015.

Burke, Robert, Lessons Learned from Anhydrous Ammonia Incident, *Firehouse Magazine*, April 1, 2017.

Burke, Robert, Crescent City Train Derailment: 40 Years Later, *Firehouse Magazine*, September 1, 2010.

Burke, Robert, The Kingman Railcar BLEVE, *Firehouse Magazine*, July 1, 1998.

Burke, Robert, Kingman, AZ, Remembers 11 Fallen Firefighters, *Firehouse Magazine*, September 1, 2008.

Burke, Robert, Remembering the Gulf Oil Refinery Fire, *Firehouse Magazine*, December 3, 2010. Visited Philadelphia on Numerous Occasions to Research the Gulf Fire with Philadelphia Firefighters and the Philadelphia Fire Department Museum.

Burke, Robert, Inside the Houston Hazmat Team, *Firehouse Magazine*, February 1, 2019. Visited Houston Hazmat Team, Road with Them for a Day, Gathered Information about Incidents from Houston Hazmat Team Members. Spotlighted the May 11, 1976 Anhydrous Ammonia Tanker Accident, Which Was Highlighted in the Article.

Burke, Robert, The Waverly Propane Explosion 25th Anniversary: What Has Changed? *Firehouse Magazine*, January 31, 2003.

Burke, Robert, Chemical Notebook: Phosphorus, *Firehouse Magazine*, November 1, 2019.

Burke, Robert, The Phenomenon of Spontaneous Combustion, *Firehouse Magazine*, October 31, 2003.

Burke, Robert, Livingston, LA, Train Derailment—36 Years Later, *Firehouse Magazine*, April 1, 2018.

Burke, Robert, Six KC, Firefighters Killed in Ammonium Nitrate Explosion, *Firehouse Magazine*, November 1, 2008. Visited the Kansas City, MO Fire Department and Interviewed Firefighters. Ray Elder of the Kansas City, MO Fire Museum Supplied Information about the Incident and We Continued to Exchange Emails and Had Telephone Conversations to Gather Further Information and Answer Questions.

Burke, Robert, Oklahoma City, Before & After the Bombing, *Firehouse Magazine*, May 1, 2018.

Burke, Robert, Weyauwega Wisconsin Propane Fire, *Firehouse Magazine*, July 1996.

Burke, Robert, New Orleans Fire Department Hazmat Response Following Katrina, *Firehouse Magazine*, February 28, 2006.

Bibliography

Butte, MT, Brief History of Butte Fire Department Stations, http://co.silverbow.mt.us/730/-Brief-History-of-Butte-Fire-Departnebt-S.

Butte-Silver Bow Public Archives and Montana PBS "Hidden Fire".

Capital Times, Madison, WI, February 9, 1968.

Chapter Four, Hazardous Materials Transportation Regulations, www.princeton.edu/~ota/disk2/1986/8636/863606.PDF.

Charleston Gazette, West Virginia, July 17, 1937.

Chemical Safety Board (CSB), Combustible Dust Safety, Advocacy Priorities, Call for Action, www.csb.gov/recommendations/mostwanted/combustibledust/.

Chemical Safety Board (CSB), Chemical Accident Investigation Report, Terra Industries Inc., Nitrogen Fertilizer Facility, Port Neal, IA, 1996, www.hsdl.org/?abstract&did=234253.

Chemical Safety Board (CSB), Herrig Brothers Farm Propane Tank Explosion, Albert City, IA, April 9, 1998.

Chemical Safety Board (CSB), Arkema Inc. Chemical Plant Fire, Crosby, TX, August 29, 2017, www.csb.gov/arkema-inc-chemical-plant-fire-/.

Chemical Safety Board (CSB), Gas Cylinder Fire and Explosions at Praxair St. Louis, MO, June 24, 2005, www.csb.gov/one-year-after-gas-cylinder-fire-and-explosions-at-praxair-st-louis-csb-issues-safety-bulletin-focusing-on-pressure-relief-valve-standards-and-good-safety-practices/.

Chemical Safety Board (CSB), Little General Store Propane Explosion, Ghent WV, January 30, 2007, www.csb.gov/little-general-store-propane-explosion/.

Chemical Safety Board (CSB), Bottled Gas Plant Fire and Explosions, Dallas TX, July 25, 2007, https://webarchive.library.unt.edu/eot2008/20080916140122/; www.sb.gov/index.cfm?folder=news_releases&page=news&NEWS_ID=388.

Chemical Safety Board (CSB), T2 Laboratories Explosion, Jacksonville FL, December 19, 2007, www.csb.gov/csb-finds-t2-laboratories-explosion-caused-by-failure-of-cooling-system-resulting-in-runaway-chemical-reaction-report-notes-company-did-not-recognize-hazards-of-chemical-process/.

Chemical Safety Board (CSB), West Fertilizer Explosion and Fire, West, TX, April 17, 2013 Ammonium Nitrate Explosion, www.csb.gov/west-fertilizer-explosion-and-fire-/.

Chemical Safety Board (CSB), Didion Milling Company Explosion and Fire, Cambria WI, May 31, 2017, www.csb.gov/didion-milling-company-explosion-and-fire-/.

Chemical Safety Board (CSB), Blowout and Fire at Pryor Trust Gas Well in Pittsburg County, OK, August 16, 2018, www.csb.gov/csb-releases-factual-update-on-blowout-and-fire-at-pryor-trust-gas-well-in-pittsburg-county-oklahoma-/.

Chemical Safety Board (CSB), Tank Fires in Deer Park, TX, March 17, 2019, www.csb.gov/statement-from-the-us-chemical-safety-board-on-recent-tank-fires-in-deer-park-tx-/.

City of Mississauga, Canada, Mississauga Train Derailment, 1979, www.mississauga.ca/portal/home?paf_gear_id=9700018&itemId=5500001

Cleveland OH, LNG Leak, Explosion, Fire, East Ohio Gas Co., October 20, 1944.

Explosion and Fire, Case Western Reserve University, Encyclopedia of Cleveland History, https://case.edu/ech/articles/e/east-ohio-gas-co-explosion-and-fire.

Chester Times, Pennsylvania, April 10, 1917.

City of Roseburg, OR, http://www.cityofroseburg.org/.

Columbus Telegram Newspaper, Columbus, NE, Explosion Wrecks Elevator, Wednesday April 8, 1981.

358 Bibliography

Columbus Times, Columbus, NE, April 8, 1981.

Corpus Christi Caller Times Newspaper, Blood and Bodies Were Scattered All Over, Wednesday April 8, 1981, www.kevinsaunders.com/photos/explosion-2/.

Corpus Christi Caller Times, Texas, February 3, 1962.

Corpus Christi Caller Times, April 6, 1981.

Cranbury Press, New Jersey, November 2, 1900.

Daily Chronicle Centralia, Washington, October 28, 1954.

Daily Courier Connellsville, Pennsylvania, April 17, 1961.

Daily Leader Pontiac, Illinois, August 7, 1978.

Daily Messenger Canandaigua, New York, December 15, 1975.

Daily Missoulian, Butte's Worse Than Awful Night, Missoula, MT, Thursday January 17, 1895.

Daviec, John, The Great Dynamite Explosions, Butte, MT, The Butte Bystander Print, 1895.

Dayton, OH, Fire Department, Miamisburg, OH, July 8, 1986. Derailment Phosphorus Fire, Visited the Dayton Department and had interview with Hazmat Chief Denny Bristow, Miamisburg, OH, July 8, 1986, Derailment Phosphorus Fire, Dayton's Hazmat Team responded on Mutual Aid to Miamisburg to assist in the train derailment and phosphorus fire.

Del Rio News-Herald, Texas, April 8, 1981.

Dykes, P. *The Day the Plant Exploded,* 2010. http://kingsportarchives.files.wordpress.com/2010/10/cay-kpt-wept-3-11.

Evening Journal, Jersey City, NJ, Thursday, May 18, 1883.

Evening Public Ledger, Philadelphia, PA, October 5, 1918.

Evening World, New York, NY, July 18, 1922.

FBI, History, World Trade Center Bombing, 1993, www.fbi.gov/history/famous-cases/world-trade-center-bombing-1993.

FBI, History, Oklahoma City Bombing, Oklahoma City, OK, April 19, 1995, Terrorist Bombing, www.fbi.gov/history/famous-cases/oklahoma-city-bombing.

FBI, The Oklahoma City Bombing 20 Years Later, https://stories.fbi.gov/oklahoma-bombing/.

Flannery, Jack, Chief, McGee, Dave, Driver, Butte MT, Fire Department, Scenes of the Explosion, DeLong & Wrenn, Butte, MT, 1895.

FEMA, U.S. Fire Administration, Technical Report Series, High Rise Office Building Fire One Meridian Plaza, Philadelphia, PA, USFA-TR-049, February 1991, www.usfa.fema.gov/downloads/pdf/publications/tr-049.pdf.

FEMA, U.S. Fire Administration, Technical Report Series, Carthage, IL USFA-TR-120, October 1997.

FEMA, U.S. Fire Administration, Technical Report Series, CSX Tunnel Fire, Baltimore, MD, USFA-TR-140, July 2001.

FEMA, Lessons Learned From West, Texas Disaster, West, TX April 17, 2013, Ammonium Nitrate Explosion, www.fema.gov/lessons-learned-west-texas-disaster.

GenDisasters, Sheridan PA Naphtha Car Explosions, May 12, 1902, http://gendisasters.com/.

French Ministry for Sustainable Development, The Bhopal tragedy Night of December 2 to 3, 1984, Bhopal, India, www.aria.developpement-durable.gouv.fr/wp-content/files_mf/FD_7022Bhopalinde_1984_ang.pdf.

Gendisasters.com. Explosions. www.gendisasters.com/.

Bibliography

Gettysburg Times, Phosphorus Truck Fire is Top News Story in 1979, December 31, 1979.

Galveston Daily News, Texas, December 29, 1977.

Grand Forks Herald, Grand Forks, ND, September 16, 1905.

Hastings, N.E., Explosion Naval Ammunition Depot, Adams County (Nebraska) Historical Society. "The Naval Ammunition Depot." From *ACHS Historical News*, Vol. 25, No. 4, and Vol. 27, No. 4, September 15, 1944. Retrieved 4 June 2014. https://adamshistory.com/?page_id=20

Hanover, Adams, *The Evening Sun*, Thursday May 26, 1994.

Hastings, Nebraska, "Gasoline & Oil Clean-up Squad," In Person Oral Interview Hastings, NE Fire Department.

Hazardous Materials Transportation Regulations Chapter 4, https://www.princeton.edu/~ota/disk2/1986/8636/863606.PDF

Health and Science Executive, Union Carbide India Ltd, Bhopal, India. December 3, 1984, www.hse.gov.uk/comah/sragtech/caseuncarbide84.htm.

Heritage Mississauga, Mississauga Train Derailment, December 10, 1979, http://heritagemississauga.com/mississauga-train-derailment/.

HFD Pictorial History 1980–2014, F. Scott Mellott.

Hindenburg Disaster Hydrogen Explosion & Fire, Lakehurst, NJ, May 3, 1937, https://hindenburg.fandom.com/wiki/The_Hindenburg_Disaster.

Independent Helena, Montana, March 12, 1927.

Indiana Evening Gazette, Pennsylvania, June 2, 1959.

Indianapolis Star, Indiana, June 27, 1911.

International Campaign for Justice in Bhopal, Bhopal, India, December 2–3, 1984, Release of Methyl Isocyanate, https://www.bhopal.net/.

Interstate Commerce Commission (ICC), www.archives.gov/research/guide-fed-records/groups/134.html

Jacoby, Anny, Marshall's Creek, PA, Nitrate Truck Explosion, June 26, 1964, www.gendisasters.com/pennsylvania/13388/marshall039s-creek-pa-nitrate-truck-explosion-june-1964.

Kevin Saunders.com, www.kevinsaunders.com/.

Kingston Daily Freeman, New York, May 3, 1946.

Kingsport Times, Tennessee, February 27, 1978.

KOTA TV, Rapid City, SD, Firefighter Killed in Blaze, September 8, 2018, Propane Tank Explosion, www.kotatv.com/content/news/Firefighter-killed-in-blaze-and-explosion-near-Tilford-492742891.html.

Logansport Pharos-Tribune, Indiana, Corpus Christi, TX Grain Elevator Blast, April 1981.

Logansport Pharos-Tribune, Indiana, December 23, 1963.

Lowell Sun, Massachusetts, May 14, 1894.

Lowell Sun, Massachusetts, June 24, 1949.

Miami Valley Disaster Services Authority, Miami Valley Disaster Services Authority After Action Report on The Miamisburg Train Derailment, Dayton, OH, September 29, 1986.

Miami Valley Fire District, Miamisburg, OH, Visited department and interviewed Battalion Chief Andy Harp. Chief Harp took me on a tour of fire stations and the incident scene supplying valuable information on the Miamisburg OH, July 8, 1986. Derailment Phosphorus Fire. After visit continued telephone and email exchanges to gather further information and answer questions. Chief Harp provided me with Miamisburg: Anatomy of a Response, author not identified, from City of Miamisburg.

360   *Bibliography*

Milwaukee Fire Historical Society, Schwab Stamp & Seal Acid Spill, Milwaukee, WI, February 4, 1903.

Milwaukee, WI Fire Department, Jim Ley, Milwaukee Fire Historical Society Milwaukee, Wisconsin's Petroleum Dispersal Unit 1, Telephone oral interview and e-mail exchanges.

Minneapolis Tribune, Minneapolis, MN, May 3, 1878.

Mississauga, Canada Fire and Emergency Services Department, Telephone Interviews and Email Exchange with Platoon Chief Stephane Malo, City of Mississauga Website, www.mississauga.ca/portal/home?paf_gear_id=9700018&itemId=5500001.

Morgan, N.J., T.A. Gillespie & Company Explosions, War History on Line, October 4, 1918, www.warhistoryonline.com/world-war-ii/heinze-reinhard-heydrich-brothers.html.

Morning Herald, Uniontown Pennsylvania, Marshall's Creek, PA Nitrate Truck Explosion, June 27, 1964.

Mt. Vernon Register-News, Illinois, August 14, 1963.

NASA, NASA Safety Center, System Failure Case Study, From Rockets to Ruins, the PEPCON Ammonium Perchlorate Plant Explosion, November 2012, Volume 6, Issue 9, Henderson NV May 4, 1988, PEPCON Explosions, www.NASA.gov.

NASA-Earth Observatory, Train Derailment and Fire, Lac Mégantic, Quebec, https://earthobservatory.nasa.gov/images/81581/train-derailment-and-fire-lac-megantic-quebec.

National Center for Biotechnology Information, The Bhopal Disaster and Its Aftermath: A Review, May 10, 2005, https://www.ncbi.nlm.nih.gov/pmc/articles/PMC1142333/.

National Geological Survey, Great Alaskan Good Friday Earthquake, Anchorage AK, March 27, 1964, https://earthquake.usgs.gov/earthquakes/events/alaska1964/.

National Hazardous Materials Fusion Center, Miamisburg, Ohio Train Derailment.

Nebraska History Museum, https://history.nebraska.gov/.

Nevada State Journal, Reno, NV, April 22, 1926.

Nevada State Journal, Reno, NV, August 8, 1959.

New Castle News, Pennsylvania, December 26, 1928.

New York Times, New York, February 18, 1882.

New York Times, New York, NY, June 6, 1892.

New York Times, New York, NY, June 27, 1964, www.nytimes.com/1964/06/27/archives/explosives-truck-blows-up-killing-6-and-freeing-snakes.html.

News Frederick, Maryland, August 5, 1897.

NTSB Investigation Report, Phosphorus Trichloride Release in Boston and Maine Yard 8 During Switching Operations, Somerville, MA, April 3, 1980.

NTSB Investigation Report, Derailment Radiation Container, Thermal, CA, January 7, 1982, www.ntsb.gov/investigations/AccidentReports/_layouts/ntsb.recsearch/Recommendation.aspx?Rec=R-83-012.

NTSB Investigation Report, Railcar Leak Nitric Acid Leak, Denver, CO, April 4, 1983, www.ntsb.gov/investigations/AccidentReports/Pages/RAR8510.aspx.

NTSB Investigation Report, CSX Freight Train Derailment and Subsequent Fire in the Howard Street Tunnel, Baltimore, MD, July 18, 2001. www.ntsb.gov/investigations/AccidentReports/Pages/RAB0408.aspx.

NTSB Investigation Report, Derailment of Canadian Pacific Railway.

Bibliography

NTSB Investigation Report, Freight Train 292-16 and Subsequent Release of Anhydrous Ammonia near Minot, North Dakota, January 18, 2002. www.ntsb.gov/investigations/AccidentReports/Reports/RAR0401.pdf

NTSB Investigation Report, Collision of Union Pacific Railroad Train with BNSF Railway Company Train, Macdona, TX, June 28, 2004, www.ntsb.gov/news/events/Pages/Collision_of_Union_Pacific_Railroad_Train_with_BNSF_Railway_Company_Train_Macdona_Texas_June_28_2004.aspx.

NTSB Investigation Report, Collision of Norfolk Southern Freight Train 192 With Standing Norfolk Southern Local Train P22 With Subsequent Hazardous Materials Release, Graniteville, SC, March 24, 2005, Chlorine Disaster, www.ntsb.gov/Investigations/AccidentReports/Pages/RAR0504.aspx.

NTSB Investigation Report, Derailment with Hazardous Materials Release, Shepherdsville, KY, January 16, 2007, Crude Oil Train Derailment & Fire, www.ntsb.gov/Investigations/AccidentReports/Reports/RAB1203.pdf.

NTSB Investigation Report, Derailment, Oneida NY Monday, March 12, 2007, Tank Car Explodes, www.ntsb.gov/investigations/AccidentReports/Reports/RAB0805.pdf.

NTSB Investigation Report, Derailment of CN Freight Train U70691-18 With Subsequent Hazardous Materials Release and Fire, Cherry Hill, IL, June 19, 2009, www.ntsb.gov/investigations/AccidentReports/Pages/RAR1201.aspx.

NTSB Investigation Report, Derailment and hazardous materials release and fire, Tiskilwa, IL, October 7, 2011, www.ntsb.gov/investigations/AccidentReports/Reports/RAB1302.pdf.

NTSB Investigation Report, BNSF Railway Train Derailment and Subsequent Train Collision, Release of Hazardous Materials, and Fire, Casselton, ND, December 30, 2013, https://ntsb.gov/news/events/Pages/2017_casselton_BMG.aspx.

NTSB Investigation Report, CSXT Petroleum Crude Oil Train Derailment and Hazardous Materials Release, Lynchburg, VA, April 30, 1014, Crude Oil Train Derailment & Fire, https://ntsb.gov/investigations/AccidentReports/Reports/RAB1601.pdf.

NTSB Investigation Report, BNSF Railway Crude Oil Unit Train Derailment, Heimdal, ND, May 6, 2015, www.ntsb.gov/investigations/AccidentReports/Reports/RAB1712.pdf.

NTSB Investigation Report, CSX Transportation, Train Derailment with Hazardous Materials Release, Hyndman PA, August 2, 2017 Train Derailment & Fire, www.ntsb.gov/investigations/AccidentReports/Pages/DCA17FR011-prelim-report.aspx.

New York Times, Explosives Truck Blows up, Killing 6 and Freeing Snakes, June 27, 1964, p. 22.

Norfolk, NE, Propane Storage Tank Fire, Visit to Fire Department and Interviews with Firefighters, December 10, 2009.

Oakland Tribune, California, August 8, 1927.

Ohio History Connection, https://ohiohistorycentral.org/w/East_Ohio_Gas_Company_Explosion.

Oklahoma State Department of Health, Terrorist Bombing, Oklahoma City, OK, April 19, 1995, www.ok.gov/health/Protective_Health/Injury_Prevention_Service/Oklahoma_City_Bombing/index.html.

Oklahoma Historical Society, Oklahoma City Bombing, www.okhistory.org/publications/enc/entry.php?entry=OK026.

362 Bibliography

Oklahoma City Fire Department, I visited the department and interviewed firefighters concerning the effects the bombing of the Murrah Federal Building had on the post bombing department. Obtained information during a previous visit to Oklahoma City in 2007 from the first chief officer to arrive on scene as he gave me a tour of the site.

Oneonta Star, New York, May, 31 1960.

OSHA, Inspection Detail, Liquid Nitrogen Asphyxiation, Springer, OK, September 1998.

New York Times, Sheridan, PA Oil Car Explosions in Railroad Yards, May 13, 1902.

PBS Hidden Fire, "The Great Disaster of 1895," Butte, MT.

Parry, Bob, Borden's Anhydrous Ammonia Explosion, Houston, TX, December 11, 1983, https://my.firefighternation.com/profiles/blogs/borden-s-explosion-dec-11-1983-houston#gref.

Philadelphia Fire Department, Visited and interviewed firefighters with the help of Battalion Chief Bill Doty and Hazmat Administrative Services Chief Mike Roeshman on all events that appear in this publication in the City of Philadelphia. Further assistance provided by Harry McGee at the Philadelphia Fire Museum.

Picatinny Arsenal, Munitions Explosion, Dover, NJ, July 10, 1926

Plattsburg Sentinel, New York, July 13, 1926.

Playground Daily News, Fort Walton Beach, Florida, May, 12, 1976.

Plattsburg Sentinel, New York, July 13, 1926.

Pocono Record, Stroudsburg, PA, June 23, 2014.

Ring, Billy, Retired, Miamisburg, OH Firefighter, Mr. Ring was at the Miamisburg, OH, July 8, 1986, Derailment Phosphorus Fire, interviewed Mr. Ring by telephone and exchanged email and phone calls obtaining additional information about the incident.

Ruston Daily Leader, Louisiana, December 23, 1977.

San Antonio Express, Texas, August 10, 1965.

Sanger, Peter, Chief Engineer, Souvenir History, Butte Fire Department, November, 1901.

Saunders, Kevin, Grain Elevator Blast, Corpus Christi, TX, April 7, 1981, www.kevinsaunders.com/. Conducted telephone interview with Mr. Saunders and in person in Grand Island, NE, 2018.

Schoeff, R.W. *Reported Grain Dust Explosion Incidents*. Manhattan: Kansas State University Extension Service, 1981.

Scott, Andrew, Marshall's Creek Blast of 64 took deadly toll on brotherhood of volunteer firemen, *Pocono Record*, June 30, 2012.

Smith, Tristan, Houston Fire Department (Images of America), *Houston Fire Museum*, May 25, 2015.

Ste. Elisabeth de Warwick, Quebec, Canada, Fire Department.

Times Shreveport-Bossier, September 18, 1984.

Titusville Morning Herald, Pennsylvania, August 6, 1897.

Titusville Morning Herald, Pennsylvania, April 29, 1963.

Thiokol Memorial Project. The Thiokol Factory Explosion Is a Largely Forgotten Tragedy, 2017. http://thiokolmemorial.org.

Tulsa Oklahoma Fire Department, Interview with hazmat team member on the phone.

Bibliography

UPI Archives, Warwick, Quebec, Canada June 27, 1993, Propane Explosion, June 21, 1993, www.upi.com/Archives/1993/06/27/Propane-tank-blast-kills-4-Quebec-firefighters-8-others-hurt/4238741153600/.

U.S. Department of State, 1993 World Trade Center Bombing, February 21, 2019, www.state.gov/1993-world-trade-center-bombing/.

U.S. Geological Survey, https://earthquake.usgs.gov/earthquakes/events/alaska1964/.

WesternMiningHistory.com, "Total Devastation: The Butte, Montana Explosion of 1895", https://westernmininghistory.com/1527/total-devastation-the-butte-montana-explosion-of-1895/.

WIBV Channel 4, Buffalo, NY, https://www.wivb.com/news/local-news/remembering-the-buffalo-propane-explosion-35-years-later/.

Wikipedia, *The Free Encyclopedia*, https://en.wikipedia.org/wiki/Hindenburg_disaster.

Winchester Star Virginia, September 4, 1984.

Wisconsin State Journal, Madison, July 22, 1947.

Woodbine GA, Thiokol Factory Solid Rocket Fuel Explosion, February 3, 1971, http://thiokolmemorial.org/the-thiokol-plant-explosion/; http://thiokolmemorial.org/the-thiokol-plant-explosion/.

Yuma Daily Sun, Arizona, December 23, 1958.

Index

acetylene 81, 267, 309
 Dallas TX July 25, 2007 Bottled Gas
  Plant Fire and Explosions 309
 San Diego, CA, August 8, 1927
  Acetylene Factory Explosion 81
Albert City, IA 270
ammonium nitrate 6, 68, 91, 110, 157, 242,
  249, 253, 255, 318, 354
 Kansas, City, MO, November 29,
  1988 242
 *Morgan NJ October 4, 1918 T.A. Gillespie
  & Company Explosions* 68
 New York City, NY, February 26,
  1993 249
 Oklahoma City, OK, April 19, 1995 255
 Port Neal, IA, December 13, 1994 253
 Roseburg OR August 8, 1959 Explosives
  Truck Explosion 110
 Texas City, TX, April 16, 1947 91
 Waco GA June 1971 Car-Truck
  Collision Sparks Dynamite
  Explosion 157
 West, TX, April 17, 2013 318
anhydrous ammonia 145, 180, 223, 228, 239,
  257, 259, 277
 Crete, NE, February 19, 1969 145
 Delaware County, PA, August 6,
  1995 259
 Houston, TX, May 11, 1976 180
 Houston, TX, December 11, 1983 223
 Minot, ND, January 18, 2002 277
 Ord, NE, July 15, 1987 239
 Orrtanna, PA, December 6, 1993 251
 Shreveport, LA, September 17, 1984
  228
Arwood, Richard 26
asphalt 100, 102
 Perth Amboy, NJ, June 24, 1949 100
Avis Rent A Truck 123

Bales, Jack 135, 140, 141, 142
benzene 5, 43
 Bradford, PA, May 14, 1894 43
 Meadville, PA, June 1892 5
 Oil City, PA, June 1892 5
 Titusville, PA, June 1892 5
Bhopal India 10, 231
boiler explosion 5, 31, 193, 289
Borden's Anhydrous Ammonia
  Explosion 223
Buffalo, NY 226
Burnside, IL 268
butadiene 159, 172
 Houston, TX, October 19, 1971, Mykawa
  Road 159
 Houston, TX, September 21, 1974
  Englewood Railyard 172
butane 104, 107, 186
 Brownfield, TX, December 23,
  1958 104
 Meldrim, GA, June 28, 1959 107

chemical explosion 75, 80, 89, 101, 124, 131,
  152, 157, 245, 310
 Carpentersville, IL, December 10,
  1969 152
 Chicago, IL, March 11, 1927 80
 Hammond, IN, December 23, 1963 124
 Houston, TX, August 10, 1965 131
 Houston TX October 23, 1989 245
 Jacksonville FL December 19, 2007 310
 New York, NY, July 18, 1922 75
 Philadelphia, PA, October 10, 1954 101
 Westport, CT, May 3, 1946 89
 Woodbine, GA, February 3, 1971 157
chemical fire 25, 56
 Memphis, TN July 5, 1979 Drexel
  Chemical Company 25
 New York, NY, November 2, 1900 56

365

Index

Chemical Safety Board (CSB) 75, 268, 270,
300, 305, 309, 310, 335, 339, 342,
344, 349
Albert City Iowa April 9, 1998 Propane
Tank Explosion 270
Burnside, IL October 2, 1997 Propane
Tank Explosion 268
Cambria WI May 31, 2017 Didion Mill
Explosion 335
Crosby TX August 29, 2017 Chemical
Plant Explosion 300
CSB Releases Call to Action on
Combustible Dust Hazards 344
Dallas TX July 25, 2007 Bottled Gas
Plant Fire and Explosions 309
Deer Park TX March 17, 2019
Petrochemical Storage Tank
Fires 342
dust incidents 2006 to 2017 349
Ghent, WV January 30, 2007 Propane
Tank Explosion 305
Jacksonville FL December 19, 2007
Chemical Plant Explosion 310
Oklahoma City, OK August 16, 2018 Oil
Well Blow Out 339
Chicago, IL 51, 80, 132
chlorine 121, 179, 190, 195, 283, 286
Brandtsville PA April 29, 1963 121
Graniteville SC March 24, 2005 286
Macdona, TX, June 28, 2004 283
Mississauga, Ontario, Canada
November 10, 1979 190
Niagara Falls, NY, December 15,
1975 179
Youngstown, FL, February 27,
1978 190
compressed gas cylinder explosion
303, 309
Dallas, TX, July 25, 2007 309
St. Louis, MO, June 24, 2005 303
corn starch 77
Pekin, IL, January 3, 1924 77
Covington, James 10, 24, 26
crude oil 5, 38, 56, 73, 102, 173, 303, 329, 331,
332, 333, 339
Casselton, ND, December 30, 2013 331
Cavan Point, NJ May 11, 1883 38
Dumas, TX Oil July 29, 1956 102
Heimdal, ND, May 6, 2015 333
Lac-Megantic, Quebec, CA July 6,
2013 329
Long Island City, NY, September 14-15,
1919 73

Lynchburg, VA, April 30, 1014 332
Oklahoma City, Ok, August 16, 2018 339
Philadelphia, PA, August 17, 1975 173
Shepherdsville, KY, January 16, 2007 303
cryogenic liquids 273
Springer, OK, September 1998 273
dust explosion 5, 71, 78, 344, 354
Cambria, WI, May 31, 2017 335
Milwaukee WI April 22, 1926 78
Pekin IL January 3, 1924 77

Drexel Chemical Company 25
Doxol Gas Distribution Plant 162
dynamite explosion 4, 45, 99, 110, 128, 158
Butte, MT, January 15, 1895 45
Marshalls Creek, PA, June 26, 1964 128
Reno, NV, August 16, 1948 99
Roseburg, OR, August 8, 1959 110
Waco, GA, June 1971 158
earthquake 125
Anchorage, AK, March 27, 1964 125

Eastman Chemical Plant 119
Englewood Rail Yard 172
ethanol 2, 22, 311, 317
Cherry Hill, IL, June 19, 2009 311
Tiskilwa, IL, October 7, 2011 317

firemen/firefighter fatalities 6, 38, 40, 51, 54,
62, 65, 73, 75, 79, 81, 83, 90, 97, 100,
101, 102, 103, 104, 105, 107, 112,
118, 119, 120, 124, 125, 131, 134,
159, 161, 168, 177, 178, 190, 227, 229,
230, 244, 249, 251, 270, 273, 307,
321, 343
firemen/firefighter injuries 5, 24, 64, 75, 82,
84, 85, 121, 133, 159, 230, 243, 248,
271, 282, 304, 312, 342
fireworks explosion 35
Chester, PA, February 16, 1882 35
flour mill 32
Minneapolis, MN, May 2, 1878 32

gasoline 73, 84, 98, 113, 118, 120, 121,
132, 230
Atlantic City, NJ, July 16, 1937 84
Auburn, NY, March 31, 1960 118
Chicago, IL, Feb 7, 1968 132
Houston, TX, February 3, 1962 121
Kansas City, KS, August 18, 1959 113
Long Island City NY September 14-15,
1919 73
Minot, ND, July 22, 1947 98

Index

Norfolk, VA, September 4, 1984 230
Philadelphia, PA, April 17, 1961 120
Gettysburg, PA 193
Ghent, WV 305
GodFather of Hazmat response xviii, 10, 11,
17, 18, 19, 20, 161
Gore, Ron xviii, 10, 11, 17, 18, 19, 20, 161
grain elevator 51, 52, 207, 213, 181, 183
Bellwood, NE, April 7, 1981 213
Chicago, IL, August 5, 1897 51
Corpus Christi, TX, April 7, 1981 207
Galveston, TX, December 29, 1977 183
Westwego, LA, December 22,
1977 181
grain dust 52, 207
grain milling 334
Cambria, WI, May 31, 2017 335
Gulf Oil Refinery Fire 173
Philadelphia, PA August 17, 1975 173
Gunpowder explosions 31, 134
Richmond, IN, April 6, 1968 134
Syracuse NY, August 30, 1841 31

Hands, Bill xix, 20, 21, 24
Hastings Nebraska "Oil and Gas Clean-up
Squad" 8
Hazmat teams 7, 8, 9, 10, 11, 20, 24
evolution 10
first career Jacksonville, FL 11
first volunteer Martin County, FL 18
Hastings, NE 8
Milwaukee, WI 9
Nebraska Fire Museum 7
pioneer Memphis, TN 24
pioneer Houston TX 20
hexane 102, 103
Dumas, TX, July 29, 1956 102
Hindenburg Disaster Lakehurst NJ May 3,
1937 83
historical hazmat incidents, selected 30,
56, 274
the 1800s 30
the 1900s 56
the 2000s 274
Hooker Chemical 179
Houston, TX xix, 20, 121, 131, 159, 172, 180,
223, 245, 297, 334
Howard Street tunnel fire 274
hurricane 291, 297, 300
Crosby, TX, August 29, 2017 300
Houston, TX, August 24, 2017 Hurricane
Harvey 297
New Orleans, LA, August 29, 200 291

hydrocarbon/petroleum/fire/explosion
64, 342
Deer Park, TX, March 17, 2019 342
Portland, OR, June 27, 1911 64
hydrogen 83
Hindenburg Disaster Lakehurst NJ
May 3, 1937 83

Indiana Storage Company 124

Jackson Fireworks Plant 35

Kansas City, MO 242
Kingman, AZ 161
Knouse Foods 251

Liquefied Petroleum Gas (LPG) 10, 24, 121,
154, 185, 262, 264, 268
Brandtsville, PA, April 29, 1963 121
Crescent City, IL, June 21, 1970 10, 154,
262, 268
Oneida, NY, Monday, March 12, 2007
307
Waverly, TN, February 24, 1978 10, 24,
185, 262, 264, 268
Livingston, LA 215

Martin County, FL 18
Meldrim Trestle Disaster 107
Memphis, TN 10, 24
methyl isocyanate 231
Bhopal, India, December 2-3, 1984 231
McCrae, Max 10, 20, 24
Miamisburg, OH 236
Milwaukee Wisconsin's Petroleum
Dispersal Unit 1, 9
Mississauga, Ontario, Canada 195
Molasses flood explosion 71
Boston, MA, January 15, 1919 71
Munitions explosion 63, 65, 67, 79, 85
Avon, CT, September 15, 1905 63
Dover, NJ July 10, 1926 79
Eddystone, PA, April 10, 1917 65
Hastings, NE, September 15, 1944 85
Morgan, NJ, October 4, 1918 67
munitions plants 4
Murrah Federal Building 255
Mykawa road 159

naphtha 54, 59
Philadelphia, PA, August 5, 1897 54
Sheridan, PA, May 12, 1902 59
NASA 240

Index

National Transportation Safety Board
(NTSB) 98, 145, 172, 181, 167, 190,
215, 217, 220, 222, 282, 283, 286,
291, 304, 305, 309, 312, 316, 318,
332, 333, 334, 338, 339
natural gas explosion 88, 134
Cleveland OH October 20, 1944 88
Richmond, IN, April 6, 1968 134
nitric acid 62, 221
Denver, CO, April 4, 1983 221
Milwaukee, WI. February 4, 1903 62

Oklahoma City, OK 254, 255
Oneida, NY 307
One Meridian Plaza high-rise fire 248

pentane 102
Dumas, TX Oil July 29, 1956 102
PEPCON 240
Philadelphia, PA 37, 54, 101, 120, 173, 248
Phillips 66 Houston Chemical
Complex 102
phosphorous 76, 191, 192, 193, 218, 236
Brownson, NE, April 2, 1978 191, 192
Gettysburg, PA, March 22, 1979 193
Miamisburg, OH, July 8, 1986 236
phosphorus trichloride 204
Somerville, MA, April 3, 1980 204
Picatinny Arsenal Dover, NJ July 10,
1926 79
Police/Law Enforcement Who Made the
Supreme Sacrifice 51, 112, 190
propane fire/explosion 105, 123, 143, 161,
195, 226, 251, 261, 270, 305, 316,
334, 337, 341
Albert City, IA, April 9, 1998 270
Buffalo, NY, December 27, 1983 226
Cleveland, OH, August 13, 1963 123
Collinsville, IL, August 7, 1978 195
Davenport, NE, June 10, 1968 143
Farmington, ME 342
Ghent, WV, January 30, 2007 305
Houston, TX, July 16, 2017 334
Hyndman, PA, August 2, 2017 337
Kingman, AZ, July 5, 1973 Doxol gas
distribution Plant 161
Mississauga, Ontario, Canada
November 10, 1979 195
Norfolk, NE, December 10, 2009 316
Schuylkill Haven, PA, June 2, 1959 105
Saint. Elisabeth de Warwick, Quebec,
Canada June 27, 1993 251
Tilford, SD, September 8, 2018 341

Weyauwega, Wisconsin, March 4,
1996 261
Pro-Serve Fire (Brooks Road) 28

radioactive 214
Thermal, CA, January 7, 1982 214
Retzloff Chemical Co. 132
Richmond, IN 134
rocket fuel, solid, explosion 119, 240
Henderson NV May 4, 1988 240
Kingsport, TN, October 5, 1960 119
Rogers, V.E. 10, 20, 161
Roseburg, OR 110

Saint. Elisabeth de Warwick, Quebec,
Canada 251
saw dust 78
Milwaukee, WI, April 22, 1926 78
Saunders, Kevin 207, 208, 210, 212, 213
Schwab Stamp & Seal 62
Shamrock Oil & Gas Corporation 102
Southwest Boulevard Fire 113
spontaneous combustion 43, 91, 181, 248
Bradford PA May 14, 1894 43
Philadelphia, PA, February 23, 1991 248
Texas City TX April 16, 1947 91
Westwego LA December 22, 1977 181
Standard Oil Works 38
Cavan Point, NJ 38

T.A. Gillespie & Company 67
A Tale of Two Hurricanes 291
terrorism 249, 255
New York City, NY, February 26,
1993 249
Oklahoma City, OK, April 19, 1995 255
Texas City, TX 91
thiokol plant 157
train derailments 107, 121, 190, 191, 215, 261,
277, 282, 303, 311, 317, 329, 331,
332, 333, 337

vinyl chloride 215
Livingston, LA, September 28, 1982 215

Washburn "A" Flour Mill 32
Waverly, TN 185
West, TX 318
Weyauwega, WI 261
World Trade Center 92, 249
New York City NY February 26,
1993 249

Yarbrough, Russell 10, 11